CONTEMPORARY ADVANCEMENTS IN MATERIALS TECHNOLOGY

Novel Solutions to Complex Problems

CONTEMPORARY ADVANCEMENTS IN MATERIALS TECHNOLOGY

Novel Solutions to Complex Problems

Edited by

Shrikaant Kulkarni, PhD

Vipul Srivastava, PhD

P. William, PhD

AAP | APPLE ACADEMIC PRESS

First edition published 2026

Apple Academic Press Inc.
1265 Goldenrod Circle, NE,
Palm Bay, FL 32905 USA

760 Laurentian Drive, Unit 19,
Burlington, ON L7N 0A4, Canada

CRC Press
2385 NW Executive Center Drive,
Suite 320, Boca Raton FL 33431

4 Park Square, Milton Park,
Abingdon, Oxon, OX14 4RN, UK

Library and Archives Canada Cataloguing in Publication

Title: Contemporary advancements in materials technology : novel solutions to complex problems / edited by Shrikaant Kulkarni, PhD, Vipul Srivastava, PhD, P. William, PhD.
Names: Kulkarni, Shrikaant, editor | Srivastava, Vipul, 1973- editor. | William, P., editor.
Description: First edition. | Includes bibliographical references and index.
Identifiers: Canadiana (print) 20240490525 | Canadiana (ebook) 20240490339 | ISBN 9781774917947 (hardcover) | ISBN 9781774917954 (softcover) | ISBN 9781003561859 (ebook)
Subjects: LCSH: Materials—Technological innovations.
Classification: LCC TA403.6 .C66 2025 | DDC 620.1/1—dc23

Library of Congress Cataloging-in-Publication Data

Names: Kulkarni, Shrikaant, editor. | Srivastava, Vipul, 1973- editor. | William, P., editor.

Title: Contemporary advancements in materials technology : novel solutions to complex problems / edited by Shrikaant Kulkarni, PhD, Vipul Srivastava, PhD, P. William, PhD.

Description: First edition. | Palm Bay, FL : Apple Academic Press, 2025. | Includes bibliographical references and index. | Summary: "This new book provides novel solutions to the problems confronted in the field of materials technology, helping to pave the way in designing new materials for the future. The authors provide a comprehensive and theoretical understanding as well as experimental findings integrated with theory for furthering research in materials science and engineering. The book looks at advances and trends, tools and technologies, and characterization of smart materials and novel materials, highlighting the correlation between structure-property-utility in the materials. The behavior of materials in general is attributed to their plethora of properties, and this volume sheds light on characteristic and unique architectures in the materials due to their structural and morphological features. Contemporary Advancements in Materials Technology: Novel Solutions to Complex Problems also deliberates on the way materials can be fabricated for acquiring optimal performance in tandem with applications. It looks at developments in smart materials for new technologies because of their exciting properties that include electronic band structure, band gap, half-metallicity, multiferroic behavior, piezoelectricity, thermoelectricity, thermodynamics, and optoelectronic behavior. These properties in materials find applications in spintronic devices, semiconducting devices, thermoelectric generators, and sustainable sources of energy. Intended for researchers, academicians, and scientists in the field of materials science and technology and computational physics, this volume will be valuable in synthesizing next-generation and novel smart materials for varied potential applications"-- Provided by publisher.

Identifiers: LCCN 2024046054 (print) | LCCN 2024046055 (ebook) | ISBN 9781774917947 (hardback) | ISBN 9781774917954 (paperback) | ISBN 9781003561859 (ebook)

Subjects: LCSH: Materials--Technological innovations.

Classification: LCC TA403.6 .C664 2025 (print) | LCC TA403.6 (ebook) | DDC 620.1/1--dc23/eng/20241219

LC record available at https://lccn.loc.gov/2024046054

LC ebook record available at https://lccn.loc.gov/2024046055

ISBN: 978-1-77491-794-7 (hbk)
ISBN: 978-1-77491-795-4 (pbk)
ISBN: 978-1-00356-185-9 (ebk)

About the Editors

Shrikaant Kulkarni, PhD
Adjunct Professor, Faculty of Business, Victorian Institute of Technology, Melbourne, Australia; Adjunct Professor, Centre of Research Outcome and Impact, Chitkara University, Punjab, India

Shrikaant Kulkarni, PhD, is currently an Adjunct Professor in the Faculty of Business at the Victorian Institute of Technology, Melbourne, Australia, as well as an Adjunct Professor at the Centre of Research Outcome and Impact, Chitkara University, Punjab, India. Dr. Kulkarni has been a senior academician and researcher for over four decades. He has delivered invited lectures and conducted sessions at national and international conferences and faculty development programs. He has guided many major and minor projects in engineering chemistry, green chemistry, nanotechnology, analytical chemistry, catalysis, chemical engineering materials, industrial organization, management, and AI. He has published over 100 research papers in national and international journals and conferences of repute. He has authored more than 50 book chapters. He has edited 25 books published by international publishers. Another ten books are in the offing. He has authored four textbooks in engineering chemistry. Dr. Kulkarni possesses MSc, MPhil, and PhD degrees in Chemistry apart from master's degrees in Economics, Business Management, and Political Science. He has expertise in materials science, green chemistry and engineering, analytical chemistry, green nanoscience, nanotechnology, AI, and computational intelligence. His expertise is also in wastewater treatment, green and analytical chemistry, and advanced areas in chemical engineering and materials science. He has published two patents and applied for two more.

Vipul Srivastava, PhD
Professor of Physics, Department of Physics, School of Chemical Engineering and Physical Sciences, Lovely Professional University, Phagwara, Punjab, India

Vipul Srivastava, PhD (Physics), is presently working as Professor of Physics in the Department of Physics in the School of Chemical Engineering and

Physical Sciences at Lovely Professional University, Phagwara, Punjab, India, where he has teaching and research responsibilities. He has a total of 21 years of research and teaching experience. He has been teaching various subjects such as condensed matter physics, semiconductors physics, digital electronics, fiber optics, engineering physics, basic physics, materials for science and engineering, at the under- and postgraduate levels. His current areas of interest are materials science, condensed matter physics, electronic structure calculations, density functional theory, optoelectronics, thermoelectrics, Heusler alloys, and perovskites materials. Presently, he is contributing in the field of theoretical physics to discover novel materials for technological applications. He has handled three research projects funded by the Government of India, with funding agencies from DST, UGC, and MPCST. He has published over 100 research articles. His h-index is 20. He has attended more than 50 conferences in India and abroad and has organized many workshops, seminars, and conferences. He is an active reviewer for over 30 indexed journals in physics, such as the *International Journal of Energy Research, Journal of Magnetism and Magnetic Materials, Journal of Physics and Chemistry of Solids, Energy Storage, Chemical Physics Letters, Ceramics International,* etc. In 2020, he was awarded research excellence award, and in 2021 he received a research appreciation award by his university. He received young scientist award from the Government of India, Department of Science and Technology, New Delhi in 2005. He was awarded his BSc (Physics) in 1995, MSc (App. Physics) in 1997, and PhD (Physics) in 2005.

P. William, PhD

Dean of Research & Development and Assistant Professor,
Department of Information Technology, Sanjivani College of Engineering,
Savitribai Phule Pune University, Pune, India

P. William, PhD, is working as Dean of Research and Development and Assistant Professor in the Department of Information Technology at Sanjivani College of Engineering, Savitribai Phule Pune University, Pune, India. He has published many papers in Scopus-indexed journals and IEEE conferences. His field of research includes natural language processing, artificial intelligence, deep learning, machine learning, soft computing, cyber security, and cloud computing. He has been associated with numerous multinational companies including IBM, TATA Consultancy Services Limited (TCS), etc. and various educational groups. A focused and hardworking professional with experience of consulting in research, innovation and development, he has served as session chair and keynote speaker at multiple Scopus-indexed international conferences. He was also appointed as a reviewer for reputed Scopus/WOS-indexed journals. He is a member of the Institute of Electrical and Electronics Engineers (IEEE) and Quality Circle Forum of India (QCFI), among other professional bodies. He received his Bachelor of Engineering in Computer Science and Engineering and Master of Technology in Computer Science and Engineering from CSVTU, Bhilai, India. He completed his PhD in the Department of Computer Science and Engineering in the School of Engineering & Information Technology, MATS University, Raipur, India.

Contents

Contributors

Shruti Bakshi
Department of Physics, School of Chemical Engineering and Physical Sciences, Lovely Professional University, Punjab, India

A. Bandyopadhyay
Department of Physics, University of Gour Banga, Malda, West Bengal, India

Dipak Bhattarai
Central Department of Physics, Tribhuvan University, Kirtipur, Kathmandu, Nepal

L. Boumia
Département Sciences de la Matière, Faculté des Sciences et Technologies, Tissemsilt University, Algérie

F. Dahmane
Laboratoire de Physique Quantique et de Modélisation Mathématique (LPQ3M), Département de Technologie, Université de Mascara, Mascara, Algérie
Laboratoire de Simulation et de Modélisation des Propriétés Magnétique des Hétéro-Structures (LPMH), Département des Sciences de la Matière, Tissemsilt University, Tissemsilt, Algeria

Y. Al-Douri
College of Health and Medical Techniques, Al-Bayan University, Baghdad, Iraq; Department of Mechanical Engineering, Faculty of Engineering, Piri Reis University, EflatunSk, No: 8, 34940 Tuzla, Istanbul, Turkey; Nanotechnology and Catalysis Research Centre, University of Malaya, 50603 Kuala Lumpur, Malaysia

Madhav Prasad Ghimire
Central Department of Physics, Tribhuvan University, Kirtipur, Kathmandu, Nepal

Navdeep Kaur
Department of Physics, School of Chemical Engineering and Physical Sciences, Lovely Professional University, Phagwara, Punjab, India

R. Khenata
Laboratoire de Physique Quantique et de Modélisation Mathématique (LPQ3M), Département de Technologie, Université de Mascara, Mascara, Algeria

Mukesh Kumar
Department of Physics, Lovely Professional University, Phagwara, India

Neha Munjal
Department of Physics, Lovely Faculty of Science and Technology, Lovely Professional University, Phagwara, India

M. Nath
Department of Physics, Deshabandhu Mahavidyalaya, Chittaranjan, West Bengal, India

Arun Negi
Department of Physics, Lovely Faculty of Science and Technology, Lovely Professional University, Phagwara, India

Sangeeta Prasher
Department of Physics, Kanya Maha Vidyalya, Jalandhar, Punjab, India

D. S. Rana
Terahertz Physics Laboratory, Department of Physics, Indian Institute of Science Education and Research, Bhopal, India

Suman Rani
Department of Physics, School of Chemical engineering and Physical Sciences,
Lovely Professional University, Punjab, India

Riyadh A. Al-Samarai
Electromechanical Engineering Department, Engineering College, University of Samarra, Samarra, Iraq

Sankar P. Sanyal
Solid State Physics Laboratory, Department of Physics, Barkatullah University, Bhopal, India

F. Semari
Laboratoire de Physique Quantique et de Modélisation Mathématique (LPQ3M),
Département de Technologie, Université de Mascara, Mascara, Algeria

Deergh Bahadur Shahi
Central Department of Physics, Tribhuvan University, Kirtipur, Kathmandu, Nepal

Bhumika Sharma
Department of Physical & Chemical Sciences, Lovely Professional University, Jalandhar, India

Ramesh Sharma
Department of Applied Science, Feroze Gandhi Institute of Engineering and Technology, Raebareli, Uttar Pradesh, India

Vilas Shelke
Rechargion Energy Pvt. Ltd, Pune, Maharashtra, India

Mandeep Singh
Department of Physics, Guru Nanak Dev University, Amritsar, India

Vipul Srivastava
Department of Physics, School of Chemical Engineering & Physical Sciences,
Lovely Professional University, Phagwara, Punjab, India

Tejas M. Tank
Department of Physics, The Patidar Gin Science College, Bardoli, India

H. S. Tiwari
Department of Pure and Applied Physics, Guru Ghasidas Vishwavidyalaya, Koni, Bilaspur, India

Deepika Tripathi
Department of Pure and Applied Physics, Guru Ghasidas Vishwavidyalaya, Koni, Bilaspur, India

Abbreviations

A	polarization strength
Å	angstrom
a.u.	atomic unit
ABS	acrylonitrile butadiene styrene
AFM	atomic force microscopy
AgNWs	silver nanowires
Al_2O_3	aluminum oxide
AZO	aluminum-doped zinc oxide
BCC	body-centered cubic
BCT	Ba $(Ca,Ti)O_3$
BET	Brunauer–Emmett–Teller
BF	$BiFeO_3$
BKT	(Bi0.5K0.5) TiO_3
BNT	$Bi0.5Na0.5TiO_3$
BT	barium titanate $BaTiO_3$
BZ	Brillouin zone
BZT	Ba $(Zr,Ti)O_3$
C	coulomb
C	capacitance
CB	conduction band
CBH	carrier barrier hopping
CeO_2	cerium oxide
CMR	colossal magnetoresistance
CNT's	carbon nanotubes
CSD	chemical solution deposition
CT	$CaTiO_3$
CVD	chemical vapor deposition
D16	shear piezoelectric coefficient
DFT	density functional theory
DM	Dzyaloshinskii-Moriya interactions
DOS	density of states
DRAM	dynamic random-access memories
DSC	differential scanning calorimetry
E	electric field

Ec	coercive field
EDX	energy dispersive x-ray analysis
Eg	band gap
Ei	external electric field components
Eloss	irrecoverable energy
emu	electromagnetic unit
Erec	recoverable energy
eV	electron volt
Exp.	experimental
F	free energy
FCC	face centered cubic
FESEM	field emission scanning electron microscope
FETs	field effect transistors
FM	ferromagnetic
FP-LAPW	full-potential linearized augmented plane-wave
FQH	fractional quantum hall
FQHE	fractional quantum hall effect
FTIR	Fourier transform infrared
FTO	fluorine doped tin oxide
FWHM	full width at half maximum
g	grains
Ga2O3	gallium(III) oxide
gb	grain boundaries
Gd: ZAO	gadolinium-doped zinc aluminum oxide
GOF	goodness of fit
GPa	gigapascal
HCs	Heusler compounds
HDPE	high-density polyethylene
HM	half-metallic
Hz	hertz
IQHE	integer quantum Hall effect
ITO	Indium tin oxide
K	electromechanical energy coupling factor
LCD	liquid crystal display
LCR	inductance (L), capacitance (C), and resistance (R)
LDPE	low-density polyethylene
LED	light emitting diode
LL	Landau level
LLDPE	linear low-density polyethylene

LN	$LiNbO_3$
LPCVD	low pressure chemical vapor deposition
LSMO	$La0.67Sr0.33MnO_3$
LSMSO	$La0.67Sr0.33Mn0.95Sn0.05O_3$
LT	$LiTaO_3$
M	magnesium
MBE	molecular beam epitaxy
MC	Monte Carlo
MD	molecular dynamics
ME	magnetoelectric coupling
MEMS	micro and nano electromechanical systems
MO	metal oxides
MOSFET	metal oxide semiconductor field effect transistor
MPa	mega pascal
MPB	morphotropic phase boundary
MR	magnetoresistance
MRAM	nonvolatile magnetic random access memories
MRI	magnetic resonance imaging
MT	muffin tin
N	Newton
NEMS	nano electromechanical systems
NM	nonmagnetic
nm	nanometers
NNs	nearest neighbors
NSPT	nonoverlapping small polaron tunneling models
NT	nanotubes
NTCR	negative temperature coefficient of resistance
OLED	organic light emitting diode
OLPT	overlapping large polaron tunneling
PBE	Perdew Burke and Ernzerhof
Pc	remanence
PDMS	poly dimethyl siloxane
PDOS	partial density of states
PEDOT	poly(3,4-ethylenedioxythiophene)
PEN	polyethylene naphthalate
PENG	piezoelectric nano generator
PET	polyethylene terephthalate
PF	power factor
Pi	polarization components

PL	photoluminescence
PLD	pulsed laser deposition
PMMA	polystyrene and polymethyl methacrylate
POPh	polyorganophosphazenes
PP	cellular polypropylene
Ps	saturation polarization
PT	lead titanate $PbZrTiO_3$
PVC	polyvinyl chloride
PVDF	poly vinyl difluoride
PVDF-TrFE	poly (vinylidene fluoride co-trifluoro ethylene)
PZT	lead zirconate titanate
PZT	$(Pb (Zr,Ti)O_3$
QAH	quantum anomalous Hall
QAHE	quantum anomalous Hall effect
QH	quantum Hall
QHA	quasi-harmonic Debye approximation
QHE	quantum Hall effect
*Q*m	mechanical quality factor
QMT	quantum mechanical tunneling
QSHE	quantum spin Hall effect
R	resistance
RE	rare Earth
ROP	ring-opening polymerization
RPM	revolutions per minute
RSM	reciprocal space mapping
RTM	resin transfer molding
SAW	surface acoustic waves
SEM	scanning electron microscopy
SHE	spin Hall effect
SOC	spin-orbit coupling
SP	Slater–Pauling
SPS	spark plasma sintering
SQUID	superconducting quantum interference device
T	applied mechanical stress
tan δ	dielectric loss
TC	ferromagnetic-paramagnetic transition temperature
Tc	Curie temperature
TCF	transparent conducting films
TCO	transparent conducting oxides

TCR	temperature coefficient of resistance
TDOS	total density of states
TEM	transmission electron microscopy
TGA	thermogravimetric analysis
TI	topological insulator
TiO_2	titanium dioxide
TN	Neel temperature
TP	metal-insulator transition temperature
TRS	time reversal symmetry
TV	television
um	micrometers
U	internal energy
UV	ultraviolet
UV-Vis	ultraviolet visible spectroscopy
V	voltage
VB	valence band
VSM	vibrating-sample magnetometry
XC	exchange-correlation
XPS	X-ray photoelectron spectroscopy
XRD	X-ray diffraction
ZnO	zinc oxide
ZnS	zinc sulphide
ZnTe	zinc telluride
ZO	oxides of zinc
ZT	figure of merit
ε0	zero frequency (dc) constant
ε∞	high-frequency dielectric constant
εr	dielectric constant
ω	frequency
*E*a	energy of activation
Z'	real part of impedance
Z"	imaginary part of impedance
*g*ij	piezoelectric voltage constant
ε	dielectric constant
η	energy storage efficiency
σ	conductivity
τ	polarization relaxation
χe	electric susceptibility
2-D	two dimension

2DEG	two-dimensional electron gas
3-D	three dimension

Preface

This book provides novel solutions to the problems confronted in the field of materials technology, which will pave the way for designing materials for the future. We have endeavored to peep into smart and novel materials in a clear and systematic manner. The materials are discussed with special reference to structure, bonding, properties, and applications. The book is comprehensive in both theoretical understanding as well as experimental findings, integrated with theory for furthering the research in materials science and engineering.

The book discusses advances, tools and technologies, characterization, and applications in frontier areas of smart materials, performance, scope, and trends of novel materials.

This book gives a deeper insight into the materials in terms of structural, physical behavior, and applications. In fact, it presents the correlation between structure–property–utility in the materials. The behavior of the materials discussed, in general, is attributed to their plethora of properties. This book sheds light on the characteristic and unique architectures in the materials due to their structural and morphological features. The book also deliberates upon the way materials can be fabricated for acquiring optimal performance in tandem with the applications. Further developments in the smart materials for upcoming technologies is because of their exciting properties, such as electronic band structure, band gap, half-metallicity, multiferroic behavior, piezoelectricity, thermoelectricity, thermodynamics, and optoelectronic behavior. These properties in materials find applications in spintronic devices, semiconducting devices, thermoelectric generators, and sustainable sources of energy.

The book is divided into three parts. Part I is devoted to smart materials for applications in frontier areas and contains four chapters. Chapter 1 discusses engineering materials, their synthesis, processing techniques, and properties. Chapter 2 deliberates on the multiferroic, ferroelectric, and piezoelectric materials in terms of their preparation, properties, and applications. Chapter 3 gives an overview of perovskites and their potential in the global market. Chapter 4 throws light on the application of Rietveld analysis as an essential tool for structural analysis of engineering materials.

Part II presents synthesis protocols for the design and development of some smart materials, their characterization, and applications in diverse areas. Chapter 5 sheds light on mechanical characteristics of ceramics, metals, and polymers for brittle and ductile materials and processing role in controlling different properties. Chapter 6 discusses physical properties of Heusler alloys in particular. Chapter 7 dwells upon Hall effects and its application in materials science. Chapter 8 studies transparent conducting oxide thin films.

Part III is devoted to novel materials for their performance, scope, and trends. Chapter 9 gives an idea about the structural, mechanical, and thermoelectric properties of lead-free halide double perovskite $Cs_2AgFeCl_{6,}$discussing the full-potential linearized augmented plane wave method. Chapter 9 discusses first principles of LDA+U prediction of electronic, magnetic, and structural characteristics of the ternary iron arsenide $CaFe_2As_2$. Chapter 10 gives a structural, electrical transport and magnetic behavior of PLD-grown La0.67Sr0.33Mn0.95Sn0.05O3 thin film. Chapter 11 gives an idea about the composition of the structural and elastic properties of boron-based III–V ternary alloys using first principles calculations. Chapter 12 deals with the mechanical properties of Rh2FeZn full-Heusler compound while Chapter 13 deals with the one-step method for synthesis of single-phase $BiFeO_3$ compound.

The book will be useful to academicians and researchers in synthesizing next generation and novel, smart materials for varied potential applications. This book is intended for researchers, academicians, and scientists in the field of materials science and technology, computational physics, etc.

PART I

Smart Materials and Techniques for Science and Technology

CHAPTER 1

Engineering Materials: Synthesis, Processing Techniques, and Properties

MUKESH KUMAR[1] and SANGEETA PRASHER[2]

[1]*Department of Physics, Lovely Professional University, Phagwara, India*

[2]*Department of Physics, Kanya Maha Vidyalaya, Jalandhar, Punjab, India*

ABSTRACT

Engineering materials are the foundation of numerous industries and modern technical breakthroughs and have always been essential to human development, security, and quality of life. These materials always open the way for new innovations and have an impact on people's lives. A material might be chosen for its strength, stiffness, hardness, toughness, wear resistance, electrical characteristics, heat or corrosion resistance, or any number of other factors, but they are all attributes. Engineering materials come in a wide variety, each with unique qualities and processing methods. These materials have unique features as well as processing options that can be used to further improve their performance. For instance, polymers can be chemically treated to boost their resilience to environmental degradation, whereas metals can be heat-treated to increase their strength. In order to benefit from the best qualities of each material or other qualities, composite materials can also be made by combining two or more distinct types of materials. The most popular engineering materials include polymers, ceramics, and composites.

Polymers are versatile materials made up of long chains of molecules bound together, which can be either natural or manufactured and can be molded into almost any desired shape. They possess a wide variety of qualities, such as being light, sturdy, and flexible, and can be strengthened with fillers to produce composite materials with increased strength and stiffness.

Contemporary Advancements in Materials Technology: Novel Solutions to Complex Problems.
Shrikaant Kulkarni, Vipul Srivastava, & P. William (Eds.)

Ceramics are another type of engineering materials that have made significant progress in recent years. They are hard and brittle nonmetals made from natural and synthetic substances commonly used for high-temperature or wear-resistant applications due to their excellent mechanical and thermal properties. Ceramics can also be made into thin films that are used as coatings on metal surfaces or as semiconductor devices in electronic circuits. New processing techniques have enabled the production of ceramic components of increasingly complex shapes and sizes while retaining their outstanding performance characteristics.

Finally, widely used in engineering applications are composite materials, which are composed of two or more different types of materials and include several distinct phases. These materials are perfect for many demanding applications because they can be engineered to optimize a set of properties. They combine the best properties of each component and create a material with excellent properties. For instance, composites with fiber reinforcement are more durable than composites with just a matrix. Significant improvements in composite manufacturing techniques over the past few years have made it possible to produce increasingly complex structures with better performance characteristics.

The present chapter has been devoted to a thorough description of the synthesis techniques, properties, and processing technology of these engineering materials.

1.1 INTRODUCTION

Materials have been ingrained so deeply in human life and culture that no aspect of life is deprived of their existence. In fact, a civilization's material technology is a measure of its advancement. It can be realized from the names of the archaeological eras, such as the Stone Age, Copper Age, and Iron Age, which are based on the discovery and application of materials in those eras. The growing societal needs lead to the production of artificial materials and advancements in synthesis and molding techniques to increase the strength and quality of these materials. With the passage of time, the development of new and improved materials with superior properties directed the scientific community to know the matter's internal structure and properties, as well as improve the processes used to create its constituent parts. On the basis of the qualitative and structural ability of industrial materials, they were termed engineering materials. Engineering materials are the materials and composites used to build systems and products for a variety of engineering applications. As their numbers grew, so did the need for classification. These materials, based on their aptitude for both structure and quality, can be divided primarily into two categories: metallic and nonmetallic. Metallic engineering materials include metals and alloys (Figure 1.1). However, this chapter primarily focuses on the characteristics, synthesis techniques, and processing methods of nonmetallic materials. The chapter will also cover various selection criteria of materials for an engineering application based on their properties, such as strength, durability, corrosion resistance, electrical conductivity, and thermal stability.

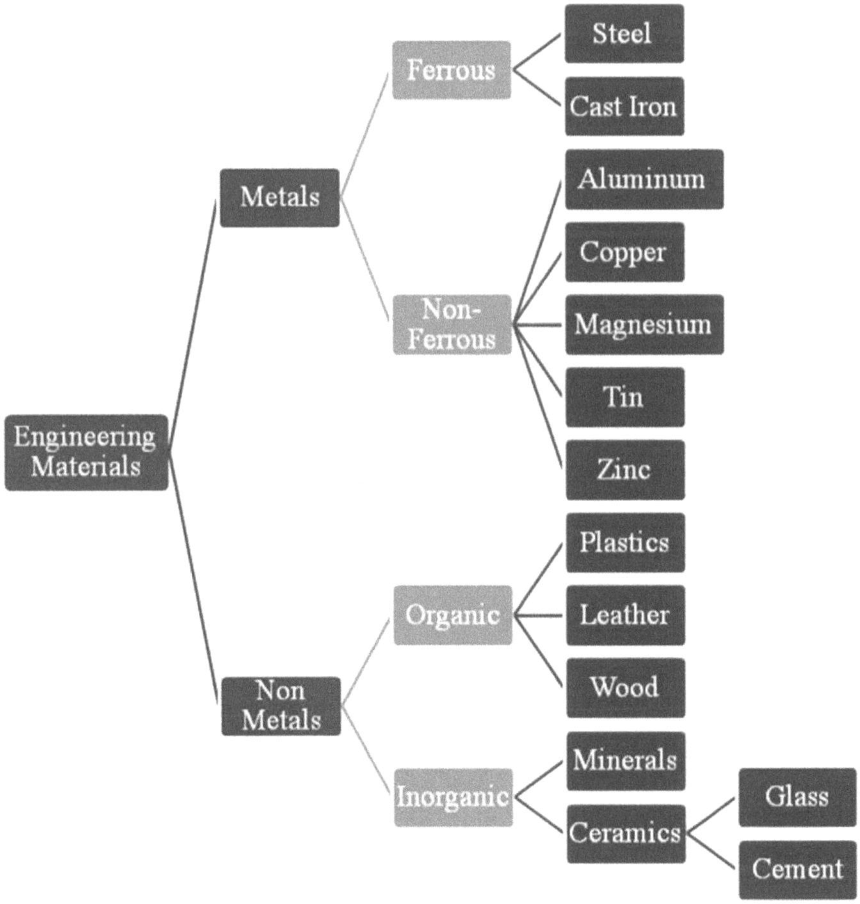

FIGURE 1.1 Classification of engineering materials.

Nonmetallic materials based on chemical makeup and atomic structure can be grouped into three main classes: polymers, ceramics, and composites. A detailed description of these materials is discussed in this chapter.

1.2 POLYMERS

Polymers, which are extremely prevalent in noncrystalline materials, are high-molecular-weight, long-chain molecules made of recurrent monomers[1] that are joined by covalent or ionic bonds. They are not always straight chains and can have short chains replaced with some side groups; such polymers are known as branched or cross-linking polymers. However, two macromolecules are held together by weak van der Waals forces.[2] These materials are light in weight, soft, and have typically low densities. They are widely used as thermal and electric insulators and are incredibly flexible.

1.2.1 CHARACTERISTICS OF POLYMERS

The phrase "material characteristics" serves as a foundation for describing a material's behavior under various circumstances and involves responses to all sorts of conditions imposed on them. These conditions may be mechanical, electrical, thermal, chemical, or radiological and help to categorize the applications of materials. The properties of polymers are affected by the degree of branching, the degree of crystallinity, the electrostatic and van der Waals attractive forces between the chains, the kind and amount of additives used, the temperature and method of processing, and the cooling rate.[3] The micelle model is used to describe the polymeric crystalline state because micelles are made up of macromolecules that are parallel to one another and form an irregular pattern that resembles crystals. The polymer is semicrystalline because an amorphous region separates these micelles from one another.

1.2.2 CLASSIFICATION OF POLYMERS

Depending on their characteristics and applications and their response to mechanical forces at elevated temperatures, polymers are usually classified into three categories.

1.2.2.1 THERMOSETS

Thermosets are typically produced by chemically reacting two or more relatively low-molecular-weight components under ambient conditions and are covalently bonded between adjacent molecular chains.[4] These bonds anchor the chains together to resist the vibrational and rotational chain motions to form a highly interlinked network when subjected to radiation or heat. These polymers undergo permanent hardness during synthesis and exhibit high rigidity and strength at room temperature. They exhibit a higher temperature resistance during heat exposure and will not creep or warp at higher temperatures.[5] At moderately high temperatures (400 K), they soften and can be molded and given any shape. However, excessive heating may cause severance of the cross-link bonds and lead to degradation. These polymers include epoxies and phenolics. Thermosets are preferred construction materials for many applications due to their long-term properties.[6]

1.2.2.2 THERMOPLASTS

Thermoplasts are pliable or moldable polymers that liquify when exposed to heat and solidify when cooled.[7] They can be repeatedly melted and solidified upon heating and cooling without undergoing any chemical changes. This property is due to their linear or branched molecular structure, which consists of long chains of polymer molecules that are not cross-linked. One of the most significant properties of thermoplasts is that they can be molded into any shape due to their low melt viscosity using a wide range of processing techniques including injection molding, blow molding, and extrusion.[8] Moreover, they can be recycled and reprocessed, making them environmentally friendly. These polymers also exhibit a wide range of mechanical, physical, and chemical properties, which can be tailored to specific applications. However, they also possess some limitations, including lower heat resistance, and their mechanical properties can be affected by temperature and environmental factors.[9,10] Furthermore, some thermoplastics can be brittle at low temperatures, which can limit their use in certain applications. A growing interest in developing sustainable thermoplasts from renewable or biodegradable sources, in response to environmental concerns, has given a push to the research in the field. These new materials are expected to replace traditional thermoplasts in various applications, thereby reducing the carbon footprint and waste generation associated with polymer production and disposal.

1.2.2.3 ELASTOMERS

Elastomers are a class of polymeric materials that exhibit great elasticity due to their unique molecular structure, which means they can undergo large deformations and recover their original shape after being stretched or compressed. They can absorb mechanical shocks without rupturing or losing their shape because of the presence of long chain cross-linked three-dimensional networks[11] that can easily slide past each other, giving the material its soft and flexible nature. The crosslinking provides the elastomer with a degree of stiffness, while the long polymer chains give the material its flexibility and ability to deform. The combination of these properties allows elastomers to exhibit excellent resilience, durability, and fatigue resistance, which make them ideal for use in high-stress applications.[12] These properties make them suitable for a wide range of applications in various industries, including automotive, aerospace, construction, and medical. Elastomers are also known for their excellent chemical inertness toward oils, solvents, and acids. Additionally, they are often used in high-temperature applications as they can withstand temperatures up to several hundred degrees Celsius without degrading. However, they exhibit low strength and poor dimensional stability[13] and are susceptible to environmental factors such as heat, humidity, and UV radiation, which can cause them to degrade over time. These polymers are difficult to process due to their low melt viscosity, which makes them difficult to mold or extrude. To address these issues, elastomers can be blended with other materials, such as reinforcing agents, plasticizers, and stabilizers, to improve their performance and durability.[14]

1.2.3 POLYMERIZATION TECHNIQUES

Polymerization is a process that involves the formation of polymers from monomers through chemical reactions. There are several techniques used for polymerization, and the choice of the technique depends on various factors such as the type of monomer, the desired polymer properties, the desired reaction conditions, and the desired polymerization process. A brief description of some of the important polymerization techniques is discussed in this section.

1.2.3.1 ADDITION POLYMERIZATION

It is a chain-growth polymerization process that involves the formation of a polymer chain by the addition of monomers to a growing polymer chain. It is characterized by the formation of new polymer chains through the addition of monomers with multiple unsaturated bonds.[15] This type of polymerization can be initiated by various means, such as heat, light, or chemical initiators, and it can be carried out in various solvents, including liquids, gases, and supercritical fluids. There are several types of addition polymerization, including the following:

- *Free-radical polymerization*: This is the most common type of addition polymerization, where free radicals are generated through initiation,[16] which then react with the monomer to form a growing polymer chain.
- *Anionic polymerization*: In this type of polymerization, the growing polymer chain is negatively charged, and the monomer is added to the chain by the removal of a leaving group, such as a halide ion.
- *Cationic polymerization*: This type of polymerization is characterized by the formation of a positively charged growing polymer chain, and the monomer is added to the chain by the removal of a proton.

Addition polymerization has several advantages, including high molecular weight control, low molecular weight distribution, and the ability to control the microstructure of the polymer.

1.2.3.2 CONDENSATION POLYMERIZATION

It is a type of step-growth polymerization process in which monomers react to form intermediate species that eventually condense into polymers. The reaction usually involves eliminating small molecules, such as water or alcohol, from the intermediate species. This type of polymerization is characterized by the formation of covalent bonds between the monomers, leading to the formation of long polymer chains. It can be classified into various categories depending upon the synthesis technology adopted:

- *Step-growth polymerization*: This is the most common type of condensation polymerization, in which the reaction between two or more monomers produces a dimer or oligomer that can then react with other monomers to form longer polymer chains.
- *Polycondensation*: This type of polymerization is characterized by the condensation of two or more monomers in the presence of a catalyst or initiator, leading to the formation of long polymer chains.
- *Intermolecular condensation polymerization*: This is a type of polymerization in which monomers react through intermolecular condensation to form a polymer chain, eliminating water and methanol

and forming strong covalent bonds between the monomer units. It is used to polymerize polyamides, polyesters, and polyurethanes.

- Intramolecular condensation polymerization: This is a type of polymerization in which monomers react through intramolecular condensation to form a polymer chain without the elimination of smaller molecules. In this process, the functional groups on the same monomer molecule react with each other to form a cyclic intermediate, which then further undergoes reactions to form the final polymer. This process results in the formation of fewer and shorter polymer chains, resulting in the formation of more compact, rigid, and less-soluble polymers.

1.2.3.3 COPOLYMERIZATION

Copolymerization is a process in which two or more different monomers are polymerized together to form a copolymer, a polymer composed of repeating units of two or more different monomers. This type of polymerization can be achieved through either addition polymerization or condensation polymerization. There are several types of copolymerization processes, including the following:

- *Random copolymerization*: In this process, different monomers are incorporated into the polymer chain in a random manner, resulting in a random distribution of different monomer units along the polymer chain.
- *Block copolymerization*: Different monomers are incorporated into the polymer chain in a sequential manner, forming blocks of one type of monomer followed by blocks of another type of monomer along the polymer chain.
- *Alternating copolymerization*: If different monomers are incorporated into the polymer chain in an alternating manner along the polymer chain, the controlled polymerization process is called alternating copolymerization.

1.2.3.4 GRAFTING

A polymer backbone is formed first in such a case and then one or more different monomers are grafted onto the backbone to form a copolymer. It offers several advantages, including the ability to tailor the properties of two polymers to meet specific requirements, the ability to improve the processing and performance of the polymers, and the ability to reduce the cost of the polymer by using less expensive monomers. It is a complex process and requires careful control of the reaction conditions and a thorough understanding of the reaction kinetics to ensure the formation of the desired copolymer.

1.2.3.5 INIFERTER-INITIATED POLYMERIZATION

This is a type of polymerization that involves the use of an iniferter molecule as an initiator for the polymerization reaction. An iniferter is a type of molecule that has both initiator and monomer functionalities, allowing it to act as both an initiator and a monomer in the polymerization reaction. The mechanism is versatile as it offers polymerization control without requiring an additional radical source. Iniferter-initiated polymerization can be used to initiate both addition polymerization and condensation polymerization reactions depending on the type of iniferter used. This type of polymerization offers several advantages, including the ability to control the molecular weight and molecular weight distribution of the polymer, the ability to control the polymer microstructure, and the ability to synthesize block copolymers and graft copolymers. Iniferter-initiated polymerization is a relatively new area of research, and much of the work has been focused on the development of new iniferters and the optimization of polymerization conditions to achieve the desired polymer properties. Despite these advances, there are still many challenges associated with iniferter-initiated polymerization, including the need for specialized equipment and the need for careful control of the reaction conditions to ensure the formation of the desired polymer. Overall, iniferter-initiated polymerization represents a promising approach for the synthesis of advanced polymers with tailored properties, and it has the potential to become a useful tool for the polymer science and engineering community in the future.

1.2.3.6 RING-OPENING POLYMERIZATION (ROP)

ROP is a type of polymerization that involves the initiation of a reaction by the opening of a cyclic monomer or prepolymer to form a linear or branched polymer. The cyclic monomer or prepolymer contains one or more cyclic bonds that are broken during the polymerization reaction, leading to the formation of a growing polymer chain. ROP can be used to polymerize a wide range of monomers, including lactones, cyclic esters, cyclic carbonates, and cyclic amides. This type of polymerization can be initiated by thermal, photochemical, or chemical means, and it can be performed in solution, in bulk, or in the presence of a solid support. ROP offers several advantages over other types of polymerization, including the ability to produce high-molecular-weight polymers with a narrow molecular-weight distribution, the ability to control the microstructure of the polymer, and the ability to produce well-defined polymers with controlled end groups. One of the key challenges associated with ROP is the need for careful control of the reaction conditions to ensure the formation of the desired polymer. This includes controlling the reaction temperature, the initiator concentration, and the monomer/initiator ratio. Despite these challenges, ROP has found a wide range of applications in various fields, including the production of biodegradable polymers, the synthesis of block copolymers, and the production of functional polymers with tailored properties. Overall, ROP represents an important tool in the polymer science and engineering community, and its continued development and optimization are expected to lead to the discovery of new and innovative applications in the future.

1.2.3.7 EMULSION POLYMERIZATION

It is a type of polymerization that involves the polymerization of monomers in a dispersed phase in an aqueous medium. This type of polymerization is characterized by the formation of small polymer particles that are dispersed in an aqueous medium, and the reaction is initiated by the addition of a suitable initiator. There are several different types of emulsion polymerization, including anionic emulsion polymerization, cationic emulsion polymerization, and nonionic emulsion polymerization, each with its own set of advantages and disadvantages. The choice of the type of emulsion polymerization depends on various factors, including the type of monomer, the desired properties of the polymer, and the desired reaction conditions. Emulsion polymerization has found a wide range of applications in the production of coatings, adhesives, and surfactants, the synthesis of block copolymers, and the production of functional polymers with tailored properties.

1.2.3.8 SUSPENSION POLYMERIZATION

It is a type of polymerization that involves the polymerization of monomers in a suspension, where the monomer is dispersed as small droplets in a continuous medium. This type of polymerization is initiated by the addition of a suitable initiator, and the reaction proceeds until the growing polymer particles become too large to remain suspended and they begin to settle to the bottom of the reaction vessel.

1.2.4 STRESS–STRAIN BEHAVIOR OF POLYMERS

The stress–strain behavior of a material is used to understand its mechanical properties and predict its behavior under different conditions. Materials exhibit either a linear or nonlinear stress–strain behavior depending on the nature of their bonds and microstructure. Metals, for example, exhibit a linear stress–strain behavior indicating a proportionality relationship between stress and strain. Polymers and composites exhibit nonlinear and more complex stress–strain behavior due to their unique microstructures and internal interactions. The behavior of these materials can be influenced by a number of parameters, including temperature, loading rate, and environment. The stress–strain curve in these materials may have multiple stages, including elastic, plastic, and brittle sections.

A number of disciplines, including engineering, materials science, and physics, depend on comprehending the stress–strain behavior of materials. It is essential for product development, material behavior prediction, and ensuring the dependability and safety of systems and structures. The mechanical characteristics of polymers are described using a number of the same metrics that are used to describe the mechanical properties of metals, such as yield and tensile strength. For many polymeric materials, the fundamental stress–strain test is employed to describe some of these mechanical characteristics. The rate of deformation (strain rate), temperature, and chemical composition of the environment all have a significant impact on the mechanical behavior of polymers. In general, polymers exhibit a nonlinear relationship between stress and strain, meaning that their behavior changes as the load increases. This behavior is due to the inherent properties of polymers, such as their long chains and low stiffness, which result in unique mechanical responses. For smaller strains, polymers exhibit elastic behavior; however, with increased strain, they develop into plastic (Figure 1.2). In the elastic region, stress changes linearly with respect to applied stress, obeying Hooke's law; however, at point b, material resistance to deformation decreases and the material can be easily deformed. It is defined as the yield stress, representing the transition from elastic to plastic behavior, and depends on the polymer crystallinity, polymer morphology, and testing conditions, such as temperature and the speed of the test. The strain hardening is defined as the region of the curve in which the stress increases rapidly with an increased amount of applied strain. In this region, the material resistance to deformation increases abruptly, and the polymer acquires a stiffer phase. Finally, the polymer breaks at a stress value higher than the yield stress, known as the tensile strength of the material. The toughness is calculated by calculating the area under the stress–strain curve of the polymer and is defined as the amount of energy required to break the polymer. In addition, the mechanical characteristics of polymers are much more sensitive to temperature changes near room temperature. As the temperature increases, the yield stress, elastic modulus (stiffness), and yield energy all decrease while the yield strain (elongation at yield) increases.

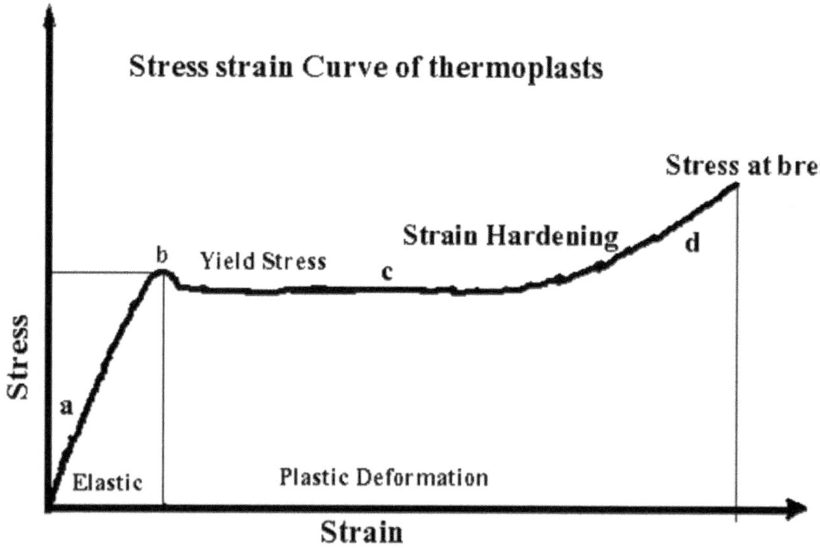

FIGURE 1.2 Stress–strain behavior of plastic polymers.

1.2.5 PROCESSING TECHNIQUES OF POLYMERS

Polymer processing refers to the techniques used to shape and mold the polymer to produce desired objects or parts from polymer materials, such as plastic and rubber. The most common processing techniques used for polymers are discussed in the following sections.

1.2.5.1 INJECTION MOLDING

This is a process in which a polymer material is melted and injected into a mold. The mold is designed in such a way that the melted polymer fills all the cavities of the mold and solidifies into the desired shape upon cooling. The injection molding process starts with the heating of the polymer to melt into fluid and then the fluid is fed into the injection molding cavity under high pressure. This helps to fill the mold evenly and eliminate any voids or air pockets. The mold is cooled to allow the polymer to solidify and take on the shape of the mold cavity. Once the polymer has cooled and solidified, the mold is opened, and the part is ejected from the mold. This method is ideal for producing parts with tight tolerances or complex shapes of small to medium-sized parts with high precision and uniformity. Additionally, it is a relatively fast and efficient process, making it suitable for largescale production runs. However, injection molding also requires a significant investment in equipment and tooling, so it is typically best suited for high-volume production runs.

1.2.5.2 BLOW MOLDING

Blow molding is a process used to produce hollow parts with a uniform wall thickness, such as containers, bottles, and tanks, from polymers. The process involves heating a tube of polymer, called a parison, until it becomes soft and pliable. Air is then blown into the parison, which expands and takes the shape of the mold. In this process, the polymer is melted and extruded into a parison to cool to a temperature where it is still flexible but not hot enough to melt again. The parison is then placed into a mold, and air is blown into it under high pressure. The air pressure causes the parison to expand and take the shape of the mold cavity. The mold is cooled to allow the polymer to solidify and retain the shape of the mold cavity, and the solidified part is ejected out from the mold.

1.2.5.3 EXTRUSION

Extrusion is a processing technique for polymers that involves forcing a melted polymer through a die to produce a continuous shape such as a tube, sheet, or profile. The melted polymer is usually fed into the extruder through a hopper, where it is melted and homogenized before being forced through the die. The polymer can then be cooled and cut into desired lengths to produce the final product. Extrusion is a versatile and efficient processing technique that can be used to produce a wide range of products, including pipes, tubing, sheets, films, profiles, and more. It can also be used to produce composite materials by incorporating fillers or reinforcements into the melted polymer. The key advantage of extrusion is its ability to produce long, continuous lengths of the product that can be cut to the desired length for use. It is also relatively fast and efficient, making it suitable for high-volume production runs. However, extrusion is not well-suited for producing parts with complex shapes or tight tolerances, and the final properties of the part may be influenced by the extrusion process, such as the processing temperature and the die design.

Additionally, the choice of polymer and the processing conditions must be carefully considered in order to produce a product with the desired properties. For example, the viscosity and thermal stability of the polymer must be taken into account to ensure a smooth and consistent flow through the die, while the cooling rate must be controlled to avoid defects such as warping or sagging.

1.2.5.4 COMPRESSION MOLDING

Compression molding is a process involving pressurizing and squeezing the polymer between two halves of a heated mold.[17] The polymer, in the form of pellets or performs, is preheated to a semimolten state, discharged into a mold, known as the press, and subjected to heat and pressure, causing the polymer to flow and take the shape of the mold cavity. Temperature and pressure are carefully controlled to ensure the desired properties of the polymer. The applied pressure helps to eliminate any air pockets or voids in the final part. The mold is then cooled under pressure to allow the polymer to solidify and retain the shape of the mold cavity. The cured and solidified material is ejected from the mold. Compression molding is ideal for producing parts with complex shapes and tight tolerances as the material is able to flow into all parts of the mold cavity. It is also well-suited for producing large parts as the material is able to fill the entire mold cavity in a single operation. However, compression molding is typically slower and less efficient than other processing techniques, and the final properties of the part may be influenced by the compression molding process, such as the curing temperature and the mold design.

1.2.5.5 THERMOFORMING

Thermoforming involves heating a thermoplastic sheet to a temperature where it becomes pliable and then drawing it over or into a mold to form a rigid or semirigid shape.[18] The material is held in the mold by vacuum pressure or mechanical clamping and then allowed to cool and solidify into desired shapes such as plastic packaging, automotive parts, and medical devices. The polymers used in thermoforming must have good thermal stability and good forming characteristics. Common polymers used in thermoforming include polyethylene, polypropylene, and acrylonitrile butadiene styrene (ABS).

1.2.5.6 ROTATIONAL MOLDING

Rotational molding, also known as rotomolding, is a manufacturing process used to produce large, hollow parts made from plastic materials. The process starts with a plastic material, usually in the form of pellets or powder, being placed into a mold. The mold is then placed in an oven and rotated around two perpendicular axes. The heat from the oven melts the plastic and causes it to spread evenly across the inside of the mold. The plastic is then evenly distributed on the inside of the mold, creating a uniform wall thickness and a smooth surface finish. The mold is then cooled, either by air or water, to allow the plastic to solidify into its final shape. Once the plastic has cooled and solidified, the mold is opened, and the part is removed. Rotational molding is commonly used to produce large, complex parts such as fuel tanks, playground equipment, agricultural containers, and outdoor furniture. The main advantage of rotational molding is the ability to produce parts with a uniform wall thickness and without seams or joints, which can make the parts stronger and more durable. Additionally, the process is well-suited for producing large, hollow parts with complex shapes. The polymers used in rotational molding are typically low-density polyethylene (LDPE), linear low-density polyethylene (LLDPE), high-density polyethylene (HDPE), polypropylene, and polyvinyl chloride (PVC).

1.3 CERAMICS

The word ceramics comes from the Greek word "keramikos", which means burnt stuff, indicating that the desirable properties of these materials are achieved through a high-temperature process called sintering. Sintering involves heating the ceramic material to a high temperature until it melts and fuses together, which results in a wear- and corrosion-resistant hard and strong material that is more durable and suitable for a wide range of applications. Ceramics refers to a broad category of nonmetallic inorganic compounds, such as clay, quartz, and feldspar, which have been used by humans for thousands of years. These materials can be shaped and hardened by heat and are characterized by their polycrystalline nature and excellent mechanical and thermal properties. These properties are affected by the purity of the raw material utilized, the amount, type, distribution, kind of impurity present, the size of the particles, and the method of processing. Ceramics are commonly used in applications that require high strength, durability, and the ability to withstand high temperatures. Some of the most common uses of ceramics are in the production of pottery, tiles for flooring and walls, bricks, and other building materials. In addition to their use in traditional applications, they are utilized in high-tech applications ranging from aerospace to healthcare, including the production of electronic components, cutting tools such as knives and drills, and refractory materials used in line furnaces. They are also used in the medical field to create artificial joints, bone replacement material, and dental implants. Ceramic components find applications in the aerospace industry for the fabrication of heat shields, engine parts, and other critical components. They secured their special place in the energy industry by creating components including capacitors, resistors, fuel cells, and solar panels that can withstand high temperatures and corrosive environments.

1.3.1 CLASSIFICATION OF CERAMICS

The standard classification of ceramics is based on their composition, properties, and intended use. Compositionally, ceramics can be divided into traditional ceramics and advanced ceramics. Traditional ceramics are made from natural raw materials such as clay, feldspar, and quartz. These materials are usually mixed with water to form a plastic paste that can be molded into a desired shape. The molded object is then dried and usually fired at high temperatures. These ceramics have been used for thousands of years and are still widely used today. Examples of traditional ceramics include pottery, bricks, and tiles. Advanced ceramics, on the other hand, are made from synthetic raw materials using advanced processing techniques and can have superior mechanical, electrical, or thermal properties. The raw materials can be oxides, carbides, nitrides, borides, or silica. It is used for a wide range of applications, including electronics, aerospace, automotive, cutting tools, engine components, semiconductors, and biomedical implants and biomedical industries. Examples of advanced ceramics include silicon nitride, alumina, magnesia, and zirconia. Alumina, magnesia, and zirconia are basically the dominant oxide ceramics used in the production of spark plugs, furnace linings, cutting tools, artificial hip joints, etc. Oxide compounds can be approximated as the closest packing of oxide ions, with the cations occupying tetrahedral or octahedral interstices. They are primarily composed of metal oxides and exhibit excellent hardness, high resistance to wear, and thermal stability, making them ideal for a wide range of applications, including structural materials such as furnace lining and heat exchangers, electronic components, catalysis, and biomedical applications including dental implants and bone substitutes for their excellent biocompatibility. These ceramics can withstand harsh chemical environments and exposure to acids and alkalis.

The other ceramics composed of two different metal oxides, called binary oxide ceramics, are utilized in catalysis, energy storage, and electronics due to their tunable properties by varying the composition of the material.[19] These materials have a large number of active sites that can facilitate chemical reactions efficiently and are also known for their high surface area and porosity. In addition to their catalytic and energy storage properties, binary oxide ceramics are also used in electronics and energy storage applications, such as batteries and supercapacitors, where their high surface area allows for increased storage capacity. Some common examples of binary oxide ceramics include zinc oxide (ZnO), titanium dioxide (TiO_2), and tin oxide (SnO_2). These materials are used in sunscreen lotions, as a UV blocker, in gas sensors, and as a photocatalyst in environmental applications, such as water purification and air pollution control due to their ability to break down organic pollutants under UV light.

Ceramics can be categorized by properties into structural ceramics, functional ceramics, and bioceramics. Structural ceramics are designed for durability, strength, wear, and are corrosion-resistant to be utilized in cutting tools and engine components. Examples of structural ceramics are silicon carbide, boron carbide, and alumina.[20] Functional ceramics are designed to have specific properties including piezoelectricity, ferroelectricity, or superconductivity. These ceramics have unique electrical, magnetic, or optical properties and are useful in a wide range of applications. Piezoelectric ceramics are used in sensors, actuators, and ultrasound transducers, while ferroelectric ceramics are used in memory devices and capacitors. Superconducting ceramics are used in magnetic resonance imaging (MRI) and power transmission. Examples of functional ceramics include lead zirconate titanate (PZT), barium titanate, and high-temperature superconductors. Bioceramics are designed for biomedical applications such as bone replacement, dental implants, and drug delivery. These ceramics are biocompatible, dimensionally stable, nontoxic, and bioinert[21] and can be used to replace or repair damaged or diseased tissues in the body. They function as a regenerative scaffold, providing a framework that sooner or later dissolves as the body reconstructs tissue. Examples of bioceramics include hydroxyapatite, bioactive glass, and ceramic-coated implants.

The other scheme of ceramic categorization is based on their applications and is titled whiteware ceramics, refractory ceramics, and electroceramics. Whiteware ceramics are used for household applications such as dinnerware, tiles, and sanitary ware. These ceramics are typically made from a mixture of clay and other materials and are fired at high temperatures to produce a hard, durable, and nonporous surface. Whiteware ceramics are usually glazed to make them more resistant to staining and scratching. They are made from materials such as porcelain, bone china, and earthenware. Refractory ceramics are used in high-temperature applications such as furnace linings, kiln furniture, and crucibles. These ceramics have high melting points, excellent thermal shock resistance, and low thermal expansion coefficients. Examples of refractory ceramics include alumina, zirconia, and silicon carbide. Electroceramics, due to their unique electrical properties such as high dielectric constant, low loss tangent, and high piezoelectricity, are used in electronic applications such as capacitors, insulators, and sensors. Examples of electroceramics include barium titanate, lead zirconate titanate (PZT), and piezoelectric ceramics.

In addition to these classification schemes, ceramics can also be classified based on other factors such as microstructure, processing method, and application. For example, ceramics can be classified as monolithic (single-phase) or composite (multiphase) based on their microstructure. Ceramics can also be classified based on the processing method used, such as powder processing, sol–gel processing, or chemical vapor deposition. Finally, ceramics can be classified based on their application, such as aerospace ceramics, energy ceramics, and environmental ceramics. These classifications provide a framework for understanding and selecting the appropriate ceramic material for specific applications based on their composition, properties, and intended use.

1.3.2 CERAMIC SYNTHESIS TECHNIQUES

Synthesizing ceramics typically involves processing raw materials into a homogeneous mixture, shaping the mixture into the desired form, and then subjecting it to high temperatures to form a solid material. There are several synthesis techniques commonly used in ceramic production, including the following:

1.3.2.1 SOLID-STATE REACTION MECHANISM

Ceramic synthesis by solid-state reaction mechanism is utilized for industrial-scale production of ceramics with a high degree of crystallinity that are known for their unique mechanical, thermal, and electrical properties, having high purity and homogeneity. It involves the mixing and grinding of different chemical compounds to achieve uniformly distributed, homogeneous mixtures which react with each other at high temperatures to form a new compound. The grinding is accomplished using a mortar and pestle or high-energy ball milling, which is critical for achieving a complete reaction. Any inhomogeneity in the mixture can result in an incomplete reaction or the formation of unwanted byproducts; hence, a solvent or a dispersant can be added during the mixing process to improve the reaction rate and product properties. The mixture is heated at a temperature high enough to initiate the reactions, such as diffusion, nucleation, and crystal growth, between the components but not high enough to cause the formation of unwanted byproducts or the degradation of the starting powders. The temperature is determined by the thermodynamic properties of the starting compounds and the desired product. The rate of the reactions involved is dependent on several factors, including the heating rate, the particle size of the starting powders, and the composition of the mixture. The final product is achieved after cooling the reactants to room temperature and removing any remaining starting materials or byproducts. The cooling rate is important because it can affect the microstructure and properties of the final product. Rapid cooling can result in the formation of a nonequilibrium phase or an amorphous material, while slow cooling can result in the formation of a crystalline material. Recent advances in the field include the use of in-situ techniques for monitoring the reaction process, such as X-ray diffraction and differential scanning calorimetry, which can provide insights into the reaction kinetics and mechanism.

1.3.2.2 SOL–GEL METHOD

The sol–gel method is a widely used and versatile technique for the synthesis of ceramic materials compared to traditional methods.[22] The method involves the preparation of a precursor solution containing metal alkoxides or metal chlorides dissolved in an appropriate solvent. The solution is then hydrolyzed by adding an appropriate hydrolyzing agent, such as water or acid, to form a colloidal suspension or sol. Subsequently, sol is allowed to undergo a gelation process whereby sol particles form a three-dimensional network or gel due to interparticle bonds. The gel is then dried and treated with a high-temperature process, known as calcination or sintering, to form the final ceramic material. The basis material, being a solution, helps in the uniform distribution of precursors and produces highly pure and homogeneous ceramics with controlled particle size and morphology, as well as nanostructured ceramics. The sol–gel method allows for the incorporation of dopants and additives into the ceramic material to tailor its properties. Some examples of ceramics synthesized by the sol–gel technique include silica, alumina, zirconia, titania, and hydroxyapatite.[23] The technique is also commonly used for the preparation of ceramic coatings on various substrates, including metals, glass, and ceramics.

1.3.2.3 HYDROTHERMAL SYNTHESIS

Hydrothermal synthesis involves the reaction of aqueous solutions of inorganic compounds, such as metal salts or metal alkoxides, in a high-pressure vessel made of stainless steel at a high temperature. The precursor solution may also contain organic additives or surfactants to control the size and shape of the ceramic particles and to modify the properties of the ceramic material. Once the solution is prepared, the vessel is sealed, and the solution is compressed at a pressure ranging from 1 to 10 MPa and usually heated above the boiling point of the solvent, typically using an autoclave or other specialized equipment. The reaction time may vary depending on the desired ceramic phase and the reaction conditions, ranging from hours to days. The mixture is then heated to promote the formation of the desired ceramic phase. The technique is often used to synthesize nanocrystalline ceramics such as zeolites, hydroxyapatite, and other complex oxides. The properties of synthesized ceramics can be tailored by adjusting the reaction conditions, such as temperature, pressure, and precursor concentration. Hence, the method is utilized to synthesize high-purity metastable ceramic phases with complex crystal structures. One of the advantages of the hydrothermal method of ceramic synthesis is that it requires only high-pressure tanks, thermal sources, and precursor solutions and is relatively simple and inexpensive. However, the requirement of high-pressure vessels and long reaction times can limit the production.

1.3.2.4 SPRAY PYROLYSIS

Spray pyrolysis is a well-established technique for producing uniform thin films with controlled thickness or powders with a high surface area and fine particle size of ceramics, valuable in various applications including electronic devices, sensors, and catalysts. It involves the controlled deposition of fine droplets of a precursor solution onto a heated substrate, typically a glass or ceramic surface using a nebulizer or spray nozzle. Subsequently, the droplets are subjected to pyrolysis to drive off any solvents or organic components and initiate the formation of the ceramic materials. The size, shape, and properties of the synthesized ceramics can be controlled by adjusting the deposition parameters, such as the precursor concentration, substrate temperature, and gas flow rate. Spray pyrolysis technology is commonly used for materials such as silicon nitride and aluminum oxide. However, the process is generally limited to smallscale production, and the quality and uniformity of the resulting material can be affected by variations in the deposition parameters. Furthermore, the high temperatures involved in the process can limit the range of substrate materials that can be used. With further research and development, spray pyrolysis may continue to provide new opportunities for the synthesis of advanced ceramic materials.

1.3.2.5 MICROWAVE-ASSISTED SYNTHESIS

Microwave-assisted synthesis is a rapidly developing innovative technology for the synthesis of complex oxide ceramics with improved phase purity, homogeneity, and crystallinity. It involves the use of microwaves to heat the precursor materials obtained by solid-state synthesis, hydrothermal synthesis, or sol–gel synthesis rather than using conventional heating methods such as a furnace or a hotplate. It results in a rapid and efficient synthesis of ceramics, which can significantly reduce the reaction time and energy requirements and improve the reaction efficiency and product quality compared to conventional heating methods. This is because microwaves can heat the reactants directly without the need for a thermal gradient to be established throughout the material. Secondly, microwave-assisted synthesis offers precise temperature control, which can lead to better control of the reaction kinetics and ultimately result in improved product quality. Examples of these materials include perovskites, spinels, and garnets, which are widely used in solid-state batteries, fuel cells, and catalysts. The various factors that can influence the success of the synthesis process involve the choice of reactants, reaction conditions, and microwave power. However, the technology can be expensive, making it less accessible to smallscale laboratories, and the yield can be influenced by the size of the microwave reactor. It also requires careful handling to avoid explosions and fire hazards.

1.3.2.6 SPARK PLASMA SINTERING

Spark plasma sintering (SPS) is an advanced processing technique used to synthesize ceramics with superior properties compared to conventional sintering methods and has gained increasing attention in recent years. It involves the application of a pulsed electric current to a powder compact in a vacuum or inert gas environment, resulting in rapid heating and consolidation of the material into a dense ceramic. The starting powdered chemical compounds are mixed in a specific ratio and then mechanically milled to achieve a fine, homogeneous powder mixture. The powder mixture is loaded into a die and compressed to a predetermined density. It is subsequently heated under a vacuum or inert gas atmosphere, and a pulsed direct current is applied to the sample. The high current and pressure cause the powders to fuse together rapidly, resulting in a dense, fully consolidated ceramic with minimal growth of grain boundaries in shorter processing times. The rapid sintering process leads to a finer and more homogeneous microstructure, resulting in superior mechanical and thermal properties. SPS can be used to synthesize structural, functional, biomedical, and nanoceramics such as hydroxyapatite, piezoelectric, and ferroelectric ceramics with improved electrical and mechanical properties. The ability to synthesize ceramics with improved mechanical, electrical, and thermal properties has the potential to impact a range of industries, from aerospace to biomedical engineering.

1.3.3 STRESS-STRAIN BEHAVIOR OF CERAMICS

Ceramics are very stiff materials with high strength, but they are prone to brittle failure and exhibit very little ductility, as represented by the point at which the material fractures. They have high ultimate strengths but low fracture strains, indicating their brittleness and low resistance to fracture. Their stress–strain behavior is governed by a combination of their microstructure, chemical composition, and processing history. The material's strength, ductility, and toughness are affected by the porosity, size, and distribution of grains, the presence of defects such as pores and cracks, the addition of certain dopants or impurities, and the sintering temperature and pressure. Increasing the grain size of a ceramic results in increasing its strength and toughness, while increasing its porosity can decrease its strength and toughness. Defects such as cracks and voids can act as stress concentrators, leading to premature failure of the material. Understanding the stress–strain behavior of ceramics is crucial for designing and engineering ceramic materials for various applications and is obtained through tensile or compressive tests, in which a ceramic sample is subjected to a gradually increasing load until it reaches its breaking point. The ceramics exhibit a linear stress–strain curve (Figure 1.3) indicating the elastic behavior of ceramics, and its slope is known as the elastic modulus, which measures the stiffness of the material. The elastic modulus of ceramics is typically very high, indicating their high stiffness. The plastic deformation of ceramics is limited due to their brittle behavior and low ductility. However, some ceramics with specific microstructures or processing conditions can exhibit some degree of plasticity.

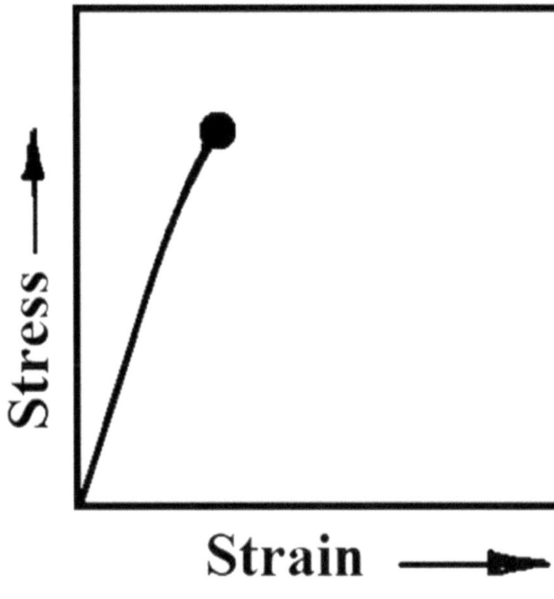

FIGURE 1.3 Stress–strain behavior of ceramics.

The tensile strength of ceramics is usually measured experimentally by three- or four-point bend tests. In the three-point bend test (Figure 1.4), the maximum stress occurs along a line on the specimen face opposite the central knife edge. The measured fracture stress is often referred to as the modulus of rupture.

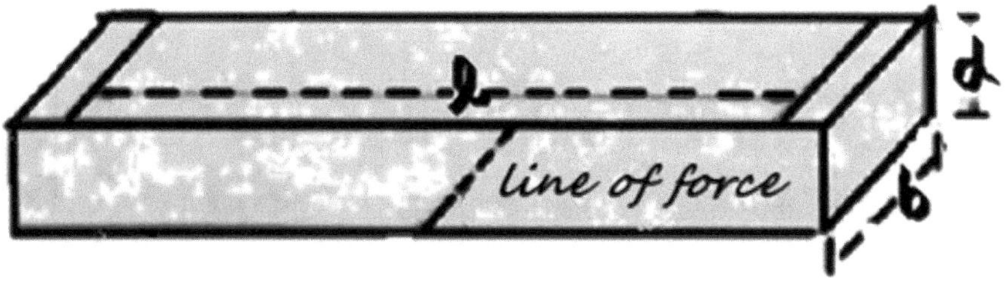

FIGURE 1.4 Three-point bend test diagram for a block having dimensions b, d, and *l* and a line of applied force.

1.3.4 PROCESSING TECHNIQUES OF CERAMICS

Ceramic processing refers to the series of steps used to produce objects from clay or other ceramic raw materials. These techniques typically include the following:

1.3.4.1 MIXING

The first step in ceramic processing is to mix the raw materials and obtain an evenly distributed and homogeneous mixture. This involves combining the clay, water, and any additional ingredients, such as feldspar, kaolin, and quartz, to form a homogeneous mixture. There are several mixing techniques used in ceramics, and some of the most common methods of mixing in ceramics include the following:

- *Pug mill mixing:* The process of mixing in a pug mill involves feeding the raw materials into the mixing chamber, where they are blended by the rotating blades or augers until a uniform mixture is achieved. This process ensures and improves the plasticity and workability of the clay, making it easier to shape and form. Pug mills can also be used to recycle scrap clay, allowing it to be reused in the production process.

- *Ball milling*: Ball milling is a mechanical process used in ceramic production to grind and blend raw materials into a fine powder. The ball mill contains steel or ceramic balls or other grinding media that, when rotated, collide with the raw materials, reducing their particle size. This process is used to create homogenous, fine powders that can then be used in the production of ceramic products, such as tiles, tableware, and sanitary ware. The process can be conducted in either a wet or dry environment depending on the desired outcome. The length of the milling process can vary depending on the desired particle size and the size of the mill used.
- *Slurry mixing*: Slurry mixing in ceramics refers to the process of blending liquid and solid materials to form a homogeneous wet mixture with the desired consistency. This mixture, known as slurry, is commonly used in the production of ceramics. The length of the mixing process and the desired particle size of the slurry can vary depending on the specific application. However, the goal is typically to create a uniform mixture with a consistent texture and viscosity.
- *Blade mixing*: Blade mixing in ceramics refers to the process of blending and homogenizing materials using a rotating blade. The blade mixer consists of a rotating blade that is located within a mixing chamber. The raw materials are fed into the chamber and blended by the rotating blade, which homogenizes the mixture and improves its plasticity and workability. It is used in conjunction with other processes, such as pugmill mixing or ball milling, to further refine the mixture and achieve the desired particle size and consistency. Blade mixing is a relatively fast and efficient process compared to other mixing methods.
- *Forming*: Once the mixture is prepared, it can be shaped into the desired form using hands and simple tools, such as a rolling pin, a knife, or a potter's wheel, such as pouring the mixture into a mold and pressing it into the desired shape or pouring a liquid mixture of clay and water into a mold and allowing it to harden.

1.3.4.2 DRYING

The drying process is a crucial step that occurs between shaping and firing. The main objective of drying is to remove the excess moisture present in the ceramic material without causing cracks or other defects. The formed piece is dried to remove any excess moisture. The water evaporates into the air as the temperature of the ceramic material increases, which results in a reduction in the moisture content of the material. As the moisture content decreases, the ceramic material shrinks, which can cause stresses to develop within the material. If these stresses are too great, cracks can form, and the final product will be damaged. The drying process can be done either naturally or in a kiln. To minimize the risk of cracking and other defects, moisture removal is taken at a controlled rate by controlling the temperature and relative humidity of the surrounding air. In general, slow drying at low temperatures and high relative humidity is preferred to minimize the risk of cracking. In addition to controlling the drying environment, the use of drying aids such as surfactants can also help to control the rate of moisture removal and reduce the risk of cracking. Surfactants are surface-active agents that can reduce the surface tension of the water and help to control the rate of evaporation. This can reduce the risk of cracking by reducing the rate at which the ceramic material shrinks.

1.3.4.3 BISCUIT FIRING

Biscuit firing is the first stage of firing in the production of ceramics. It refers to the process of heating a formed ceramic piece, also known as a "biscuit," to a temperature that is sufficient to remove all moisture and organic matter and to partially sinter the clay particles together. Biscuit firing takes place at a temperature between 900 and 1100°C. The goal of biscuit firing is to produce a strong, stable ceramic piece that is ready for further processing, such as glazing or additional firing. The duration of the biscuit firing process depends on the size and shape of the piece, as well as the desired temperature and the type of clay used. Biscuit sets the foundation for the final properties and appearance of the ceramic product. Improper biscuit firing can result in cracking, warping, or other defects that can affect the final outcome of the piece.

1.3.4.4 GLAZING

Glazing is used to add a protective and decorative layer to the surface of a ceramic piece. It involves applying a thin coating of a mixture called a "glaze" to the surface of the piece, which is then fired at high temperatures to produce a hard, glass-like coating. Glazing can provide protection to the underlying ceramic material as well as enhance its appearance and durability. The glaze mixture typically consists of a combination of minerals, such as feldspar and kaolin, along with other materials such as colorants, opacifiers, and fluxes. The mixture is mixed into a slurry, which is then applied to the surface of the ceramic piece. Glazing can be done by brushing, pouring, dipping, or spraying the glaze slurry onto the surface of the piece.

1.3.4.5 GLAZE FIRING

Glaze firing refers to the process of firing a ceramic piece after it has been coated with a glaze mixture. It involves heating the piece to high temperatures, typically between 800 and 1200°C, in a kiln. The high temperature causes the glaze to melt and fuse to the surface of the ceramic piece, forming a hard, glass-like coating that is both protective and decorative. The firing process also causes chemical reactions to occur within the glaze, which can result in changes in color, texture, and other properties. The temperature and duration of the firing, as well as the atmosphere within the kiln, can greatly affect the final outcome of the piece. For example, a slow ramp-up to the firing temperature and a controlled cooling rate can help to reduce the risk of thermal shock and minimize cracking in the glaze. Additionally, the type of fuel used in the kiln and the presence of any elements in the atmosphere, such as carbon or sulfur, can affect the final color and appearance of the glaze.

1.3.4.6 COOLING

The cooling process in ceramics refers to the gradual decrease in temperature of a ceramic material as it transitions from a molten state to a solid state. The cooling rate, or the rate at which the temperature decreases, can have a significant impact on the microstructure, mechanical properties, and overall quality of the final ceramic product. Slow cooling can result in more uniform crystal structures and reduced residual stress, while rapid cooling can result in greater porosity and lower strength. The cooling rate, or the rate at which the ceramic is cooled, is carefully controlled to avoid thermal shock, cracking, or other defects. There are several cooling methods used in ceramics, including slow, rapid, or controlled cooling. Slow cooling involves gradually cooling the ceramic at a controlled rate in a kiln or furnace. This method results in a homogeneous microstructure and improved mechanical properties. During rapid cooling, ceramics are cooled quickly, either by immersing them in a cool bath or blowing cool air on them, resulting in high residual stresses and an altered microstructure. Controlled cooling is completed either by using a combination of slow and rapid cooling or by using a specific cooling profile.

1.4 COMPOSITES

Composite materials emerged in the mid-20th century as promising engineering materials with new opportunities for modern technology. Materials consisting of two or more components, with distinct properties and different components, can be referred to as composite materials. A few typical examples of composite materials are wood, clad metals, fiberglass, reinforced plastics, and cemented carbides. The constituents retain their original properties in composites and are usually described as a matrix, a continuous phase of composites and reinforcing material embedded as a discontinuous phase in the matrix. The components are selected, considering their mechanical and other properties, to meet specific application requirements. The reinforcement material is significantly stiffer and stronger than other composite components and serves as a primary load-bearing component. The reinforcement may be in the form of particles, short fibers, or continuous fibers. Some of the reinforcement materials utilized in composites are glass, aramid or kevlar, carbon, boron, or silicon carbide, for they exhibit lightweight properties, high modulus or strength, and are resistant to high temperatures. The matrix material suspends and binds the reinforcement material and hardens to control the final shape. Compared to reinforcement materials, the matrix materials are relatively weak and lack stiffness.[24] The matrix materials include polymers, minerals, or metals. Combining several components produces unique properties and a high strength–weight ratio of the composite (Figure 1.5). Another advantage of composite materials is that they can be manufactured in complex shapes and offer flexibility in design. It is also often assumed that the elastic resistance is uniform in the whole composite, resulting in differences in the distribution of stress and variations in local strain. There are many kinds of composite materials, including carbon-reinforced fiber plastic, glass fiber-reinforced aluminum, composites with carbon nanotubes, etc. Other types of composites include metal-matrix composites and ceramic-matrix composites, which are widely used in engineering applications.

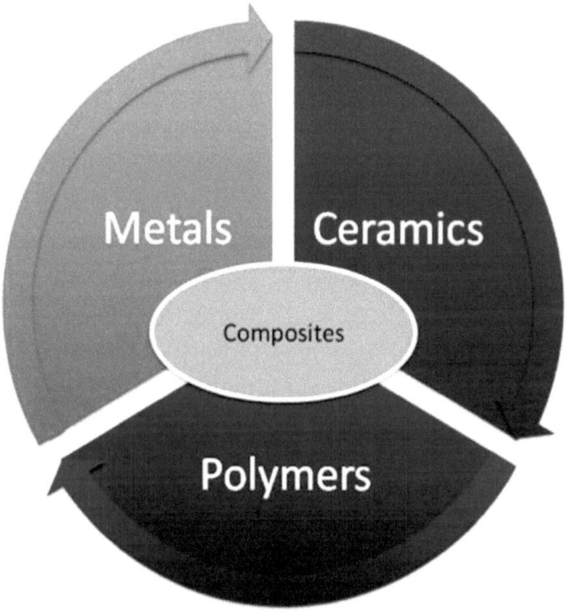

FIGURE 1.5 Evolution of various types of composites.

1.4.1 CHARACTERISTICS OF COMPOSITES

Composites are characterized by a reinforcement-resin mixture in a quasi-bidimensional material by the fiber mass fraction, defined as the ratio of the mass of reinforcement to the total mass of the composite, the fiber volume fraction, and the mass density of a ply, calculated as the ratio of the total mass to the volume of the composite. The other parameter utilized to characterize the composite is ply thickness. It is defined as the mass of the reinforcement material distributed per unit area of the composite. If h is defined as the ply thickness, then

$$h = total\,volume\,of\,the\,composite \times \frac{mass\,of\,fiber}{Fiber\,of\,volume \times density\,of\,fiber}$$

1.4.2 CLASSIFICATION OF COMPOSITES

Functional composites and structural composites are two broad categories of composite materials that differ in their primary purpose and design.

Structural composites are primarily designed for mechanical performance, while functional composites are designed for specific functional properties beyond mechanical performance.

1.4.2.1 STRUCTURAL COMPOSITES

Structural composites are designed primarily to provide high strength, resistance to deformation, stiffness, and durability to structures, such as buildings, bridges, aircraft, vehicles, and construction industries. The structural composite constituents are high-strength fibers, such as carbon fibers, glass fibers, or aramid fibers, embedded in a matrix material, such as epoxy resin. These composites are optimized for their mechanical properties, and their design is focused on ensuring that the material can withstand mechanical loads without failure.

1.4.2.2 FUNCTIONAL COMPOSITES

Functional composites, on the other hand, are designed to provide specific functional properties, such as thermal conductivity, electrical conductivity, or energy absorption. They are used in a wide range of applications, such as thermal management systems, electromagnetic shielding, and energy storage. They can be made from a wide variety of materials, including metals, ceramics, polymers, and composites. They are designed to have a unique combination of properties that cannot be achieved with individual materials, and they often incorporate multiple layers or regions with different compositions to achieve their desired function.

1.4.3 STRESS–STRAIN BEHAVIOR OF COMPOSITES

The composites exhibit a brittle structure and a nonlinear stress–strain curve before reaching the breaking stress depending on the constituents, volume fraction, orientation, arrangement of the fibers, and particle size of the matrix (Figure 1.6). These curves are different from those of individual materials and show maximum stress with the flexural strength of the material indicating a strong adhesive force between the fibers and matrix. Variations in elongation and tensile strength of the composites are due to polar attractive forces and van der Waals bonding interactions between the fillers and matrix.[25]

Most of the composites exhibit a yield drop and are fatigue-resistant. Their stress–strain behavior can also be tailored to achieve desired mechanical properties. For example, changing the fiber volume fraction, orientation, and arrangement can alter the stiffness, strength, and ductility of the composite. Similarly, changing the matrix material or modifying the interfacial properties between the fibers and matrix can also influence the stress–strain behavior of composites. The curve can be divided into three stages, known as elastic, plastic, and failure stages. In the elastic stage, the composite material behaves like a linear elastic material, and the stress–strain curve is linear. The slope of the curve represents the composite's stiffness and increases with an increase in fiber volume fraction. In the plastic stage, the composite material starts to deform plastically, and the slope of the curve decreases. This stage is characterized by a gradual increase in strain with a minimal increase in stress. The plastic behavior of composites is dependent on the type of matrix material, the volume fraction and orientation of the fibers, and the interfacial properties between the fibers and matrix. In the failure stage, the composite material experiences a significant drop in stress, leading to failure. The failure of composites can occur due to several reasons, such as fiber fracture, delamination, or matrix cracking. Several studies have investigated the stress–strain behavior of composites. For example, a study by Dwivedi et al.[26] investigated the effect of fiber volume fraction and fiber orientation on the mechanical properties of carbon fiber-reinforced polymer composites. The study found that an increase in fiber volume fraction resulted in higher stiffness and strength, while an increase in fiber orientation angle reduced the stiffness and strength. Another study by Huseynov et al.[27] and Wang et al.[28] investigated the effect of matrix material on the mechanical properties of glass fiber-reinforced polymer composites. The study found that the type of matrix material had a significant impact on the stress–strain behavior of the composites, with epoxy matrix composites having higher stiffness and strength than polyester matrix composites.

1.4.4 PROCESSING TECHNIQUES OF COMPOSITES

Composite processing refers to the methods used to fabricate composite materials into the desired shape and size. It depends on the specific properties desired in the final product, the size and shape of the part, and the production volume. Some of the common composite processing methods are detailed in the following section.

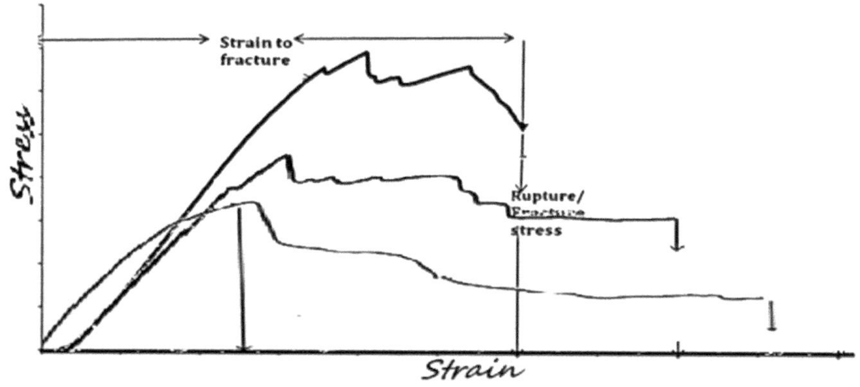

FIGURE 1.6 Stress–strain behavior of different composites.

1.4.4.1 HAND LAY-UP

This is a simple and low-cost method of producing composite parts by reinforcing fiber layers, such as glass or carbon, in a polymer matrix (such as epoxy) and placing them by hand in a mold. The mold is then subjected to heat, ultraviolet radiation, and pressure to cure the composite, causing the polymer matrix to solidify and bond the layers of reinforcement fibers together. This process is often used to manufacture large, complex-shaped composite parts for various applications such as the aerospace, sporting goods, and automotive industries. It is a relatively simple and low-cost process, making it an attractive option to be used to produce a wide range of composite parts of different shapes and sizes. However, the technique is a manual process, which means that the quality and consistency of the final product can be dependent on the operator's skill and attention to detail. This can result in inconsistencies in the final product and limited precision in terms of thickness and uniformity.

1.4.4.2 VACUUM INFUSION

This method is similar to hand lay-up, but instead of manually saturating the reinforcement material, the resin is drawn into the mold by vacuum pressure through a permeable surface. The process starts by laying out the reinforcement fibers in a mold, followed by the placement of a permeable material, such as a mesh or breathing bag, over the fibers. The resin is then poured over the permeable material, and the mold is sealed. A vacuum pump is then used to create a vacuum within the mold, which draws the resin through the permeable material and into the reinforcement fibers, wetting them completely. The mold is then subjected to heat and pressure to cure the composite, causing the polymer matrix to solidify and bond the layers of reinforcement fibers together. This process is known for its efficient and consistent distribution of resin, resulting in a high-quality composite part with a low void content.[29] It is often used in high-volume production runs for applications such as the marine, wind energy, and transportation industries.

1.4.4.3 COMPRESSION MOLDING

This method involves heating a preimpregnated (prepreg) reinforcement material in a mold and applying pressure to compact and cure the material. In this technique, reinforcement fibers and resin are placed into a mold and subjected to heat and pressure. The mold is heated to the temperature required to activate the curing process of the resin, and pressure is applied to consolidate the reinforcement fibers and distribute the resin evenly throughout the mold. The mold is then cooled, allowing the composite to solidify and bond the fibers together.[30] Compression molding is known for producing composite parts with high strength and uniform thickness, making it well-suited for high-volume production of parts that require consistent quality. The process is commonly used for producing a wide range of composite products, such as automotive parts, electrical components, and consumer goods.[31] Its main advantage is that it can produce high-quality parts in large quantities, which makes it an economical option for mass production of composites.

1.4.4.4 RESIN TRANSFER MOLDING (RTM)

RTM is a composite material processing technique that involves injecting a liquid resin into a closed mold containing reinforcement fibers under pressure. The process starts by placing the reinforcement fibers into a mold, followed by the placement of a sealed, inflatable bag over the fibers. The resin is then injected into the mold through ports under pressure, and the resin is distributed evenly throughout the reinforcement fibers through the action of the bag. The mold is then subjected to heat and pressure to cure the composite, causing the polymer matrix to solidify and bond the reinforcement fibers together. RTM is well-suited for producing high-quality composite parts with low void content, uniform thickness, and precise detail.[32] It is often used for producing complex-shaped parts for high-performance applications, such as in the aerospace and sporting goods industries. The process has the advantage of producing parts with precise and consistent quality in a relatively short cycle time, making it well-suited for high-volume production of composite parts.

1.4.4.5 PULTRUSION

Pultrusion is a composite material processing technique that involves pulling reinforcement fibers through a bath of polymer resin and then into a heated die, where the resin cures and solidifies, bonding the fibers together. The fibers are typically continuous strands, such as fiberglass or carbon, which are pulled through the resin bath and into the die, where they are shaped into the desired profile. The cured composite part is then cooled and removed from the die. Pultrusion is a continuous process that allows for the production of long, continuous composite parts with a uniform cross-section and is used for producing building materials, structural profiles, and reinforcing bars, among others. Its main advantage is that it allows for the efficient production of long, uniform composite parts with a consistent cross-section, making it an economical option for mass production of composites.

1.4.4.6 FILAMENT WINDING

Filament winding is a composite material processing technique where reinforcement fibers, such as fiberglass or carbon, are wound onto a mandrel in a specific pattern to form a composite structure.[33] The fibers are impregnated with resin, either by being dipped into a resin bath or by having resin applied to them as they are being wound. The mandrel is then rotated, and the fibers are wound onto it in a predetermined pattern, determined by a computer-controlled winding machine. The composite part is then cured, bonding the fibers and resin together to form a solid structure. Filament winding is utilized in producing pipes, pressure vessels, and sporting goods. Its main advantage is that it allows for the production of composite parts with a high degree of precision and uniformity as the winding pattern can be controlled by a computer.

1.4.4.7 INJECTION MOLDING

It is a manufacturing process where molten material, typically a thermoplastic or thermoset polymer, is injected into a mold cavity, where it cools and solidifies to form a solid part. Injection molding can also be used for composite materials, where reinforcement fibers, such as fiberglass or carbon, are added to the polymer matrix to improve its mechanical properties. The reinforcement fibers are typically chopped and mixed with the polymer resin before being injected into the mold. The mold is then closed and subjected to pressure, which helps to consolidate the reinforcement fibers and distribute the resin evenly throughout the mold. The composite part is then cured, bonding the fibers and resin together to form a solid structure. Injection molding is well-suited for high-volume production of composite parts with complex shapes and tight tolerances as the process allows for precise control over the fiber placement and resin distribution. Its main advantage is that it allows for the efficient production of high-quality composite parts with complex shapes in a single operation, making it an economical option for mass production of composites.

1.5 CONCLUSIONS

Engineering materials have impacted the daily lives of humans by enabling technologies ranging from automobiles to lasers and medical equipment utilized in pathological labs and surgeries. The chapter focuses on the classification, characterization, processing technology, and mechanical properties of these materials. Polymers are lightweight, extremely flexible, and low-density materials widely used as thermal and electric insulators. Their properties are affected by the degree of branching, crystallinity, the kind and amount of additives used, temperature and method of processing, and cooling rate. Depending on their characteristics and applications and their response to mechanical forces at elevated temperatures, polymers are usually classified into three categories. Ceramics find applications in the aerospace industry for the fabrication of heat shields, engine parts, and other critical components. Except for these applications, they exhibit piezoelectricity, ferroelectricity, or superconductivity, have unique electrical, magnetic, or optical properties, and are useful in a wide range of applications. The study will lead to the development of new and improved materials for various industries and optimize the performance of existing materials to increase the lifespan and reliability of products where these materials are utilized. This study is also a source of benefit to advance technology, improve performance and product lifespan, and create economic and environmental benefits. The knowledge gained from studying materials can be used to solve real-world problems.

KEYWORDS

- **polymers**
- **ceramics**
- **composites**
- **sintering**
- **mechanical and structural properties**
- **stress–strain behavior**

REFERENCES

1. Carothers, W. H. Studies on Polymerization and Ring Formation. I. An Introduction to the General Theory of Condensation Polymers. *J. Am. Chem. Soc.* **1929,** *51* (8), 2548–2559. https://doi.org/10.1021/ja01383a041.
2. Piccolo, A.; Conte, P. Molecular Size of Humic Substances, Supramolecular Associations Versus Macromolecular Polymers. *Adv. Environ. Res.* **2000,** *3* (4), 508–521.
3. Tan, L. J.; Zhu, W.; Zhou, K. Recent Progress on Polymer Materials for Additive Manufacturing. *Adv. Funct. Mater.* **2020,** *30* (43), 2003062. https://doi.org/10.1002/adfm.202003062.
4. AlMaadeed, M. A. A.; Ponnamma D.; El-Samak, A. A. Chapter 1 - Polymers to Improve the World and Lifestyle: Physical, Mechanical, and Chemical Needs. In *Polymer Science and Innovative Applications*; AlMaadeed, M. A. A., Ponnamma, D., Carignano, M. A., Ed.; Elsevier, 2020; pp 1–19. ISBN: 9780128168080, https://doi.org/10.1016/B978-0-12-816808-0.00001-9.
5. Peterson, A.; Östergren, I.; Lotsari, A.; Venkatesh, A.; Thunberg, J.; Ström, A.; Rojas, R.; Andersson, M.; Berglund, L. A.; Boldizar, A.; Müller, C. Dynamic Nanocellulose Networks for Thermoset-like yet Recyclable Plastics with a High Melt Stiffness and Creep Resistance. *Biomacromolecules* **2019,** *20* (10), 3924–3932. https://doi.org/10.1021/acs.biomac.9b00993
6. Kulkarni, G. S. Chapter 1 - Introduction to Polymer and Their Recycling Techniques. In *Plastics Design Library, Recycling of Polyurethane Foams*; Thomas, S., Rane, A. V., Kanny, K., Abitha, V. K., Thomas, M. G., Ed.; William Andrew Publishing, 2018; pp 1–16. ISBN 9780323511339, https://doi.org/10.1016/B978-0-323-51133-9.00001-2.

7. Scott, S. M.; Ali, Z. Fabrication Methods for Microfluidic Devices: An Overview. *Micromachines* **2021,** *12*, 319. https://doi.org/10.3390/mi12030319
8. Patil, A.; Patel, A.; Purohit, R. An Overview of Polymeric Materials for Automotive Applications. *Mater. Today Proc.* **2017,** *4* (2), 3807–3815, https://doi.org/10.1016/j.matpr.2017.02.278.
9. Boubakri, A.; Haddar, N.; Elleuch, K.; Bienvenu, Y. Impact of Aging Conditions on Mechanical Properties of Thermoplastic Polyurethane. *Mater. Design* **2010,** *31* (9), 4194–4201. https://doi.org/10.1016/j.matdes.2010.04.023
10. Boubakri, A.; Guermazi, N.; Elleuch, K.; Ayedi, H. F. Study of UV-Aging of Thermoplastic Polyurethane Material. *Mater. Sci. Eng.* A **2010,** *527* (7–8), 1649–1654. https://doi.org/10.1016/j.msea.2010.01.014.
11. Martinez, R. V.; Glavan, A. C.; Keplinger, C.; Oyetibo, A. I.; Whitesides, G. M. Soft Actuators and Robots that Are Resistant to Mechanical Damage. *Adv. Funct. Mater.* **2014,** *24*, 3003–3010. https://doi.org/10.1002/adfm.201303676
12. Dang, C.; Peng, F.; Liu, H.; Feng, X.; Liu, Y.; Hu, S.; Qi, H. Facile Solvent-Free Synthesis of Multifunctional and Recyclable Ionic Conductive Elastomers from Small Biomass Molecules for Green Wearable Electronics. *J. Mater. Chem. A* **2021,** *9* (22), 13115–13124. http://dx.doi.org/10.1039/D1TA01659H
13. Tjong, S. C.; Ke, Y. C. Fracture Toughening Behavior of Elastomer Modified Poly-phenylene Ether/Polyamide Blends. *Euro. Polym. J.* **1998,** *34* (11), 1565–1570. https://doi.org/10.1016/S0014-3057(98)00012-3.
14. McKeen, L. W. 3 - Plastics Used in Medical Devices. In *Plastics Design Library, Handbook of Polymer Applications in Medicine and Medical Devices*; Modjarrad, K., Ebnesajjad, S., Ed.; William Andrew Publishing, 2014; pp 21–53. https://doi.org/10.1016/B978-0-323-22805-3.00003-7.
15. Dainton, F. S.; Ivin, K. J. Some Thermodynamic and Kinetic Aspects of Addition Polymerisation. *Quart. Rev. Chem. Soc.* **1958,** *12* (1), 61–92. http://dx.doi.org/10.1039/QR9581200061
16. Colombani, D.; Chaumont, P. Addition-Fragmentation Processes in Free Radical Polymerization. *Prog. Polym. Sci.* **1996,** *21* (3), 439–503. https://doi.org/10.1016/0079-6700 (95)00024-0.
17. Greene, J. P. 16 - Compression Molding. In *Plastics Design Library, Automotive Plastics and Composites*; Greene, J. P., Ed.; William Andrew Publishing, 2021; pp 265–278. https://doi.org/10.1016/B978-0-12-818008-2.00015-5
18. Riley, A. 14 - Plastics Manufacturing Processes for Packaging Materials. In *Packaging Technology*; Emblem, A.; Emblem, H., Eds.; Woodhead Publishing, 2012; pp 310–360. https://doi.org/10.1533/9780857095701.2.310
19. Chavali, M. S.; Nikolova, M. P. Metal Oxide Nanoparticles and their Applications in Nanotechnology. *SN Appl. Sci.* **2019,** *1*, 607. https://doi.org/10.1007/s42452-019-0592-3
20. Grigoriev, S. N.; Fedorov, S. V.; Hamdy, K. Materials, Properties, Manufacturing Methods and Cutting Performance of Innovative Ceramic Cutting Tools−A Review. *Manufacturing Rev.* **2019,** *6*, 19. https://doi.org/10.1051/mfreview/2019016
21. Raghavendra, S. S.; Jadhav, G. R.; Gathani, K. M.; Kotadia, P. Bioceramics in Endodontics - A Review. *J. Istanb. Univ. Fac. Dent.* **2017,** *51* (3 Suppl 1), S128–S137. DOI: 10.17096/jiufd.63659; PMID: 29354316; PMCID: PMC5750835.
22. Choi, A. H.; Conway, R. C.; Cazalbou, S.; Ben-Nissan, B. 3 - Maxillofacial Bioceramics in Tissue Engineering: Production Techniques, Properties, and Applications. In

Biomaterials, Fundamental Biomaterials: Ceramics; Thomas, S., Balakrishnan, P., Sreekala, M. S., Eds.; Woodhead Publishing, 2018; pp 63–93. https://doi.org/10.1016/B978-0-08-102203-0.00003-2

23. Macan, J.; Gajović, A.; Ivanković, H. Porous Zirconium Titanate Ceramics Synthesized by Sol–Gel Process. *J. European Ceram. Soc.* **2009,** *29* (4), 691–696. https://doi.org/10.1016/j.jeurceramsoc.2008.07.001
24. Hussain, M.; Imad, A.; Nawab, Y.; Saouab, A.; Herbelot, C.; Kanit, T. Effect of Matrix and Hybrid Reinforcement on Fibre Metal Laminates under Low–Velocity Impact Loading. *Compos. Struct.* **2022,** *288*, 115371. https://doi.org/10.1016/j.compstruct.2022.115371.
25. Cécile, A. C.; Chazot, A.; Hart, J. Understanding and Control of Interactions between Carbon Nanotubes and Polymers for Manufacturing of High-Performance Composite Materials. *Compos. Sci. Technol.* **2019,** *183*, 107795. https://doi.org/10.1016/j.compscitech.2019.107795.
26. Dwivedi, V.; Mandla, P.; Gupta, D.; Chandel, R.; Sharma, A. Simulation of Tensile Behaviour of Carbon Fiber Reinforced Polymer Matrix Composite. *Mater. Today Proc.* **2023.** https://doi.org/10.1016/j.matpr.2023.01.422.
27. Huseynov, O.; Hasanov, S.; Fidan, I. Influence of the Matrix Material on the Thermal Properties of the Short Carbon Fiber Reinforced Polymer Composites Manufactured by Material Extrusion. *J. Manuf. Process.* **2023,** *92*, 521–533. https://doi.org/10.1016/j.jmapro.2023.02.055
28. Wang, A.; Liu, X.; Yue, Q.; Xian, G. Tensile Properties Hybrid Effect of Unidirectional Flax/Carbon Fiber Hybrid Reinforced Polymer Composites. *J. Mater. Res. Technol.* **2023**. https://doi.org/10.1016/j.jmrt.2023.03.078
29. Sunilpete, M. A., Cadambi, R. M. Development of Cost Effective Out-of-Autoclave Technology – Vacuum Infusion Process with Tailored Fibre Volume Fraction. *Mater. Today Proc.* **2020,** *21* (2), 1293–1297. https://doi.org/10.1016/j.matpr.2020.01.165.
30. Rajeshkumar, G.; Balaji, B. A.; Seshadri, S. A. 10 - Compression and Injection Molding Techniques. In *Composites Science and Engineering, Cellulose Fibre Reinforced Composites*; Ramnath, R. A., Sanjay, M. R., Siengchin, S., Fiore, V., Eds.; Woodhead Publishing, 2023; pp 165–181. https://doi.org/10.1016/B978-0-323-90125-3.00018-5
31. Park, C. H.; Lee, W. I. 3 - Compression Molding in Polymer Matrix Composites. In *Composites Science and Engineering, Manufacturing Techniques for Polymer Matrix Composites (PMCs)*; Advani, S. G., Hsiao, K. T., Ed.; Woodhead Publishing, 2012; pp 47–94. https://doi.org/10.1533/9780857096258.1.47.
32. Bodaghi, M.; Costa, R.; Gomes, R.; Silva, J.; Correia, N.; Silva, F. Experimental Comparative Study of the Variants of High-Temperature Vacuum-Assisted Resin Transfer Moulding. *Compos. A Appl. Sci. Manuf.* **2020,** *129*, 105708. https://doi.org/10.1016/j.compositesa.2019.105708.
33. Azeem, M.; Ya, H. H.; Alam, M. A.; Kumar, M.; Stabla, P.; Smolnicki, M.; Gemi, L.; Khan, R.; Ahmed, T.; Ma, Q.; Sadique, M. R.; Mokhtar, A. A.; Mustapha, M. Application of Filament Winding Technology in Composite Pressure Vessels and Challenges: A Review. *J. Energy Storage* **2022,** *49*, 103468. https://doi.org/10.1016/j.est.2021.103468

CHAPTER 2

Multiferroics, Ferroelectric, and Piezoelectric Materials

MANDEEP SINGH

Department of Physics, Guru Nanak Dev University, Amritsar, India

ABSTRACT

The piezoelectric materials are those which develop surface charges (acquire electric dipole moment) of opposing polarities across them upon application of external stress. A typical example is a quartz (often used for transducer applications) crystal. High-end piezoelectrics find applications in micro and nanoelectromechanical systems (MEMS and NEMS) as well as in sensors and actuators. The stress response of piezoelectrics is typically linear in nature. The ferroelectrics differ from piezoelectrics in the sense that they possess electric dipole moment even in the unstrained state. Barium titanate (BT), lead titanate (PT), and lead zirconate titanate (PZT), etc., are some typical examples of ferroelectrics. In addition to piezoelectric applications, they are useful in multilayer capacitors and energy storage applications. The stress response of any given material is inherently related to its crystal structure. From crystal structural view point, non-existence of inversion center is the key to piezoelectricity (and also ferroelectricity). Out of a total of 32 point groups (to which a material may belong), piezoelectricity is supported by only 20 of them. 10 out of these 20 point groups (supporting piezoelectricity) support ferroelectricity. In nutshell, all ferroelectrics are also piezoelectrics but not vice versa. A single ferroelectric order (temporary or permanent) is what characterizes both piezoelectric and ferroelectric materials. Multiferroics are special kind of materials which exhibit co-existence of more than one ferroic (out of magnetic (anitferromagnetic/ferromagnetic),

Contemporary Advancements in Materials Technology: Novel Solutions to Complex Problems.
Shrikaant Kulkarni, Vipul Srivastava, & P. William (Eds.)

ferroelectric and ferroelastic) states/orders. The potential for existence of coupling among different ferroic orders in ferrites makes them suitable for exciting applications, such as tunable μ-wave phase shifter, dynamic random access memories (DRAM), spintronic devices, and magnetic field sensors.

2.1 PEROVSKITES AND THEIR CRYSTAL STRUCTURE

The word perovskite refers to those compounds which have their chemical composition identical to $CaTiO_3$ (CT). Historically, CT was the first perovskite to be discovered (in 1839 by Gustav Rose). The term "Perovskite" was coined by him in honor of Count Lev Aleksevich Von Perovski, who was a Russian mineralogist.[1] The general formula of perovskites is thus ABO_3 where A, B, and O represent larger in size bivalent metal cation, relatively smaller in size tetravalent metal cation and oxygen anion, respectively. The oxide perovskites (with tetragonal unit cell) are base materials for well-known piezoelectric and ferroelectric materials (both lead and lead-free). The crystal structure of oxide perovskites is quite stable in a very broad range of temperature. They are hence useful for a number of dielectric, piezoelectric, ferroelectric, and magnetic applications.[2–5]

Figure 2.1 shows the representative crystal structure of an ideal centro-symmetric (cubic $BaTiO_3$) perovskite. The corner positions (A-sites) are usually occupied by cations belonging to alkali or rare-earth group of elements. The A-site cations are surrounded by 12 oxygen anions. The body center position (B-site) is usually occupied by Ti (4+) cation. Eight oxygen anions octahedrally surround B-site cations. The face center positions are occupied by oxygen anions.

The corner-sharing octahedra are quite flexible. As a consequence, a number of different cations can be accommodated at the A and B-sites in the perovskite structure. The flexibility leads to tetragonal, rhombohedral, ortho-rhombic, or monoclinically distorted crystal structures in perovskites. The tetragonal distortion (Ti^{4+} displaced from the body center) is most significant from the piezoelectric/ferroelectric/multiferroic properties point of view and is shown in Figure 2.2.[6,7]

2.2 PIEZOELECTRICS

These are the materials which develop electrical voltage in response to external pressure and vice versa. From a symmetry perspective, crystals can be classified into 32 point groups. Out of a total of 32 point groups (to which a material may belong), piezoelectricity is supported by only 20 of them.

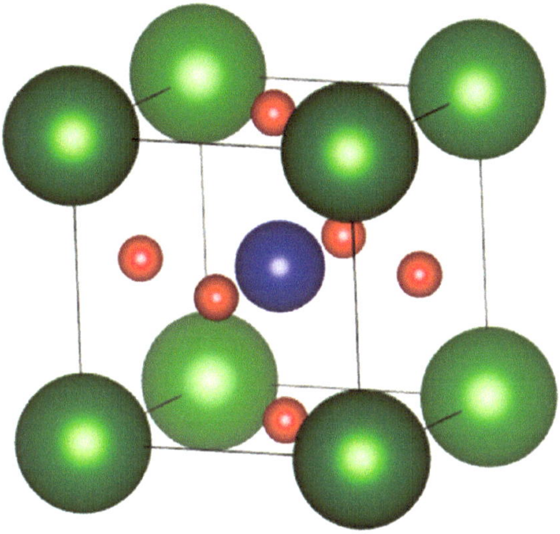

FIGURE 2.1 The centrosymmetric $BaTiO_3$ unit cell. The green (at corner), blue (at body center), and red balls (face centers) represent Ba^{2+}, Ti^{4+}, and O^{2-} ions, respectively.

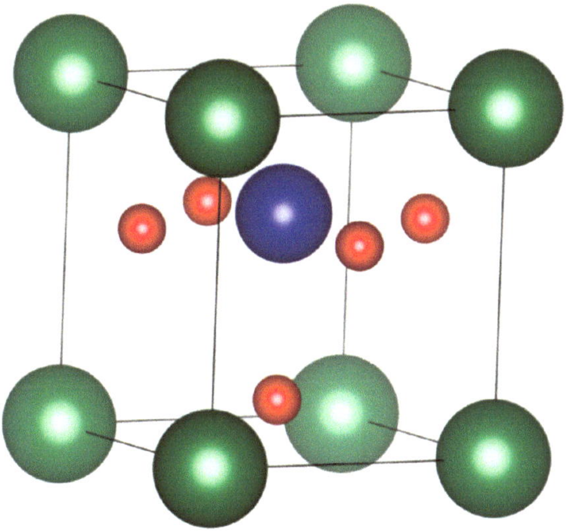

FIGURE 2.2 Distorted cubic (tetragonal $PbTiO_3$) unit cells showing displaced B-site cation from center. The shift of the Ti^{4+} cation (blue ball) away from body center is exaggerated for clarity. The actual fractional co-ordinates of Pb^{2+} (green balls), Ti^{4+} (blue ball), and O^{2-} (red balls), respectively are (0, 0, 0) and (0.5, 0.5, 0.541) and ((0.5, 0, 0.6208), (0.5, 0.5, 0.1148)).

2.2.1 HISTORY OF PIEZOELECTRICS

The piezoelectric effect was discovered in the year 1880 by Pierre and Jacques Curie (brothers) in quartz (SiO_2) single crystal. They found that electrical voltage is developed across quartz crystal upon application of pressure. The word piezo is rooted in Greek which means pressure. The converse piezoelectric effect (electric field-induced mechanical strain) is also displayed by piezoelectric materials. Gabriel Lippman discovered this phenomenon in 1881. Ultrasonic technology (SONAR) was invented nearly 30 years after discovery of ferroelectricity. $BaTiO_3$ (BT) and $PbZrTiO_3$ (PZT) are two historically important piezoelectrics to be discovered in addition to various ternary perovskite ($Pb(Zn_{1/3}Nb_{2/3})O_3$, $Pb(Mg_{1/3}Nb_{2/3})O_3$, $Pb(Co_{1/3}Nb_{2/3})O_3$, etc.) and non-perovskite ($LiTaO_3$ (LT) and $LiNbO_3$ (LN) and) piezoelectric systems. In addition, composite piezoelectric systems (with polyvinyl difluoride (PVDF)) and lead-free piezoelectric systems, such as (Bi, Na) TiO_3 (BNT), and (Na, K) NbO_3 (NKN) are also gaining significant attention.[8,9]

2.2.2 SYNTHESIS METHODS/TECHNIQUES

The experimental techniques which are used for synthesizing piezoelectric materials are common to those which are used for ferroelectric and multiferroic material synthesis. Solid-state route is the traditional method of choice for the synthesis of piezoelectric/ferroelectric/multiferroic materials. This method involves stoichiometric mixing and ball milling (in usually propanol medium) of either metal oxides or their carbonate powders. A high-temperature treatment (termed calcination, temperature ≈ 900–1100°C) is usually given to the dried mixture in order to allow cationic interdiffusion.[10] The heat-assisted cation interdiffusion leads to phase formation. Calcination treatment is often followed by sintering treatment (temperature ≈ 1100–1300°C). The purpose of the sintering treatment is to enhance sample density without causing phase modification. Simplicity and low cost are the main advantages of solid-state route. In addition, a variety of chemical methods (also termed liquid phase methods) are also increasingly employed for the synthesis task. Among the chemical synthesis methods, autocombustion methods, sol–gel method and co-precipitation method are typically used.[11] The chemical methods offer advantages such as (1) lower temperature of synthesis, (2) better stoichiometry control, (3) better size and compositional homogeneity, and (4) better purity over solid-state route. Sekhar et al. have listed some piezoelectric materials which are recently developed using additional techniques of synthesis. These include (1) Bridgeman technique, (2) Dice and fill technique, (3) molten salt technique, (4) solution coating technique, (5) electro spinning technique, and (6) polymeric precursor techniques.[12]

2.2.3 ORIGIN OF PIEZOELECTRICITY: MICROSCOPIC MECHANISM

Born in the year 1933 was the first to propose the mechanism of piezoelectricity in crystals. As per him, the net displacement of cations and anions from their mean position is responsible for piezoelectricity. However, his theory was unable to account for the observed disagreement between the calculated and experimentally determined values, particularly in crystals having both covalent and ionic bonds. Presently, the piezoelectric response of not only simple but complex solid solution (belonging to perovskite family) piezoelectrics is best described using theoretical calculations based on first-principles approach. The slope of strain (determined from free energy, $F = U–P_i E_i$) vs the electric field is used to determine various piezoelectric coefficients (d_{ij}). U (internal energy) and P_i (polarization components) are determined from calculations based on first principles. Here, E_i are external electric field components. The theory proposed by Landau, Ginzburg, and Devonshire can also adequately describe piezoelectricity in ferroelectrics. This theory ascribes piezoelectricity to electrostriction resulting from spontaneous polarization. In materials like $SrTiO_3$, the piezoelectricity is hypothesized to be of quantum origin. In some of the materials (like $AgNa (NO_2)_2$), multiple processes (electronic, ionic, and orientational polarizations) can combine to give piezoelectricity. The above-mentioned mechanisms of piezoelectricity are termed intrinsic. In addition, extrinsic factors (such as domain wall displacement) may also be important in some materials. In materials (such as PZT), the piezoelectric response is dramatically enhanced in compositions nearer to morphotropic phase boundary (MPB).[13]

2.2.4 POLARIZATION STRESS RELATIONSHIP

Polarization components (P_i), piezoelectric modulus components (d_{ijk}, which is a third rank tensor) and applied mechanical stress components (X_{jk}, which is a second-rank tensor) are related as:

$$P_i = d_{ijk} X_{jk} \tag{2.1}$$

Ideally, a second- and third-rank tensor in three-dimensions will have a total of 9 and 27 components. However, the stress tensor (X_{jk}) is symmetric with respect to its indices "j" and "k." As a consequence, modulus tensor (d_{ijk}) also becomes symmetric with respect to its indices "j" and "k." This symmetry restricts the possibility of maximum number of components to 18 from 27. The symmetry of "j" and "k" indices in X_{jk} and d_{ijk} implies that the tensor relation (2.1) can be written in the simple matrix from as follows:

$$P_i = d_{ij} X_j \tag{2.2}$$

In this representation, P_i and X_j can be regarded as three-dimensional vectors and d_{ij} as the 6 × 3 matrix has 18 components (maximum in crystals of lowest symmetry).[14] Out of these 18 elements, crystal symmetry will further restrict the number of independent components. A very nice matrix representation of total number of independent components (of d_{ij} matrix) for different point groups is given by Kim et al.[15] The point group "1" (lowest symmetry crystal) has 18 independent components. On the other hand, the point groups $\bar{4}3$ m and 23 (highest symmetry crystals) have only "1" independent component. The most interesting point group (from piezoelectric/ferroelectric/multiferroic perspective), that is, 4 mm has three independent components which are $d_{31} = d_{32}$, d_{33} and $d_{15} = d_{24}$.

2.2.5 FIGURES OF MERIT OF A PIEZOELECTRIC MATERIAL/SENSOR

The set of d_{ij}'s (piezoelectric modulus) for a given crystal symmetry along with some other factors are collectively termed figures of merit. Some of the important ones have been described below.

2.2.5.1 d_{11}

This parameter (and also d_{22} and d_{33}) is used to characterize piezoelectric effect in the longitudinal (also called L-effect) direction. This effectively means that the electrical polarization in the crystal is induced in the direction of applied mechanical stress. Thus, a modulus component with identical indices (11, 22, 33, etc.), describes piezoelectric effect along one of the three orthogonal directions (i.e., x, y, z).

2.2.5.2 d_{12}

This parameter (and also d_{13}, d_{31}, d_{21}, d_{32}) is used to characterize the piezoelectric effect in the transverse (also called T-effect) direction. In the transverse case, electric polarization is developed across axis 1 (x-axis) upon the application of mechanical stress along a perpendicular axis 2 (y-axis). These elements therefore characterize the piezoelectric response in a direction perpendicular to the direction of stress application.

2.2.5.3 d_{14}

This parameter is used to characterize piezoelectric response arising because of shear stresses. This effectively means that the piezoelectric response is developed in direction 1 (x-axis) when two opposing forces (acting at opposite faces) are applied in direction 2 (y-axis). The kind of piezoelectric response is typically observed in ferroelectric ceramics and it is of great engineering utility.

It is noteworthy that d_{ij}, P_i, and X_i have got the dimensions of C/N, C/m^2, and N/m^2.

2.2.5.4 PIEZOELECTRIC VOLTAGE CONSTANT (g_{ij})

It (g_{ij}) is defined as the ratio of electric field developed across a piezoelectric to applied mechanical stress. Alternatively, it is the ratio of mechanical strain developed inside a piezoelectric to applied electric displacement. Numerically,

$$g_{ij} = \frac{E_i}{T_j} = \frac{S_i}{D_j} \tag{2.3}$$

Here, the subscript "i" and "j" represent the direction of response generation and cause application, respectively. It is related to d_{ij} as

$$g_{ij} = \frac{d_{ij}}{\varepsilon T_{ii}} \tag{2.4}$$

Here, "ε" represents material permittivity. It is a very important parameter to determine the suitability of a piezoelectric for energy harvesting and sensing applications.[16,17]

2.2.5.5 ELECTROMECHANICAL ENERGY COUPLING FACTOR (K)

It is a measure of the efficiency of conversion of mechanical/electrical energy to electrical/mechanical energy in a piezoelectric. It is given by the following expression:

$$K = \sqrt{\frac{E_{m-e}}{E_m}} = \sqrt{\frac{E_{e-m}}{E_e}} \tag{2.5}$$

Here E, e, m represent, energy, electrical, and mechanical, respectively. The energy transformation paths are represented by m-e and e-m. Higher K values are desirable.

2.2.5.6 DIELECTRIC CONSTANT (ε_r)

Numerically it is the ratio of material permittivity to the permittivity of free space. Higher the value of ε_r, better are the insulating properties of the given piezoelectric material. However, very high value of ε_r leads to decreased sensitivity of the piezoelectric sensor.

2.2.5.7 DIELECTRIC LOSS (tan δ)

It is the energy which is lost because of the heating of dielectrics. For high-performing piezoelectrics, tanδ should be small.

2.2.5.8 MECHANICAL QUALITY FACTOR (Q_m)

It is defined as:

$$Q_m = \frac{E_{m-res}}{E_{m-cons}} \tag{2.6}$$

Here “res” and “cons” stand for the mechanical energy stored in a resonantly vibrating crystal and the energy consumed per cycle. For high-performing piezoelectric materials, high Q_m values are desirable.[18,19]

2.2.6 APPLICATIONS OF PIEZOELECTRICS

The applications of piezoelectric materials are dependent upon the nature (direct, inverse, etc.) of piezoelectric phenomenon. The phenomenological classification is as follows:

2.2.6.1 APPLICATIONS BASED ON DIRECT PIEZOELECTRIC EFFECT

The functioning of devices, such as electric power generators, deformation sensors, acoustic receivers, microphones, and adapters, etc., is based upon direct piezoelectric phenomenon.

2.2.6.2 APPLICATIONS BASED ON INVERSE PIEZOELECTRIC EFFECT

Devices, such as piezoelectric motors, ultrasonic/sound radiators, and voltage and current piezoelectric transformers exploit inverse piezoelectric phenomenon for their functioning.

2.2.6.3 APPLICATIONS BASED ON PIEZOELECTRIC RESONANCE

It is the natural frequency of oscillation of a piezoelectric at which it is (1) vibration amplitude and (2) energy inter-conversion (electrical to mechanical and vice versa) efficiency are both maximum. Devices such as piezoelectric filters, a wide variety of resonant sensors (such as humidity, pressure, chemical composition, and temperature, etc.), voltage and current piezoelectric transformers and resonance frequency balancers are piezoelectric resonance phenomenon based.

2.2.6.4 APPLICATIONS BASED ON ELECTROSTRICTION

The external electric field-driven expansion and/or contraction of a piezoelectric/solid-state material is termed electrostriction. The deformation is the consequence of slight atomic displacements of their constituent ions (in opposite directions) due to force exerted by the electric field. The functioning of nonlinear acoustic devices, electrically controllable filters, and non-hysteresis μ-positioners is based upon the electrostriction phenomenon.[14]

2.2.7 PIEZOELECTRIC MATERIALS

Broadly, the piezoelectric materials may be divided into three categories (1) lead based, (2) non-lead based, and (3) composites. However, some biological and polymer materials also exhibit piezoelectric properties. The lead-based piezoelectrics though possess best piezoelectric/ferroelectric properties but their usage is recently discouraged on account of toxicity of lead.

2.2.7.1 LEAD BASED PIEZOELECTRICS

Commercially and due to complications involved in synthesizing high-quality single crystals, lead titanate ($PbTiO_3$ (PT)) in its pure form is never used as a piezoelectric. However, some of its solid solutions (with other materials) exhibit brilliant piezoelectric properties. It therefore serves as the base material for lead-based piezoelectrics. It is ferroelectric at room temperature and transforms from tetragonal (lattice parameters a = 3.902 Å & c = 4.156 Å, c/a = 1.063, space group P4mm) to a non-ferroelectric cubic phase (space group *Pm-3m*) at ≈493°C [9, 20]. d_{33}, d_{31}, and d_{15} values ≈117 pC N^{-1}, −25 pC N^{-1}, and 65 pCN^{-1} for PT have been reported by Gavrily et al.[21]

Lead zirconate titanate ($PbZr_{1-x}Ti_xO_3$, PZT) is the most significant PT-based piezoelectric/ferroelectric material. It was discovered at the Tokyo Institute of Technology in 1950s. PZT became a popular piezoelectric material after the extensive work of Bernard Jaffe.[22] It is electromechanical coupling coefficient and piezoelectric constant values are ≈ 0.69 and 360 pC N^{-1}, respectively.[23] It has a Curie temperature ≈350°C, and as a consequence, it also serves as a high-temperature piezoelectric.[24] In addition to being stable in humid and chemical environments, it is also mechanically robust. It has been used as (1) a sensor in reactive environments and (2) energy harvesting requiring large loads applications.[25] It exhibits enhanced piezoelectric response at MPB. PZT in pure form is hardly used for commercial applications. Often dopants are used to modify its response. The nature of dopants leads to its categorization into hard and soft PZT.

There are also some lead-based ternary piezoelectric materials which were developed by various commercial companies in the 1960s. Some of these materials are solid solutions of PZT with $PbZn_{1/3}Nb_{2/3}O_3$, $PbSb_{1/2}Sn_{1/2}O_3$, $PbCo_{0.5}W_{0.5}O_3$, $PbMg_{0.5}W_{0.5}O_3$, $PbMn_{1/3}Sb_{2/3}O_3$, $PbCo_{1/3}Nb_{2/3}O_3$, $PbMn_{1/3}Nb_{2/3}O_3$, $PbNi_{1/3}Nb_{2/3}O_3$, $PbMg_{1/3}Nb_{2/3}O_3$, etc. The piezoelectric performance of these mentioned ternary materials is much superior to binary PZT.

2.2.7.2 NON-LEAD PIEZOELECTRICS

The first ever lead-free perovskite piezoelectric/ferroelectric to be discovered by scientists (during Second World War) was $BaTiO_3$ (BT).[22] Initially, it was used for high capacitance ($\varepsilon_r \approx 1100$)[26] as well as for SONAR and phonographic needle applications.[27] It was later realized (when electrically poled) to have piezoelectric characteristics. The piezoelectric constant, voltage constant, and Curie temperature for pure BT are quite low and are ≈ 191 pCN^{-1}, 11.4×10^{-3} V m N^{-1}, and 130°C, respectively. Grain size reduction was realized to lead to a dramatic improvement in the piezoelectric characteristics of BT.[28] A d_{33} value ≈460 pCN^{-1} was observed in BT grown via the hydrothermal method (grain size ≈1–2 μm).[26] Suitable A and B-site substitution is another way to improve the piezoelectric performance of BT. Using this approach, d_{33} value ≈ 620 pCN^{-1} was demonstrated in BZT–BCT system.[29]

$Bi_{0.5}Na_{0.5}TiO_3$ (BNT) is another important lead-free piezoelectric. It was discovered in 1960s by Smolensky et al.[30–32] Its room temperature crystal structure is rhombohedral (lattice parameters α = 3.98 Å and α = 89.67°).[33] In polarized (at 40 kV/cm electric field) BNT ceramics, $d_{31} \approx$ 15pCN^{-1}, $d_{33} \approx$ 70pCN^{-1}, $K_{31} = 0.10$, $\varepsilon_r \approx 300$, and $\tan\delta \approx 0.011$ have been reported at 200°C.[34] Another important lead-free piezoelectric is $Bi_{0.5}\,K_{0.5}TiO_3$ (BKT) (also discovered by Smolenskii et al.). At room temperature, the crystal structure is tetragonal[35] and T_c (Curie temperature) ≈380C[36] and lattice constants with $a = 3.913$ Å and $c = 3.993$ Å.[37] Hiruma et al. reported $K_{33} \approx 0.28$ and $d_{33} = 69.8$ pCN^{-1} for BKT.[38] Other modifications of BNT and BKT (using various dopants) have been discussed in detail by Quan et al.[30] $Na_xK_{1-x}NbO_3$ (NKN) is another lead-free (hence nontoxic) piezoelectric. However, its piezoelectric coefficient ($d_{33} \approx$ 100–400 pCN^{-1}) as well as electromechanical coupling constant ($K_p \approx 0.36$, "p" represents planar) are much smaller in comparison to PZT-based piezoelectrics.[39] Several studies (using different dopants) to enhance its performance have been reported.[40–46]

Another lead-free non-perovskite piezoelectric is $LiNbO_3$ (LN) with $T_c > 1000$°C.[47–49] Its high-temperature stability allows it to be used as a transducer at very high temperatures. However, its d_{33} (≈6 pCN^{-1}) is quite small as compared with PZT.[37] AlN_3 (crystal structure being wurtzite), is another lead-free non-perovskite piezoelectric with d_{33} ≈5 pC N−1.[50] T_c > 1000°C, moisture stability, and its better processing compatibility with semiconductor metal oxides are what makes it useful in microelectromechanical (MEMS) devices.[51]

$BiFeO_3$ (BF) is a very important perovskite from multiferroic perspective. BF-based materials such as (1) BF–$SrTiO_3$ (BF–ST), and (2) BF-BNT–BKT, and (3) BF-BNT, etc., have also been extensively investigated for enhancing their piezoelectric performance.[52–57] $d_{33} \approx 69$ pCN^{-1} was observed for 0.625 BF–0.375 ST coupled with unstable piezoelectric response with respect to time.[53] Fujii et al.[58] reported the piezoelectricity of a BNT–BF system. A maximum $d_{33} \approx 93$ pCN^{-1} was observed in 0.1 BF–0.9 BNT.[58] The piezoelectric response though was observed to be asymmetric (due to defect dipoles generating strong internal bias) in a wide composition range thereby hindering its application utility. The best value of $d_{33} \approx 402$ pCN^{-1} and $T_c \approx 454$°C was reported in $0.67Bi_{1.05}Fe_{0.97}Ga_{0.03}O_3$ −0.33 BT sample by Lee et al. (in 2015).[59]

2.2.7.3 POLYMER PIEZOELECTRICS

Piezoelectricity in polymers (suitable from the perspective of viable applications) was discovered by Kawai et al.[60] in 1960s. They can be both semicrystalline and amorphous. PVDF (Poly Vinyledene fluoride, discovered in 1969) along with its copolymers represent an example of semicrystalline polymers. They consist of an amorphous matrix containing suspended minute ferroelectric crystals.[61] Application of electric field reorients them leading to enhanced piezoelectric properties.

Among polymer materials, PVDF has a relatively higher value of piezoelectric coefficient along with its chemical stability, lower acoustic impedance, and brilliant electromechanical response.[62] Its d_{33} value is ≈ −24 to −34 pCN^{-1} in the poled state.[63] Its negative d_{33} implies that in contrast to expansion in normal piezoelectrics (upon electric field application, it compresses. Another polymer piezoelectric in a copolymer of PVDF is Poly (vinyledene fluoride co-trifluoroethylene) abbreviated PVDF-TrFE. It is exceptional in the sense that it does not require poling to behave as a piezoelectric. As far as fabrication and piezoelectric applications are concerned, it enjoys an edge over PVDF.[62,64]

Polyamides (the most well-known is nylon) are another type of polymers which exhibit piezoelectric response. Nylon, though known since 1940s;

however, its piezoelectric properties were discovered by Kawai et al. in 1970s.[65,66] Nylon-11 (developed by Wu et al.[67] using solution casting technique and applying uniaxial stretching) observed its piezoelectric constant ≈7.2 pCN^{-1}.

Polyimides (which are amorphous) represent another family of piezoelectric polymers which possess brilliant dielectric, mechanical, and thermal properties.[68] Even though PVDF possesses better piezoelectric properties, polyimides stand out owing to their ability to retain piezoelectric properties even up to temperatures as high as 15°C. On the contrary, piezoelectric performance of PVDF deteriorates beyond 70°C.[69,70] The piezoelectric mechanism of polyimides is drastically different from semicrystalline ones. The stability of polyimides at high temperatures lends them utility in MEMS devices like pyroelectric and tactile sensors operating at high temperatures.[68,69] A $d_{33} \approx 250$ pCN^{-1} (at 150°C) was observed by Yang et al.[71] developed in a copolyimide. This is significant as compared with the value observed in PVDF.

Besides these, there are some piezoelectric polymers called ferroelectrets (cellular polypropylene (PP and polyorganophosphazenes (POPh) are typical examples.) The electrets consist of metastable dipoles formed either by trapping of electrostatic charges or aligned (temporarily) molecular dipoles.[72] Kirjavainen[73] was the one who discovered PP. In a single film, d_{33} value as high as 1200 pCN^{-1} has been observed.[74,75] POPh exhibit piezoelectric behavior and are nothing but linear organopolymers. Their mesophase accounts for their piezoelectric response. It has been observed that the addition of salts (such as $LiSO_3CF_3$) to their structure can enhance their piezoelectric behavior.[76] A d_{31} as high as ≈150 pCN^{-1} in has been observed in 10% salt-added POPh (much larger than PVDF).[77] Habib et al. have extensively reviewed various polymer piezoelectrics.[22]

2.2.7.4 BIOLOGICAL PIEZOELECTRICS

The piezoelectricity in biological materials is thought to have physiological significance.[78] Experimentally it is confirmed that growth of bone and remodeling are linked to piezoelectricity in the bone.[79] It has been proposed that piezoelectricity in the elastin of human lung's plays a key role in binding atmospheric oxygen with hemoglobin.[80] In the human nervous system, the sensing of external stimulation is also believed to involve some kind of piezoelectricity.[81] The piezoelectric properties of proteins (like collagen, actin, elastin, and myosin, etc.) have been associated with their constituents.

These constituents include (1) electrical dipoles, (2) amino acids, and (3) the packing of peptide chains.[82–84]

Cellulose (also its derivatives) is an abundant biological material with interesting piezoelectric properties. It is a linear polymer which contains >10,000 repeating glucose units. The units are linked to each other by β-(1–4)-glycosidic bonds. In one of the cellulose-based biomaterial EAPap (electroactive paper based on cellulose), $d_{31} \approx 1425$ pCN^{-1} has been observed.[85] The understanding of amino acids (proteins building blocks) piezoelectric properties is of fundamental importance. Piezoelectric properties of all other amino acids (except methionine) have been experimentally confirmed.[86,87] Among amino acids which are naturally occurring, the piezoelectric coefficients of glycine and hydroxyproline have been determined to be highest.[88] d_{16} (shear piezoelectric coefficient) ≈ 178 pCN^{-1} has been reported for an amino acid β-glycine.[84] The second highest value of shear piezoelectric coefficient (d_{16}) ≈ −28 pCN^{-1} has been observed for amino acid Hydroxy-L-proline.[83] A number of proteins are also known to exhibit piezoelectric behavior. The piezoelectric properties of collagen are most widely investigated.[89] Piezoelectric coefficient in collagen as high as ≈12 pCN^{-1} has been observed.[90] Elastin (found in walls of blood vessels, lungs, etc.) is another widely investigated protein which is fibrous in nature. The measured value of its piezoelectric coefficient is ≈ 0.1 pCN^{-1}[80,91] In the egg white, lysozyme is a globular protein present in a major amount. Using PFM, its piezoelectric coefficient (longitudinal) has been measured ≈19.3 pCN^{-1}.[92] This value is quite high considering the material to be biological. Besides these, piezoelectricity has also been observed in tissues/organs of living beings as well as viruses.[78]

2.2.7.5 COMPOSITE PIEZOELECTRICS

Composites are a combination (mixture) of more than one piezoelectric material. The combinations are formed to enhance the overall piezoelectric response of the composite material. Their properties can be further enhanced by altering (1) their relative proportion, (2) their way of connectivity, and (3) structure, etc. There are 10 ways of classifying di-phasic piezoelectric composites as per the pattern ((0–0), (0–1), (0–2), etc.) of their connectivity.[93] Six more patterns of connectivity ((1–0), (2–0), (3–0), etc.) were suggested by Pilgrim et al.[94] Out of these, (0–3) pattern is advantageous on account of its ease of fabrication and ability to scale up. It usually comprises ceramic particles (regarded zero-dimensional) dispersed within a polymer matrix which is connected in three-dimensions.[95]

PVDF-PZT (first reported by Kitayama et al.[96]) is a very common (0–3) type piezoelectric (polymer-ceramic) composite. The stiffness coefficient of this composite is much smaller as compared with PZT whereas the piezoelectric constant is slightly smaller.[97] A composite formed out of electrospun nanofibers of PVDF-TrFE/BT mixture dispersed in PDMS (poly dimethylsiloxane) exhibits excellent piezoelectric response. This composite (containing 15% BT by weight) was found to generate a voltage ≈25 V at ≈0.6 Hz walking frequency when placed inside human shoes.[98] Similarly, from piezoelectric nanogenerator (PENG) perspective, composites such as PMN-PT with polyimide and PDMS (M = magnesium), ZnO-PVDF (ZnO = zinc oxide), and Sm-PMN-PT (Sm = samarium) in PVDF are quite promising.[25] The piezoelectric material/device aspects (nature of material, synthesis, and optimization techniques) on energy harvesting, nanogenerator, and active vibration flow and noise control applications have been discussed in a number of reviews.[99–103]

2.3 FERROELECTRICS

A ferroelectric (unlike piezoelectric) can possess non-zero electric dipole moment (below a critical temperature) without being subjected electric field. Under the impact of an oscillating electric field, a ferroelectric will exhibit a hysteresis loop (not displayed by piezoelectrics). Crystalline ferroelectrics can be both in the single crystalline or polycrystalline state. Out of these, the widely used ones are in polycrystalline form. The ferroelectrics respond nonlinearly and irreversibly to an external electric field. This is in contrast to a linear and reversible response of piezoelectrics. 10 (out of 21) non-centrosymmetric crystals display ferroelectric response. They are polar materials in which spontaneous polarization can have two equilibrium orientations (along and against the field). They were termed "ferroelectrics" owing to similarities between their electrical response and the magnetic response of ferromagnetics.

2.3.1 IMPORTANT CONCEPTS RELATED TO FERROELECTRICITY

Some important concepts related to ferroelectricity are discussed below.

2.3.1.1 SPONTANEOUS POLARIZATION

Technically, polarization ($\vec{P}$) of a dielectric material is defined as the net electrical dipole moment developed in a material per unit volume. It is measured in units of C/m^2 ($\mu C/cm^2$, is a more practical unit). The word spontaneous refers to the emergence of electric polarization even without electric field stimulus (accompanies a centrosymmetric to non-centrosymmetric crystal structure transition at T_c). In general, electric polarization can be due shift in the center of gravity of (1) atoms and ions, (2) a change in ion orientation, and (3) due to displacement of interfacial charges. All of these types of polarizations are a function of the applied field frequency. The interfacial polarization is dominant at a lowest frequencies where the atomic polarization is dominant at the highest frequencies.[104]

2.3.1.2 HYSTERESIS (P-E) LOOP

Ferroelectric characteristics of ferroelectric samples (usually in the form of a flat cylindrical disc with metal electrodes on both sides) are measured by applying an oscillating electric ($\vec{E}$, measured in kV/cm) and measuring the resulting sample response ($\vec{P}$). The measurements are done using an instrument called a ferroelectric tester (based on Sawyer the Tower circuit). It turns out that the plot of $\vec{P}$ (along the y-axis) vs $\vec{E}$ (along the x-axis,) is nonlinear and irreversible (as shown in Figure 2.3). The important sample parameters coercive field (E_c, intersection with x-axis), remnance (P_c, intersection with y-axis), and saturation polarization (P_s, maximum obtainable polarization) are shown in Figure 2.3 as well.[104–106] The shape of the loop can be square (Figure 2.3), gradually rounded (Figure 2.4, left), or pinched (Figure 2.4, right). The presence of square loops in a sample implies sudden switching of the polarization state of the sample (in a small field range around E_c). This kind of P–E loop is expected in undoped and defect-free samples (with very low levels of leakage current). In such samples, all of the electric dipoles will be almost alike and they will therefore exhibit an almost simultaneous switch of their polarization with the applied field. However, the doped samples (containing structural defects, oxygen vacancies, and compositional inhomogeneity) exhibit a rather rounded (not leaky) P–E loop. Such samples consist of electric dipoles (not exactly alike) which exhibit an extended/gradual response to the applied electric field. The pinched loop is a typical characteristic of antiferroelectric samples.[34,107–110]

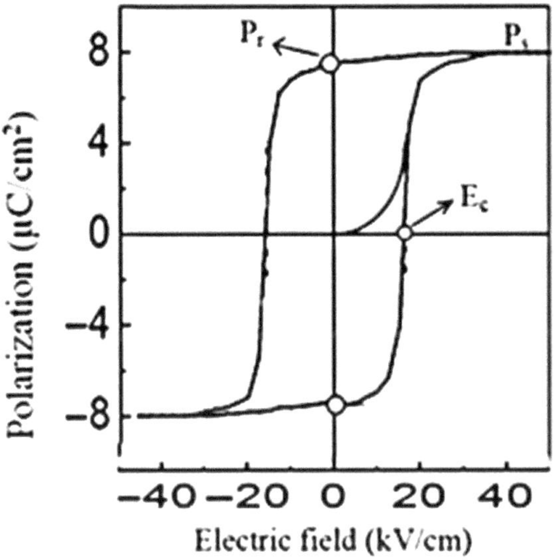

FIGURE 2.3 The square P–E loop depicting coercive field (E_c), remnance (P_c), and saturation polarization (P_s).

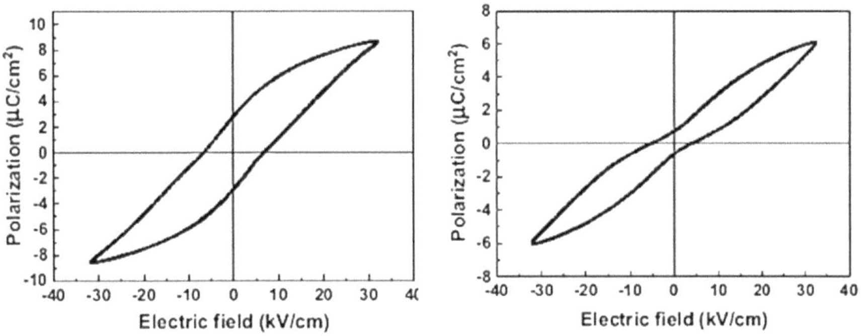

FIGURE 2.4 Rounded (left) and pinched (right) P–E loops. The loops correspond to $Ba_{0.7}Ca_{0.3}TiO_3$ (left) and $Ba_{0.6}Ca_{0.4}TiO_3$ (right) samples.

2.3.1.3 *P-E LOOP AND ENERGY STORAGE*

Ferroelectric materials having characteristics such as (1) low value of dielectric loss, (2) fatigue resistance, and (3) intermediate breakdown field strength have potential for energy storage applications. As a capacitor (dielectric filled) is subjected to charge–discharge cycles, some part of the energy spent can be recovered (E_{rec}) whereas some of the energy cannot be recovered (E_{loss}). E_{rec}, is taken as the area under the discharge curve and y-axis whereas E_{loss} is the are under the P–E loop. These two energies are systematically shown in the quarter P–E loop in Figure 2.5. The efficiency of energy stored is defined as

$$\eta = \frac{E_{rec}}{E_{rec} + E_{loss}} \quad (2.7)$$

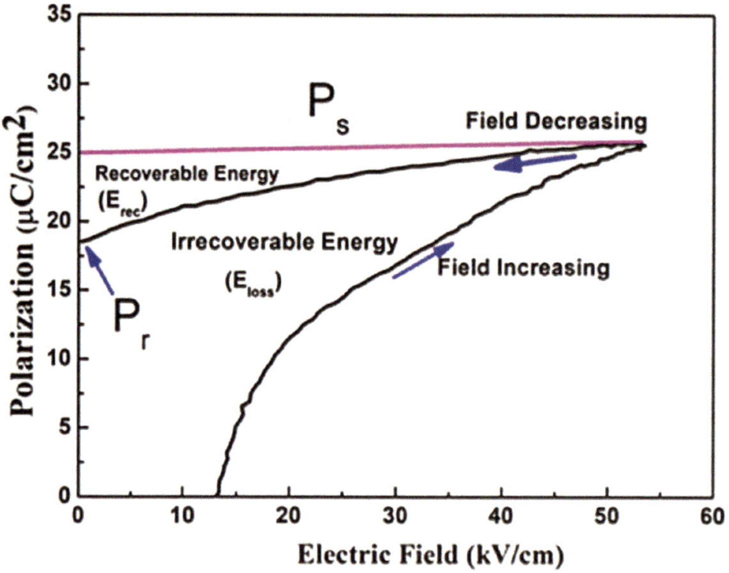

FIGURE 2.5 The figure shows E_{rec} (blue square boxes, area under discharge curve), E_{loss} (pink solid fill, area under the loop) and other ferroelectric parameters.

It is clear (from Figure 2.5) that η will increase with an increase in E_{rec} which will increase with an increase in P_s and a decrease in P_r (as this will increase the area between the discharge curve and the y-axis). Therefore, to realize higher efficiency (thereby high density of energy storage), a ferroelectric with a large P_s, low P_r and a very high value of electric breakdown strength is desirable.[111]

2.3.1.4 RELAXATION

A ferroelectric material acquires polarization under the influence of an applied electric field. After the removal of the field, the material tries to switch back to its unpolarized (though may not be complete due to finite P_r). The process of returning to the original unpolarized state from a polarized one is termed relaxation. The characteristic time over which this polarization relaxation can take place is represented by "τ." In ferroelectrics, relaxation time can vary from a fraction of picoseconds to 100's of nanoseconds.[112]

2.3.1.5 DEBYE RELAXATION, COLE–COLE RELAXATION, NORMAL AND RELAXOR FERROELECTRICS

Physicist Peter Debye in 1913 introduced this model of relaxation of polarization. This model assumes only a single relaxation time (τ). His approach led to the following equations for the real (ε') and imaginary (ε'') part of the dielectric constant at frequency "ω"[112,113]:

$$\varepsilon' = \varepsilon_\infty + \frac{\varepsilon_s - \varepsilon_\infty}{1+\omega^2\tau^2} \tag{2.8}$$

$$\varepsilon'' = \frac{(\varepsilon_s - \varepsilon_\infty)\omega\tau}{1+\omega^2\tau^2} \tag{2.9}$$

Here, ε_s and ε_∞ represent zero frequency (dc) and high-frequency dielectric constants. A ferroelectric with only one relaxation time is called a normal ferroelectric. If we eliminate "ω" between eqs "2.8" and "2.9," we get

$$\left(\varepsilon' - \frac{(\varepsilon_s - \varepsilon_\infty)}{2}\right)^2 + (\varepsilon'')^2 = \left(\frac{\varepsilon_s - \varepsilon_\infty}{2}\right)^2 s \tag{2.10}$$

Equations represent a circle (with ε' on x-axis and ε'' on y-axis) with center at $\left(\frac{\varepsilon_s - \varepsilon_\infty}{2}, 0\right)$. The plot of ε' vs ε'' (referred to as Cole–Cole plot) is shown in Figure 2.6. The plot has been mathematically generated assuming conductivity $\sigma = 0$, $\varepsilon_s = 1000$, $\varepsilon_\infty = 50$ and $\tau = 10^{-9}$s. The frequency has been varied from 10 to 9×10^{10} (in Hertz). The plot is a semicircle in the first quadrant and the center on the x-axis is clearly evident.

If the material also has finite conductivity, this will result in the modification of "7" to the following equation for ε''.

$$\varepsilon'' = \frac{(\varepsilon_s - \varepsilon_\infty)\omega\tau}{1+\omega^2\tau^2} \tag{2.11}$$

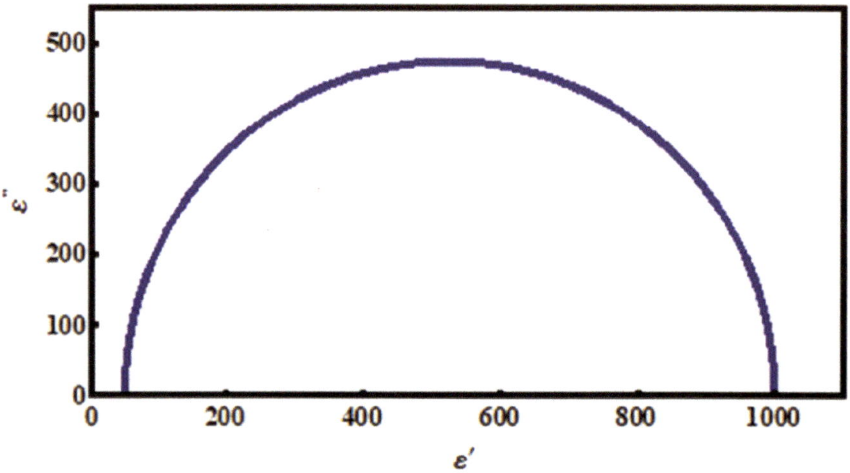

FIGURE 2.6 The Cole-Cole plot corresponding to a ferroelectric exhibiting Debye relaxation. $\varepsilon_s = 1000$, $\varepsilon_\infty = 50$ and $\tau = 10^{-9}$s have been assumed.

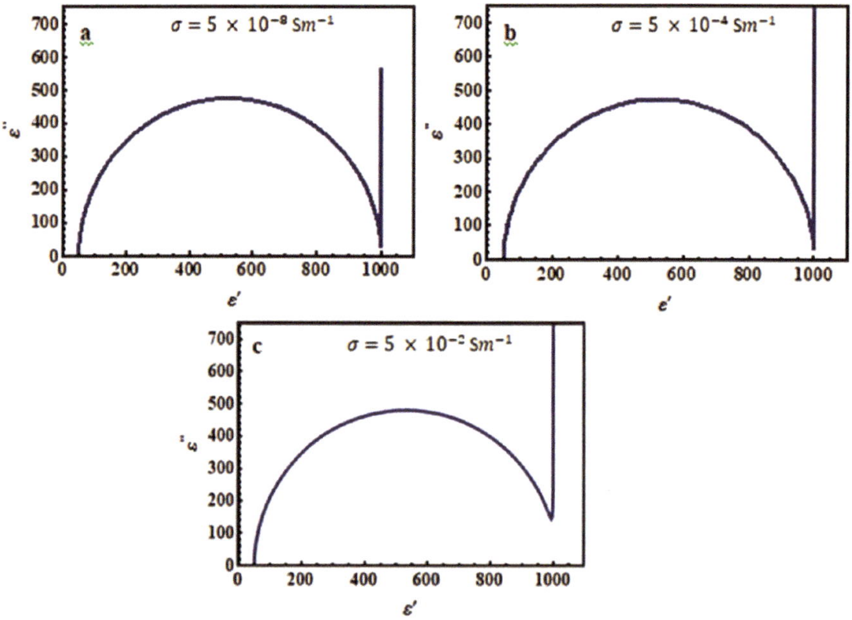

FIGURE 2.7 The Cole-Cole plot corresponding to a ferroelectric exhibiting Debye relaxation with (a) $\sigma = 5 \times 10^{-8}$ Sm^{-1}, (b) $\sigma = 5 \times 10^{-4}$ Sm^{-1} and (c) $\sigma = 5 \times 10^{-2}$ Sm^{-1}. $\varepsilon_s = 1000$, $\varepsilon_\infty = 50$ and $\tau = 10^{-9}$s have been assumed.

Where "σ" is the material conductivity. The Cole–Cole plots for a material with finite conductivity are shown in Figure 2.7. In case, the material exhibits two different relaxation times (typically related to grain, grain boundary, etc.), the Cole–Cole plots appear as shown in Figure 2.8. The generation parameters have been mentioned in the figure captions.

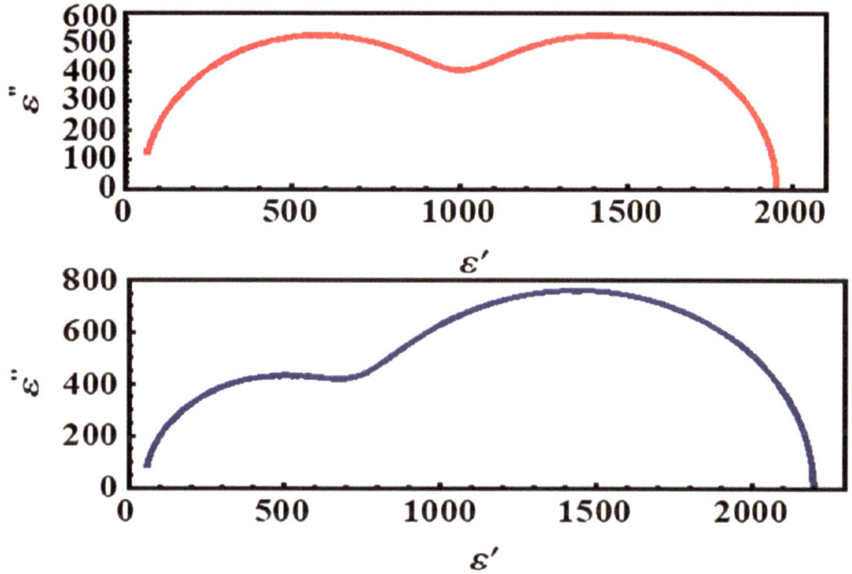

FIGURE 2.8 The Cole-Cole plot corresponding to two different relaxation times with same (red) and different (blue) radii. For red curve $\varepsilon_s = 1000$, $\varepsilon_\infty = 50$ $\tau_1 = 2\ 10^{-9}s$ and $\tau_2 = 10^{-10}$s and for blue curve $\varepsilon_{s1} = 1500$, $\varepsilon_{s2} = 750$, $\varepsilon_\infty = 50$, $\tau_1 = 2\ 10^{-9}$s and $\tau_2 = 10^{-10}$s.

However, Debye eqs 2.8 and 2.9 do not adequately describe the relaxation process for a vast majority of ferroelectric/dielectric materials. This is because in such materials, the description of experimental data requires the assumption of distributed relaxation times. A ferroelectric material exhibiting distributed relaxation times is called a relaxor ferroelectric. The relaxation phenomena in such materials can be understood by considering Cole and Cole dispersion relations.[114] As per these dispersion relations, ε' and ε'' are given as

$$\varepsilon' = \varepsilon_\infty + \frac{(\varepsilon_s - \varepsilon_\infty)(1 + (\omega\tau)^{1-\alpha}\ \mathrm{Sin}\left[\frac{\pi\alpha}{2}\right])}{1 + 2(\omega\tau)^{1-0.1}\mathrm{Sin}\left[\frac{\pi\alpha}{2}\right] + (\omega\tau)^{2(1-\alpha)}} \tag{2.12}$$

$$\varepsilon'' = \frac{(\varepsilon_s - \varepsilon_\infty)(\omega\tau)^{1-\alpha} \quad \mathrm{Cos}\left[\frac{\pi\alpha}{2}\right]}{1 + 2(\omega\tau)^{1-0.1}\mathrm{Sin}\left[\frac{\pi\alpha}{2}\right] + (\omega\tau)^{2(1-\alpha)}} \tag{2.13}$$

Once again, if we eliminate "ω" between eqs "2.10" and "2.11," we get

$$\left(\varepsilon' - \frac{(\varepsilon_s - \varepsilon_\infty)}{2}\right)^2 + \left(\varepsilon'' + \frac{(\varepsilon_s - \varepsilon_\infty)}{2}\mathrm{Tan}\left[\frac{\pi\alpha}{2}\right]\right)^2 = \left(\frac{\varepsilon_s - \varepsilon_\infty}{2}\ \mathrm{Sec}\left[\frac{\pi\alpha}{2}\right]\right)^2 \tag{2.14}$$

Equation 2.11 once again represents a circle with center at $\left(\frac{\varepsilon_s - \varepsilon_\infty}{2}, -\frac{(\varepsilon_s - \varepsilon_\infty)}{2}\mathrm{Tan}\left[\frac{\pi\alpha}{2}\right]\right)$. Clearly, the center is not at the x-axis but in the fourth quadrant. The Cole–Cole plots have been mathematically generated assuming conductivity $\sigma = 0$, $\varepsilon_s = 1000$, $\varepsilon_\infty = 50$ and $\tau = 10^{-9}$s at $\alpha = 0$, 0.2 and 0.4 and are shown in Figure 2.9. The frequency has been varied from 10 to 9×10^{10} (in Hertz). The plots are semicircular in nature with their centers gradually depressed along the y-axis (except for $\alpha = 0$ at which Cole–Cole and Debye dispersion relations are identical) with increasing value of "α." In several materials, Davidson–Cole and Havriliak–Negami dispersion relation gives a better fit to the experimental data.[115,116]

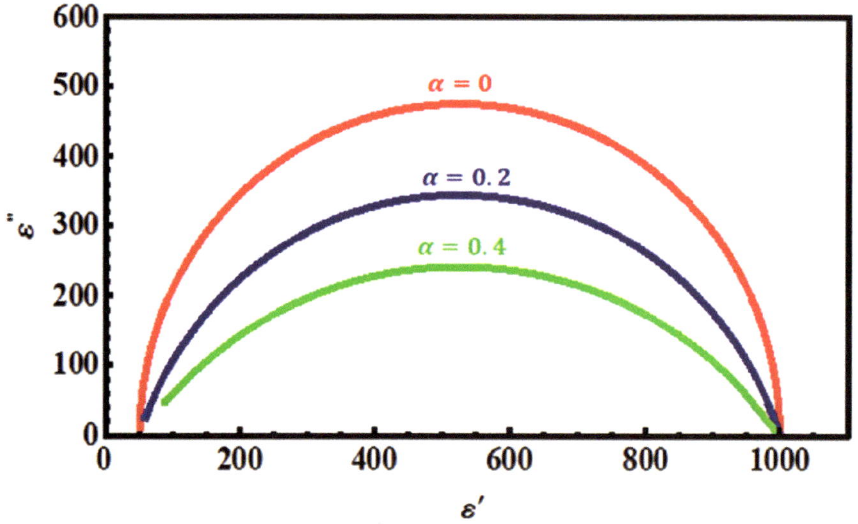

FIGURE 2.9 The Cole-Cole plots (red for $\alpha = 0$, blue for $\alpha = 0.2$, and green for $\alpha = 0.4$) corresponding to a ferroelectric exhibiting non-Debye relaxation. $\varepsilon_s = 1000$, $\varepsilon_\infty = 50$ and $\tau = 10^{-9}$s have been assumed.

2.3.1.6 ORIGIN OF FERROELECTRIC PHASE TRANSITION

The soft mode theory explains ferroelectric phase transition in perovskites. As per this theory, the lowest frequency phonon (polar) mode is softened during cooling and its frequency approaches zero at T_c (transformation temperature from centrosymmetric to non-centrosymmetric structure). This centrosymmetry breaking instability is generally termed second-order Jahn-Teller effect.[117–122] The phase transition can be (1) first order, (2) second order, and (3) diffuse. The differences between these transitions (change of electric polarization) are qualitatively shown in Figures 2.10a and 2.10b. It is clear that first-order transition is abrupt. The second-order transition takes place in a small range of temperatures around T_c whereas the diffuse transition takes place over a very broad temperature range around T_c. Further, in the first- and second-order transition, no ferroelectric region (no tailing above T_c) of the sample survives above T_c whereas for the diffuse transition; the sample above T_c is still a mixture (apparent from tailing above T_c) of ferroelectric (nano) regions and paraelectric regions. A normal ferroelectric exhibits first- or second-order order transition whereas a relaxor exhibits diffuse transition.

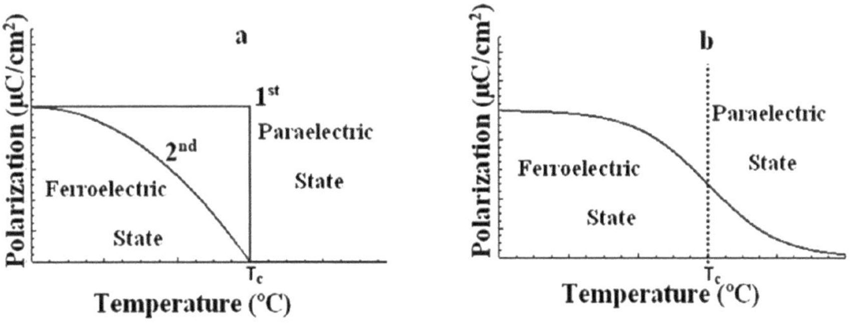

FIGURE 2.10 (a) Difference between a 1st order and 2nd order and (b) a diffuse ferroelectric phase transition at T_C.

However, not only the polarization changes dramatically at T_c, but also does the real part of the dielectric constant. The dielectric constant vs temperature profile corresponding to a normal ferroelectric transition (in $Ba_{0.95}Ca_{0.05}TiO_3$ ceramics, left) and relaxor ferroelectric transition (in $Pb_{0.9}Sm_{0.10}Zr_{0.405}Ti_{0.495}Fe_{0.10}O_3$, right) are shown in Figure 2.11.[108,123]

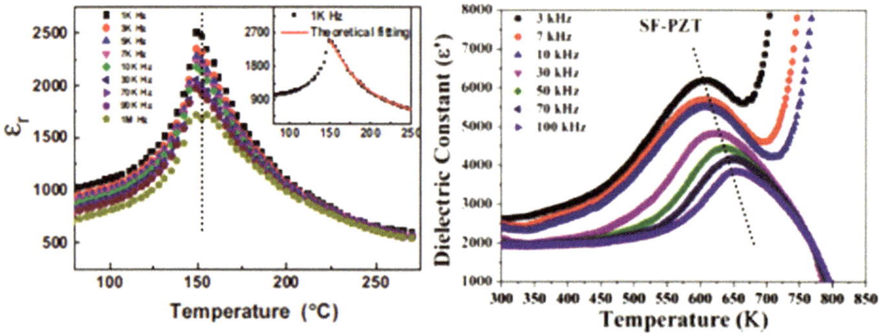

FIGURE 2.11 The figure shows sharp transition in a normal ferroelectric (in $Ba_{0.95}Ca_{0.05}TiO_3$, left) and broad transition in a relaxor ($Pb_{0.9}Sm_{0.10}Zr_{0.405}$ $Ti_{0.495}$ $Fe_{0.10}O_3$, right) ferroelectric.

2.3.1.7 CURIE–WEISS AND MODIFIED CURIE–WEISS LAW

As per Curie–Weiss law, the electric susceptibility (χ_e) of a ferroelectric material above Curie temperature (T_c) is given by the following expression[124]:

$$\chi_e = \frac{A}{T - T_C} \tag{2.15}$$

Here "T" is the absolute temperature and "A" is just a constant. However, relaxors and ferroelectrics exhibiting diffuse behavior obey a modified form of Curie–Weiss law given by the following expression.[125–127]

$$\frac{1}{\varepsilon_r} - \frac{1}{\varepsilon_C} = B(T - T_C)^{\gamma} \tag{2.16}$$

Here ε_r, ε_C, γ, B are dielectric constant, dielectric constant at Curie temperature and numbers, respectively. Usually, $\gamma = 1$ for normal ferroelectrics and $\gamma = 2$ for relaxors and between "1" and "2" for ferroelectrics displaying diffuse phase transition. Aanchal et al.[109] applied eq 2.14 to study ferroelectric transition in $Ba_{1-x}Ca_xTiO_3$ system. The value of γ was found to vary between "1.22" and "1.53."[109]

2.3.1.8 COMPLEX IMPEDANCE (MODULUS) SPECTROSCOPY

The electrical properties of polycrystalline ferroelectrics/piezoelectrics in a non-destructive manner are characterized using complex impedance (modulus) spectroscopy (CIS, $Z^*(\omega) = Z'(\omega) - j\,Z''(\omega)$. The measurements are made over a sample (in the form of a sandwich between two metallic electrodes) for a range of frequencies at different temperatures. The dominating factors contributing to sample impedance (grains/grain boundaries/electrodes) can be easily deciphered using this technique.[128–131] The expressions for "Z'" and "Z''" (with impedance contributions from grains and grain boundaries) for a ferroelectric sample exhibiting Debye behavior are given by

$$Z'(\omega) = \frac{R_g}{1+(\omega R_g C_g)^2} + \frac{R_{gb}}{1+(\omega R_{gb} C_{gb})^2} \tag{2.17}$$

$$Z''(\omega) = \frac{\omega_g R_g{}^2 C_g}{1+(\omega R_g C_g)^2} + \frac{\omega_{gb} R_{gb}{}^2 C_{gb}}{1+(\omega R_{gb} C_{gb})^2} \tag{2.18}$$

where ω, R, and C stand for frequency, resistance, and capacitance, respectively. The subscripts "g" and "gb" stand for grains and grain boundaries. A different set of expressions for samples exhibiting distributed relaxation times is used.[132] The plots (termed Nyquist plots) between "Z'" and "Z''" look similar to Figures 2.6–2.9. The Nyquist plot (experimental and fitted) corresponding to $Ba_{0.96}Ca_{0.04}TiO_3$ sample at two different temperatures is shown in Figure 2.12.[132] The data were fitted assuming both the grain and grain boundary contributions. The decreasing circular radius with temperature is indicative of NTCR (negative temperature coefficient of resistance) behavior in $Ba_{0.96}Ca_{0.04}TiO_3$ sample. The complex electric modulus ($M^*(\omega) = j\omega C_0 Z^*(\omega)$) is further helpful in determining whether short-range or long-range carrier movement dominates the relaxation process. The relaxation time is inversely related to the frequency at which the imaginary part of electric modulus is maximum. A measurement of relaxation time (τ) at different temperatures and exploitation of the relation $\tau = \tau_0 e^{\frac{E_a}{kT}}$ (by performing a linear fit to ln τ vs $\frac{1}{T}$ data, other symbols having their usual meaning), can be used to calculate the energy of activation (E_a).[132]

2.3.1.9 AC CONDUCTION

The frequency (ω)-dependent ac conductivity (σ_{ac}) usually follows the following expression:

$$\sigma_{ac} = \sigma_{dc} + A\,\omega^n \tag{2.19}$$

where "A" (measure of polarization strength) and "n" ($0 < n < 1$, measure of lattice carrier interaction) are temperature-dependent constants and σ_{dc} is dc conductivity of the sample. The value of the "n" and its variation with temperature is related to the nature of conduction mechanism (CBH (carrier barrier hopping), QMT (quantum mechanical tunneling), OLPT (overlapping large polaron tunneling) or NSPT (non-overlapping small polaron tunneling) models). If in a sample, n = 0.8 and is either temperature independent or varies (increases) only partially with temperature, the QMT model is applicable. If "n" decreases with increasing temperature, the QMT model is applicable. If "n" is a function of both frequency and temperature and exhibits minima with temperature, OLPT model is applicable. Finally, if "n" increases with temperature, the NSPT model is applicable.[133] The graphical plots (mathematically generated) of eq 2.19 are shown in Figure 2.13.

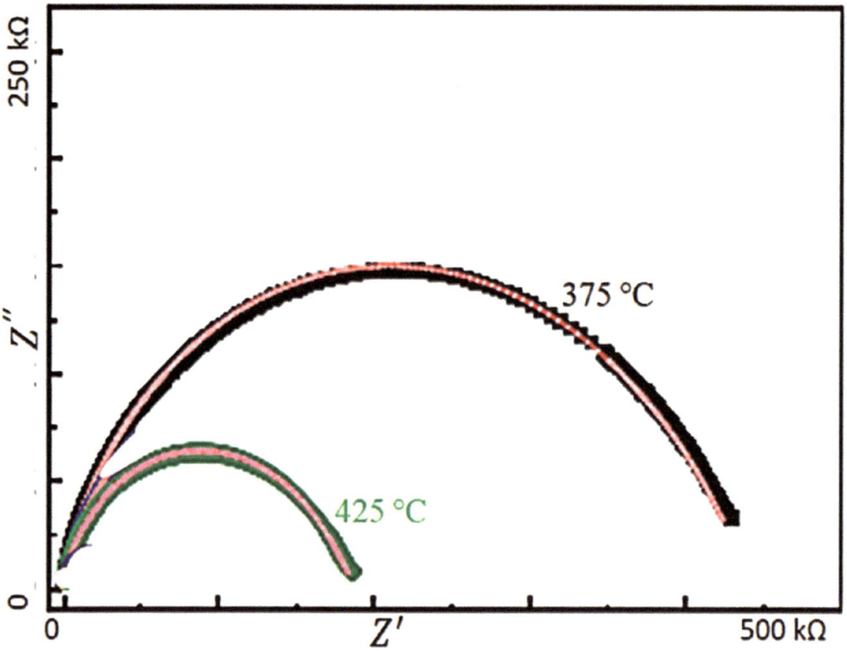

FIGURE 2.12 Nyquist plots for $Ba_{0.96}Ca_{0.04}TiO_3$ at 375 °C and 425 °C.

2.3.1.10 SOME FERROELECTRICS AND THEIR CHARACTERISTICS

Typically, all materials (discussed in the section) can be used for ferroelectric applications. The ferroelectric properties of some important ferroelectric materials are given in Table 2.1.

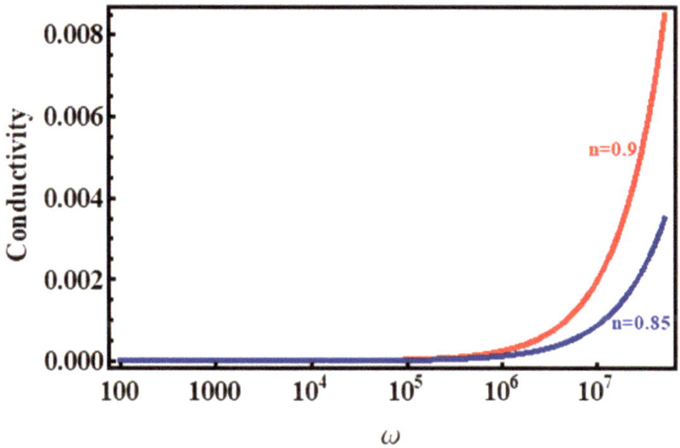

FIGURE 2.13 The mathematically generated ac conductivity vs. frequency plots. σ_{dc} = 5 10^{-7} Sm^{-1} and A = 10^{-9} have been assumed.

TABLE 2.1 Ferroelectric Parameters of Some Typical Ceramics. Abbr. used PZT (Pb(Zr,Ti) O_3, BF ($BiFeO_3$), BT ($BaTiO_3$), BZT (Ba(Zr,Ti)O_3), BCT (Ba(Ca,Ti)O_3), BNT ($Bi_{0.5}Na_{0.5}$) TiO_3, BKT (($Bi_{0.5}K_{0.5}$)TiO_3.

Sr. No.	Material/composition	P_r µC/cm²	E_c (×10) kV/cm	T_c (°C)	d_{33} × 10 (pCN^{-1})	Ref.
1	PZT	10–40	5–7	≈400	–	134
2	$Hf_xZr_{1-x}O_2$	10–40	80–200	0–500	–	134
3	$Al_xSc_{1-x}N$	80–110	200–500	>600	–	134
4	BNT	38	–	≈540	9.1	135
5	$Bi_{0.5}(Na_{1-x-y}K_xLi_y)TiO_3$	40.2	2.47	–	23.1	135
7	BZT-0.5BCT	14.5	0.168		56–62	135
8	0.68BF-0.22PT-0.1BT	60	–	≈600	100	136
12	$0.7Bi_{1.02}(Fe_{0.96}Mg_{0.02}Ti_{0.02})O_3$-0.3BT	35.75	–	497	19.8	136
17	0.90BNT–0.05BT–0.05BKT	22.8	2.7	–	–	137
18	0.94BNT–0.06BT	27.4	2.9	–	–	137

2.4 MULTIFERROICS

Materials exhibiting co-existing contrasting ferroic properties (any two among ferromagnetism ferroelectricity and ferroelasticity) are termed multiferroics. The phenomenon was predicted theoretically in 1894 by Curie.[138] They are potentially useful for innovative technological applications such as magnetic field sensors, energy efficient devices spintronic and memory devices and actuators, etc.[139–146] A brief historical account of multiferroics is presented below.

2.4.1 HISTORY OF MULTIFERROICS

The phenomenon was predicted theoretically in 1894 by Curie.[138] Magnetoelectric (linear) effect in Cr_2O_3 from symmetry considerations was predicted in 1959 by Dzyaloshinskii.[147] Astrov confirmed it experimentally in 1961.[148] Cr_2O_3 was later found unsuitable for real-life applications. In the early 1960s, the two most extensively studied multiferroic systems (i.e., $BiFeO_3$ and $RMnO_3$ (R=Y, Sc, In, etc.,)) these days, were already identified.[149–152] However, giant magnetoelectric (linear) effect was first observed by Ascher et al.[153] in $Ni_3B_7O_{13}I$ in the year 1966. In this material, the multiferroic state could be hysteretically switched using an electric or magnetic field. Simultaneous electric and magnetic ordering was observed by Schmid[144] in boracite in 1994. He coined the term multiferroic thereby complementing Aizu's earlier *ferroic* materials classification.[154] $BiFeO_3$ was extensively studied by Smolenskii et al.[146,155,156] In 2003, large values of ferroelectric and magnetic parameters were reported in $BiFeO_3$ thin films by Wang et al.[157] The pace of research in multiferroics was once again enhanced with the observation of strong magnetoelectric (ME) coupling in $TbMn_2O_5$[158] and $TbMnO_3$.[159]

2.4.2 WHY ONLY FEW MULTIFERROICS EXISTS?

Only 13 (among 122 Shubnikov–Heesch point groups) point groups permit the co-existence of electric and magnetic orders. As a result, the number of potential multiferroic materials becomes severely limited.[144] The property-based hierarchical relationship in oxide materials has been nicely depicted pictorially by Bea et al. and Eerenstein et al.[139,141]

2.4.3 d^0– d^N CONFIGURATION ISSUE

The contrasting d-shell electronic configuration for (1) ferroelectrics (Ti^{4+} in d^0 state, empty d-orbital) and (2) ferromagnetics (B-site cation in d^n state, partially filled d-orbital) were pointed out by Hill.[3] Therefore, a material exhibiting multiferroicity must have transition metal cation in both d^0 and d^n state. This requirement is inherently contradictory. She termed this as d^0– d^n problem. As a consequence, the naturally occurring multiferroics are rare. The potential multiferroic candidates are either ferrimagnetic or weakly ferromagnetic (which are electrically insulating) materials.[5,160,161]

2.4.4 ORIGIN OF MAGNETIC ORDERING IN PEROVSKITES

Magnetic is introduced into perovskites by partially replacing, the Ti^{4+} (d^0) cations with cations belonging to the transition group (d^n). The presence of surrounding oxygen anions (which are octahedrally arranged) results in the splitting of the d-orbital into two groups. One group is termed e_g (containing $d_{x^2-y^2}$ and $d_{3z^2-r^2}$ orbitals) and the other t_{2g} (containing d_{xy}, d_{yz}, and d_{zx} orbitals). The second group has lower energy in comparison to the first. The oxygen anion-mediated parallel/orthogonal coupling between spin magnetic moments of two nearest B-site magnetic cations results in diverse materials which are magnetically ordered. Quantum mechanical exchange interactions (super-exchange/double-exchange, etc.) play a key role in the magnetic behavior of perovskites. Double-exchange usually imparts ferromagnetic character whereas super-exchange can impart both ferromagnetic and antiferromagnetic character to doped perovskites. The binary nature of super-exchange interaction is dependent upon whether the d-orbital is (1) half-filled or (2) below half-filled.[121,122,162]

2.4.5 MECHANISMS RESPONSIBLE FOR MULTIFERROIC BEHAVIOR

Some mechanisms responsible for multiferroic behavior are discussed below.

2.4.5.1 A-SITE LONE-PAIR DRIVEN

This mechanism relies upon the availability of non-bonding lone pair of electrons with A-site cation. The distribution of these electrons about A-site is non-isotropic. The spatial symmetry is broken as a consequence, which results in ferroelectricity. $BiFeO_3$ is an example of such a material. The 6s electron pair (of Bi^{3+} cation) is not involved in any kind of bonding. A local dipole is created as a result of the movement of this lone pair toward FeO_6 octahedra (away from Bi^{3+}). This dipole (oriented in the[111] direction) induces ferroelectricity in the system. The magnetic behavior is a consequence of the antiferromagnetic coupling of Fe^{3+} cations.[160]

2.4.5.2 GEOMETRY DRIVEN

Space-filling effects and constraints imposed by geometry cause structural deformation in certain perovskite materials. This results in a shift of ions, hence causing geometric ferroelectricity. This kind of ferroelectricity is typically observed in hexagonal manganites ($RMnO_3$, R may any of Y, Dy, Sc). In these materials, deformation & tilting of MnO_5 bipyramids cause a shift of rear-earth cations which therefore leads to the appearance of ferroelectricity. Similarly, the cooperation of two nonpolar lattice modes leads to ferroelectric polarization in the case of $Ca_3Mn_2O_7$. The interaction between the canted magnetic moments of the system with this ferroelectric polarization leads to multiferroic behavior.[160]

2.4.5.3 CHARGE ORDERING DRIVEN

Periodic superstructures may be formed if the valence electron distribution around a host ion is non-uniform. The formation of such superstructures because of the periodic arrangement of Fe^{2+} and Fe^{3+} ions has been put forward in $LuFe_2O_4$. It has been further suggested that spontaneous polarization results due to the alternating Fe^{2+}/Fe^{3+} layers (with a cation ratio of either 1:2 or 2:1). Likewise, $Pr_{1-x}Ca_xMnO_3$ (a mixed manganite) is also known to possess ferroelectricity due to charge ordering.[99,163,164]

2.4.5.4 SPIN-DRIVEN MECHANISMS

The transfer of non-centro symmetry to electric lattice from magnetic lattice due to spin–charge interaction leads to the development of polarization. The nature of magnetically driven ferroelectrics (termed improper) is drastically different from the displacive ferroelectrics. The magnetic coupling is not possible in displacive ferroelectrics.[160] The following mechanisms lead to multiferroicity wherein electric polarization sets in due to magnetic ordering:

a) Exchange striction.
b) Inverse Dzyaloshinskii–Moriya (DM) interactions.
c) Spin-dependent p–d hybridization.

2.4.6 TYPES OF MULTIFERROICS

Based on the origin of multiferroicity, multiferroics can be grouped into two categories. Microscopically, oxygen anion-mediated interactions between magnetic moments of localized electrons (in partly filled "d" or "f" shells of transition metal or rare-earth cations) are responsible for the birth of magnetism.

2.4.6.1 TYPE-I MULTIFERROICS

In this category of multiferroics, the origin of (1) ferroelectric and (2) magnetic ordering is not the same. They possess good ferroelectric properties with their ferroelectric/magnetic transition temperatures being above ambient. The ferroelectric behavior in them can arise due to any of the geometric/lone pair/charge ordering effects. The two ordered states are usually weakly coupled. $YMnO_3$ ($T_c \approx 914$ K, $P \approx 6$ μC cm^{-2}, $T_N \approx 76$ K) and $BiFeO_3$ ($T_c \approx$ 1110 K, ($P \approx 90$ μC cm^{-2}, $T_N \approx 643$ K) are typical examples.[4]

2.4.6.2 TYPE-II MULTIFERROICS

In Type-2 multiferroics, ferroelectricity is driven by magnetic ordering. Since one is driving the other, the coupling strength between two ordered states is quite strong. However, there is a major drawback. Very weak ferroelectric polarization ($\approx 10^{-2}$μC/cm^2) is observed in these materials which is hard to measure. They can be further categorized into two groups. The first group consists of materials such as $TbMnO_3$. In the first group, a magnetic spin spiral is responsible for ferroelectricity. The second group consists of materials like Ca_3CoMnO_6 having collinear magnetic structure. Exchange-restriction drives ferroelectricity in these materials.[4]

2.4.7 LINEAR MAGNETOELECTRIC (ME) COUPLING COEFFICIENT

It is the coupling (one impacts the other) between different ordered states which makes multiferroics exciting. Linear ME coupling coefficient is a measure of the strength of ME coupling in a multiferroic. As per Landau, in a magnetoelectric (which is homogeneous) system, contribution to free energy (F) due to ME is given by

$$FM(E,H) = -\alpha_{ij} E_i H_j \tag{2.20}$$

In this expression, α_{ij} (a tensor) is termed the linear ME coefficient. E_i and H_j are the electric and magnetic field components along different coefficients. There is an upper limit to α_{ij} given by

$$\alpha_{ij} = \sqrt{\frac{\varepsilon_{ij}\mu_{ij}}{c_0^{\,2}}} \tag{2.21}$$

Here c_0 is the speed of light and ε_{ij}, μ_{ij} represent dielectric permittivity and magnetic permeability. The linear magnetoelectric coefficient can be measured using (1) static, (2) quasi-static, and (3) dynamic techniques. Using the dynamic technique, the magnetoelectric coefficient is determined from the following equation:

$$\alpha_{ME} = \frac{\delta V}{\delta H\ t} \tag{2.22}$$

Here δV represents voltage developed across a multiferroic sample of thickness (t) upon application of "δ" alternating magnetic field. However, in addition to eq 2.22, the following expressions are also used to measure coupling strength.

$$\eta(in\ \%) = \frac{X(x) - X(0)}{X(0)} \times 100 \tag{2.23}$$

Here $X(x)$ represents any one of the physical properties such as dielectric constant (ε), ferroelectric polarization (P, both measured in the presence of magnetic field), and magnetic susceptibility (χ), magnetization (M, both measured in the presence of electric field). $X(0)$, represents the same property measured in the absence of the corresponding field. It is measured in units of $mVcm^{-1}Oe^{-1}$ (CGS).[120,165] A small value of α_{ME} is observed in bulk materials. For example, $\alpha_{ME} \approx 55.4$ $mVcm^{-1}Oe^{-1}$ has been reported in bulk $SrCo_2Fe_{24}O_{41}$.[166] Relatively higher values of α_{ME} are observed in thin film materials. For example, $\alpha_{ME} \approx 3000$ $mVcm^{-1}Oe^{-1}$ has been reported in $BiFeO_3$ thin films.[167] Very high value of α_{ME} has been reported in multilayer structures. For example, $\alpha_{ME} \approx 4.30 \times 10^5$ $mVcm^{-1}Oe^{-1}$, has been observed in Ta/Cu/Mn–Ir/Fe-Co-Si-B/AlN multilayer structure.[168]

2.4.8 REALIZING MULTIFERROICITY IN PRACTICE

The rarity of magnetic ferroelectrics is essentially related to d^0–d^n condition (put forward by Nicola Hill) which inhibits the co-existence of magnetic and ferroelectric ordered states in a single material. The following four approaches can be used to synthesize multiferroics.

1. By making magnetic and polar occupy different positions (Wyckoff) in the unit cell as in $BiFeO_3$ and $EuTiO_3$.
2. By synthesizing a double perovskite structure having two sub-lattices: one ferroelectric and the other magnetic.
3. By synthesizing magnetic perovskites where spiral magnetic ordering drives ferroelectric behavior.
4. By forming composites, where one of the materials is a strong ferroelectric and the other exhibits magnetic ordering.[161]

2.4.9 *DIFFERENT TYPES OF MULTIFERROIC MATERIALS*

A brief summary of different types of multiferroic materials is given below.

2.4.9.1 A-SITE DRIVEN FERROELECTRIC MULTIFERROICS

It was shown using first-principles calculations that ferroelectricity can be induced in materials having A-site located lone-pair electrons (as in $PbTiO_3$).[161] Such materials are termed as A-site driven. The magnetic ion replacement at the B-site can induce magnetic. $BiMnO_3$[169] and $BiFeO_3$[170] are typical examples. $BiFeO_3$ is a G-type antiferromagnetic[171] having rhombohedrally distorted crystal structure with T_c and $T_N \approx 1103$ K and 643 K, respectively.[139,172] Studies on multiferroic $BiFeO_3$ have resulted have clarified canting-related ferromagnetic ordering. The absence of magnetic ordering in bulk $BiFeO_3$ has been explained in terms of the presence of a cycloidal spin structure. However, $BiFeO_3$ thin films do exhibit magnetic ordering.[170] Theoretically, it has been established that tilting of oxygen octahedra in $BiFeO_3$ thin films leads to spin canting and therefore magnetic ordering.[173,174] Experimental confirmation of canting has been done using the polarized small-angle neutron scattering method.[175] High leakage current (attributable to oxygen vacancies) severely restricts $BiFeO_3$ from practical applications.[176] Therefore, aliovalent doping strategies are used to limit leakage current in $BiFeO_3$.[177] $BiMnO_3$ is bismuth-based on another perovskite. It is unique in the sense that it exhibits true ferromagnetism. All other bismuth-based oxides exhibit either antiferromagnetic or canted ferromagnetic behavior.[178] Its crystal structure is monoclinically distorted with magnetic and ferroelectric transition temperatures $\approx$ 100 and 450 K.[179,180] However, it cannot be prepared at ambient pressure usingthe conventional solid-state reaction method. Specialized high-temperature and high-pressure methods are needed for its synthesis.

2.4.9.2 B-SITE DRIVEN FERROELECTRIC MULTIFERROICS

A potential ferroelectric multiferroic which is B-site driven is $EuTiO_3$. Its magnetic structure is G-type antiferromagnet at temperatures below 5.3 K.[181–183] and possesses a cubic crystal structure (space group *Pm- m*). It is a quantum paraelectric antiferromagnet in which ferroelectricity is strain-induced.[184,185] Ferroelectricity is principally forbidden due to its cubic structure. However, it displays antiferrodistortive phase transition (like strontium titanate). This kind of transition is believed to have a very strong association with antiferroelectric behavior.[186] The first-principles calculation predicted three possible antiferrodistortive phases.[184] Ferroelectricity has been observed in doped $EuTiO_3$ samples such as $Eu_{0.8}Y_{0.2}MnO_3$[188] and $Eu_xSr_{1-x}MnO_3$.[189,190] Magnetodielectric studies have confirmed strong ME coupling in $EuTiO_3$ and systems based upon it.[191]

2.4.9.3 DOUBLE AND COMPLEX PEROVSKITE

Complex perovskite have $AB'_{1-x}B_xO_3$ as their compositional formula. The x = 0.5 materials are termed double perovskites.[192] Lu_2MnCoO_6,[193] Bi_2NiMnO_6,[194] Y_2NiMnO_6,[195] and Bi_2FeCrO_6[196] are typical double perovskites multiferroics. The arrangement of B-site cations in all these materials is rock-salt type. However, only Bi_2FeCrO_6 (in the form of thin films) displays both ferroelectric and magnetic transitions above room temperature and is therefore multiferroic.[197] Double perovskites (such as $Pb(Fe_{2/3}W_{1/3})O_3$, $PbFe_{0.5}Nb_{0.5}O_3$, $PbFe_{0.5}Ta_{0.5}O_3$, etc.) exhibit relatively higher ferroelectric and magnetic transition temperatures.[197–201] They are ferroelectric relaxors and are G-type antiferromagnets. Among these, solid solutions of $Pb(Fe_{2/3}W_{1/3})O_3$ with $PbTiO_3$ are most exciting for multiferroic applications. Room temperature weak ferromagnetism has been reported 0.8 $Pb(Fe_{2/3}W_{1/3})O_3$ -0.2 $PbTiO_3$ thin films[202] and $(Pb(Fe_{2/3}W_{1/3})O_3)_{1-x}$ – $(PbTiO_3)_x$ ceramics (x varying from 0.10 to 0.30).[203,204]

2.4.9.4 MANGANITES

The B-site cation (in ABO_3 compositional formula) manganites is occupied by manganese. Some of the manganites display very exciting magnetoelectric properties. For example, $TbMnO_3$[159] displays magnetic field-controlled ferroelectric polarization whereas $HoMnO_3$ displays electric field-controlled magnetic polarization.[205] However, these compounds usually display small ferroelectric polarization. Promising multiferroic properties have been observed in $CaMn_7O_{12}$.[206]

2.4.9.5 PEROVSKITE COMPOSITES

Magnetoelectric/multiferroic behavior has also been observed in composites of (1) ferromagnetic and (2) ferroelectric materials. These composites comprise combinations of materials like $CoFe_2O_4$ and $NiFe_2O_4$ (which are ferromagnetic) with excellent ferroelectrics such as $BaTiO_3$, $PbTiO_3$, and $Pb(Zr,Ti)O_3$, etc.[207,208] Such combinations offer higher technical viability on account of their (1) simplicity, (2) scalability, and (3) affordable cost.[209] $BaTiO_3/CoFe_2O_4$ heterostructures thin films supported on $SrTiO_3$ were prepared by Zheng et al.[208] using pulsed laser deposition. Saturation polarization ≈ 23 mC/cm^2 and saturation magnetization ≈ 300 emu/cc were observed.

KEYWORDS

- **piezoelectrics**
- **ferroelectrics**
- **multiferroics**
- **perovskites**
- **impedance spectroscopy**
- **dielectrics**

REFERENCES

1. Kay, H. F.; Bailey, P. C. A Structure and Properties of $CaTiO_3$. *Acta Crystallographica* 1957, *10* (3), 219–226.
2. Fiebig, M.; Lottermoser, Th.; Fröhlich, D.; Goltsev, A. V.; Pisarev, R. V. Observation of Coupled Magnetic and Electric Domains. *Nature* **2002,** *419*, 818–820.
3. Hill, N. A. Why Are There so Few Magnetic Ferroelectrics? *J. Phys. Chem. B* **2000,** *104* (29), 6694–6709.
4. Khomskii, D. Classifying Multiferroics: Mechanisms and Effects. *Physics* **2009,** *2*, 20.

5. Khomskii, D. I. Multiferroics: Different Ways to Combine Magnetism and Ferroelectricity. *J. Magn. Magn. Mater.* **2006,** *306* (1), 1–8.
6. Momma, K.; Izumi, F. VESTA 3 for Three-Dimensional Visualization of Crystal, Volumetric and Morphology Data. *J. Appl. Crystallogr.* **2011,** *44*, 1272–1276.
7. Jona, F.; Shirane, G. *Ferroelectric Crystals*; Dover Publications. Inc.: New York.
8. Uchino, K. et al. Science and Technology. *Adv. Mater.* **2015,** *16*, 046001.
9. Ahmed, S.; Akdogan, E. K. (Eds.). *Piezoelectric and Acoustic Materials for Transducer Applications*; Springer, 2008.
10. Sanyal, B.; Eriksson, O. (Eds.). Advanced Functional Materials: A Perspective from Theory and Experiment: Volume 2. In: *Science and Technology of Atomic, Molecular, Condensed Matter & Biological Systems*; Elsevier Science Ltd., 2012.
11. Tian, H. (Ed.). *Perovskite Materials, Devices and Integration*; IntechOpen, 2020.
12. Sekhar, M. C.; Veena, E.; Kumar, N. S.; Naidu, K. C.; Mallikarjuna, A.; Basha, D. B. A Review on Piezoelectric Materials and Their Applications. *Cryst. Res. Technol.* **2022,** 2200130.
13. Damjanovic, D. Ferroelectric, Dielectric and Piezoelectric Properties of Ferroelectric Thin Films and Ceramics. *Rep. Progress Phys.* **1998,** *61*, 1267.
14. Poplavko, Y., Functional Dielectrics for Electronics. *Piezoelectricity* **2020,** 161–216. DOI:10.1016/B978-0-12-818835-4.00005-5.
15. Kim, H.-P.; Kang, W.-S.; Hong, C.-H.; Lee, G.-J.; Choi, G.; Ryu, J.; Jo, W. *Advanced Ceramics for Energy Conversion and Storage, Elsevier Series in Advanced Ceramic Materials*; 2020; pp 157–206.
16. https://www.americanpiezo.com/knowledge-center/piezo-theory/piezoelectric-constants.html.
17. Xu, T.-B. Structural Health Monitoring (SHM) in Aerospace Structures; Woodhead Publishing, 2016; pp 175–212.
18. Gao, S.; Ao, H.; Jiang, H. Properties and Performance of General Piezoelectric Materials on a Novel Cantilevered Energy Harvester. *IOP Conf. Ser.: Material Science and Engineering* **2019,** *562*, 012098.
19. Carlsson, L.; Crane, R. L.; Uchino, K. *Comprehensive Composite Materials*; Elsevier, 2000.
20. Nelmes, R. J.; Kuhs, W. F. The Crystal Structure of Tetragonal $PbTiO_3$ at Room Temperature and at 700 K. *Solid State Commun.* **1985,** *54* (8), 721–723.
21. Gavrilyachenko, V. G.; Fesenko, E. G. Piezoelectric Effect in Lead Titanate Single Crystals. *Soviet Phys. Crystallogr. USSR* **1971,** *16* (3), 549.
22. Habib, M.; Lantgios, I.; Hornbostel, K. A Review of Ceramic, Polymer and Composite Piezoelectric Materials. *J. Phys. D: Appl. Phys.* **2022,** *55*, 423002.
23. Low, T. S.; Guo, W. Modeling of a Three-Layer Piezoelectric Bimorph Beam with Hysteresis. *J. Microelectromech. Syst.* **1995,** *4*, 230.
24. Butt, Z.; Pasha, R. A.; Qayyum, F.; Anjum, Z.; Ahmad, N.; Elahi, H. Generation of Electrical Energy Using Lead Zirconate Titanate (PZT-5A) Piezoelectric Material: Analytical, Numerical and Experimental Verifications. *J. Mech. Sci. Technol.* **2016,** *30*, 3553–3558.
25. Zhu, X. *Piezoelectric Ceramic Materials: Processing, Properties, Characterization & Applications (Materials Science and Technologies)*; Nova Science Publishers, 2010.
26. Hao, J.; Li, W.; Zhai, J.; Chen, H. Progress in High-Strain Perovskite Piezoelectric Ceramics. *Mater. Sci. Eng. R* **2019,** *135*, 1–57.
27. Uchino, K. *Advanced Piezoelectric Materials Science and Technology*; Woodhead Publishing Series in Electronic and Optical Materials, 2010; pp 1–85.

28. Gao, J.; Xue, D.; Liu, W.; Zhou, C.; Ren, X. Recent Progress on $BaTiO_3$-Based Piezoelectric Ceramics for Actuator Applications. *Actuators* **2017,** *6* (3), 24.
29. Karaki et al. *Jpn. J. Appl. Phys*. 46 L97.
30. Quan, N. D.; Bac, L. H.; Van Thiet,D.; Hung, V. N.; Dung, D. D. Current Development in Lead-Free $Bi_{0.5}$ $(Na,K)_{0.5}TiO_3$-Based Piezoelectric Materials. *Adv. Mater. Sci. Eng.* **2014, 2014,** Article ID 365391.
31. Smolensky, G. A.; Isupov, V. A.; Agranovskaya, A. I.; Krainic, N. N. New Ferroelectrics with Complex Compounds: IV. *Fizika Tverdogo Tela* **1960,** *2*, 2982–2985.
32. Jones, G. O.; Thomas, P. A. Investigation of the Structure and Phase Transitions in the Novel A-Site Substituted Distorted Perovskite Compound $Na_{0.5}Bi_{0.5}TiO_3$. *Acta Crystallographica B* **2002,** *58*, 168–178.
33. Isupov, V. A. Ferroelectric $Na_{0.5}Bi_{0.5}TiO_3$ and $K_{0.5}Bi_{0.5}TiO_3$ Perovskites and Their Solid Solutions. *Ferroelectrics* **2005,** *315* (1), 123–147.
34. Jaffe, B.; Cook, W. R..; Jaffe, H. *Piezoelectric Ceramics*. Academic Press: London, 1971.
35. Smolenskii, G. A.; Agranovskaya, A. I. Dielectric Polarization of a Series of Compounds of Complex Composition. *Fizika Tverdogo Tela* **1959,** *1*, 1562–1572.
36. Razumovskaya, O. N.; Kuleshova, T. B.; Rudkovskaya, L. M. Reactions of Formation of $BiFeO_3$, $K_{0.5}$ $Bi_{0.5}TiO_3$, and $Na_{0.5}Bi_{0.5}TiO_3$. *Neorganicheskie Materialy [Inorganic Materials]* **1983,** *19*, 113–115.
37. Ivanova, V. V.; Kapishev, A. G.; Venevtsev, Y. N.; Zhdanov, G. S. X-ray Determination of the Symmetry of Elementary Cells of the Ferroelectric Materials $(K_{0.5}Bi_{0.5})TiO_3$ and $(Na_{0.5}Bi_{0.5})TiO_3$ and of High-Temperature Phase Transitions in $(K_{0.5}Bi_{0.5})TiO_3$. *Izvestiya Akademii Nauk SSSR, Seriya Fizicheskaya* **1962,** *26*, 354–356.
38. Yoshii, K.; Hiruma, Y.; Nagata, H.; Takenaka, T. Electrical Properties and Depolarization Temperature of (Bi1/2Na1/2)TiO^3- (Bi1/2K1/2)TiO_3 Lead-Free Piezoelectric Ceramics. *Jpn J. Appl. Phys.* **2006,** *45* (part 1, no. 5B), 4493–4497.
39. Sezer, N.; Koç, M. A Comprehensive Review on the State-of-the-Art of Piezoelectric Energy Harvesting. *Nano Energy* **2021,** *80*, 105567.
40. Wang, Y.; Wu, J.; Xiao, D.; Zhu, J.; Jin, Y.; Zhu, J.; Yu, P.; Wu, L.; Li, X. Microstructure, Dielectric, and Piezoelectric Properties of (Li, Ag, Ta) Modified $(K_{0.50}Na_{0.50})$ NbO_3 Lead-Free Ceramics with High Curie Temperature. *J. Appl. Phys.* **2007,** *102* (5), 054101.
41. Guo, Y.; Kakimoto, K.; Ohsato, H. $(Na_{0.5}K_{0.5})NbO_3$–$LiTaO_3$ Lead-Free Piezoelectric Ceramics. *Mater. Lett.* **2005,** *59* (2–3), 241–244.
42. Zhang, S.; Xia, R.; Shrout, T. R.; Zang, G.; Wang, J. Piezoelectric Properties in Perovskite 0.948 $(K_{0.5}Na_{0.5})NbO_3$–0.052$LiSbO_3$ Lead-Free Ceramics. *J. Appl. Phys.* **2006,** *100* (10), 104108.
43. Du, H.; Tang, F.; Liu, D.; Zhu, D.; Zhou, W.; Qu, S. The Microstructure and Ferroelectric Properties of $(K_{0.5}Na_{0.5})NbO_3$–$LiNbO_3$ Lead-Free Piezoelectric Ceramics. *Mater. Sci. Eng. B* **2007,** *136* (2–3), 165–169.
44. Guo, Y.; Kakimoto, K.; Ohsato, H. Dielectric and Piezoelectric Properties of Lead-Free $(Na_{0.5}K_{0.5})NbO_3$–$SrTiO_3$ Ceramics. *Solid State Commun.* **2004,** *129* (5), 279–284.
45. Zhang, S.-T.; Kounga, A. B.; Aulbach, E.; Granzow, T.; Jo, W.; Kleebe, H. J.; R¨odel, J. Lead-Free Piezoceramics with Giant Strain in the System $Bi_{0.5}Na_{0.5}TiO_3$–$BaTiO_3$–$K_{0.5}Na_{0.5}NbO_3$. I. Structure and Room Temperature Properties, *J. Appl. Phys.* **2008,** *103* (3), 034107.
46. Wang, R.; Xie, R. J.; Hanada, K.; Matsusaki, K.; Kawanaka, H.; Bando, H.; Sekiya, T.; Itoh, M. Enhanced Piezoelectricity Around the Tetragonal/Orthorhombic Morphotropic

Phase Boundary in (Na,K)NbO_3–$ATiO_3$ Solid Solutions. *J. Electroceram.* **2008,** *21* (1–4), 263–266.
47. Tittmann, B. R.; Parks, D. A.; Zhang, S. O. High Temperature Piezoelectrics—A Comparison. In: *13th International Symposium on Nondestructive Characterization of Materials (NDCM-XIII)*, 20–24 May **2013,** Le Mans, France.
48. Islam, Md. S. Piezoelectric and Dielectric Properties of $LiNbO_3$, PMN-PT, and PZT-5A Materials, at Cryogenic Temperatures, 2018. https://era.library.ualberta.ca/items/610f0c17–6653–44b7–873a-fd1ef5daf65e/download/c75f50c6–5e11–420f-bc4e-8d63 f1c970ef.
49. Lithium Niobate—Boston Piezo-Optics Inc. www.bostonpiezooptics.com/lithium-niobate.
50. Guy, I. L.; Muensit, S.; Goldys, E. M. Extensional Piezoelectric Coefficients of Gallium Nitride and Aluminum Nitride. *Appl. Phys. Lett.* **1999,** *75* (26).
51. Swart, J. W. (Ed.). *Solid State Circuits Technologies*; Intechopen, 2010. https://doi.org/10.5772/6887.
52. Kim, S.; Nam, H.; Calisir, I. Lead-Free $BiFeO_3$-Based Piezoelectrics: A Review of Controversial Issues and Current Research State. *Materials* **2022,** *15* (13).
53. Makarovic, M.; Bencan, A.; Walker, J.; Malic, B.; Rojac, T. Processing, Piezoelectric and Ferroelectric Properties of (x)$BiFeO_3$- (1−x)$SrTiO_3$ Ceramics. *J. Eur. Ceram. Soc.* **2019,** *39*, 3693–3702.
54. Yan, F.; Bai, H.; Ge, G.; Lin, J.; Shi, C.; Zhu, K.; Shen, B.; Zhai, J.; Zhang, S. Composition and Structure Optimized $BiFeO_3$-$SrTiO_3$ Lead-Free Ceramics with Ultrahigh Energy Storage Performance. *Small* **2022,** *18*, 2106515.
55. Lu, Z.; Wang, G.; Li, L.; Huang, Y.; Feteira, A.; Bao, W.; Kleppe, A.K.; Xu, F.; Wang, D.; Reaney, I. M. In Situ Poling X-ray Diffraction Studies of Lead-Free $BiFeO_3$-$SrTiO_3$ Ceramics. *Mater. Today Phys.* **2021,** *19*, 100426.
56. Wu, M.; Fang, L.; Li, G.; Elouadi, B. Dielectric and Ferroelectric Properties of (1−x) $BiFeO_3$-x$Bi_{0.5}Na_{0.5}TiO_3$ Solid Solution. *Ferroelectrics* **2015,** *478*, 18–25.
57. Zhou, C.; Liu, X.; Li, W.; Yuan, C. Structure and Piezoelectric Properties of $Bi_{0.5}Na_{0.5}TiO_3$-$Bi_{0.5}Ka_{0.5}TiO_3$-$BiFeO_3$ Lead-Free Piezoelectric Ceramics. *Mater. Chem. Phys.* **2009,** *114*, 832–836.
58. Fujii, I.; Ito, Y.; Suzuki, T.; Wada, T. Ferroelectric and Piezoelectric Properties of $(Bi_{1/2}Na_{1/2})TiO_3$-$BiFeO_3$ Ceramics. *J. Mater. Res.* **2016,** *31*, 28–35.
59. Lee, M. H.; Kim, D. J.; Park, J. S.; Kim, S. W.; Song, T. K.; Kim, M. H.; Kim, W. J.; Do, D.; Jeong, I. K. High-Performance Lead-Free Piezoceramics with High Curie Temperatures. *Adv. Mater.* **2015,** *27*, 6976–6982.
60. Heiji Kawai, The Piezoelectricity of Poly (Vinylidene Fluoride). *Jpn. J. Appl. Phys.* **1969,** *8*, 975.
61. Bauer, S.; Bauer, F.. Piezoelectric Polymers and Their Applications. In: *Piezoelectricity: Springer Series in Materials Science*, Vol. 114; Springer: Berlin, Heidelberg, 2008. https://doi.org/10.1007/978-3-540-68683-5_6.
62. Sappati, K. K.; Bhadra, S. Piezoelectric Polymer and Paper Substrates: A Review. *Sensors* **2018,** *18* (11), 3605.
63. Gomes, J.; Serrado Nunes, L.; Sencadas, V.; Lanceros-Mendez, S. Influence of the β-Phase Content and Degree of Crystallinity on the Piezo- and Ferroelectric Properties of Poly (Vinylidene Fluoride). *Smart Mater. Struct.* **2010,** *19*, 065010.
64. Harrison, J.; Ounaies, Z. Piezoelectric Polymers; NASA/CR-2001–211422 ICASE Report No. 2001–43.

65. Anwar, S.; Amiri, M. H.; Jiang, S.; Abolhasani, M. M.; Rocha, P. R. F.; Asad, K. Piezoelectric Nylon-11 Fibers for Electronic Textiles, Energy Harvesting and Sensing. *Adv. Funct. Mater.* **2021,** *31*, 2004326.
66. Kawai, H. Electrostriction and Piezoelectricity of Elongated Polymers Films Oyo Buturi. **1970,** *39*, 413.
67. Wu, J.; Fu, Y.; Hu, G.-H.; Wang, S.; Xiong, C. Effect of Stretching on Crystalline Structure, Ferroelectric and Piezoelectric Properties of Solution-Cast Nylon-11 Films. *Polymers* **2021,** *13* (13), 2037.
68. Simpson, J.; Ounaies, Z.; Fay, C. Polarization and Piezoelectric Properties of a Nitrile Substituted Polyimide. *Mater. Res. Soc. Symp.—Proc.* **1997,** *459*, 59–64.
69. Atkinson, G. M.; Pearson, R. E.; Ounaies, Z.; Park, C.; Harrison, J. S.; Midkiff, J. A. Piezoelectric Polyimide Tactile Sensors. In: *Biennial University/Government/Industry Microelectronics Symposium—Proceedings*; Published—1 Sep 2003; pp 308–311.
70. Sherman, J. D.; Elloian, J.; Jadwiszczak, J.; Shepard, K. L. On the Temperature Dependence of the Piezoelectric. *ACS Appl. Polym. Mater.* **2020,** *2* (11), 5110–5120.
71. Yang, L.; Chi, S.; Dong, S.; Yuan, F.; Wang, Z.; Lei, J.; Bao, L.; Xiang, J.; Wang, J. Preparation and Characterization of a Novel Piezoelectric Nanogenerator Based on Soluble and Meltable Copolyimide for Harvesting Mechanical Energy. *Nano Energy* **2020,** *67*, 104220.
72. Usher, D.; Cousins, K. R.; Zhang, R.; Ducharmec, S. The Promise of Piezoelectric Polymers Timothy. *Polym. Int.* **2018,** *67*, 790–798.
73. Kirjavainen, K. Electromechanical Film and Procedure for Manufacturing Same, US Patent US465**4546,** 1987. https://patents.google.com/patent/US4654546A/en.
74. Mohebbi, A.; Mighri, F.; Ajji, A.; Rodrigu, D. Cellular Polymer Ferroelectret: A Review on Their Development and Their Piezoelectric Properties. *Adv. Polym. Technol.* **2018,** *37*, 468–483.
75. Hamdi, O.; Mighri, F.; Rodrigue, D. Piezoelectric Cellular Polymer Films: Fabrication, Properties and Applications. *AIMS Mater. Sci.* **2018,** *5* (5), 845–869.
76. Kulichikhin, V.; Semakov, A.; Tur, D. New Flexible Piezoelectrics and Actuators Based on Polyorganophosphazenes. *Sens. Actuators A* **2016,** *252*, 48–53.
77. Furukawa T. *Phase Trans.* **1989,** *18*, 143–211.
78. Lay, R.; Deijsabc, G. S.; Malmstř, J. The Intrinsic Piezoelectric Properties of Materials—A Review with a Focus on Biological Materials. *RSC Adv.* **2021,** *11*, 30657.
79. Šutka, A.; Sherrell, P. C.; Shepelin, N. A.; Lapčinskis, L.; M̄alnieks, K.; Ellis, A. V. *Adv. Mater.* **2020,** *32*, 2002979.
80. Jiang, P.; Yan, F.; Nasr Esfahani, E.; Xie, S.; Zou, D.; Liu, X.; Zheng, H.; Li, J. *ACS Biomater. Sci. Eng.* **2017,** *3*, 1827–1835.
81. Jiang, H. Y.; Yen, F.; Huang, C. W.; Mei, R. B.; Chen, L. *AIP Adv.* **2017,** *7*, 045215.
82. Kim, D.; Han, S. A.; Kim, J. H.; Lee, J. H.; Kim, S. W.; Lee, S. W. *Adv. Mater.* **2020,** *32*, 1906989.
83. Guerin, S.; Syed, T. A. M.; Thompson, D. *Nanoscale* **2018,** *10*, 9653–9663.
84. Guerin, S.; Stapleton, A.; Chovan, D.; Mouras, R.; Gleeson, M.; McKeown, C.; Noor, M. R.; Silien, C.; Rhen, F. M. F.; Kholkin, A. L.; Liu, N.; Soulimane, T.; Tofail, S. A. M.; Thompson, D. *Nat. Mater.* **2018,** *17*, 180–186.
85. Yun, G. Y.; Kim, H. S.; Kim, J.; Kim, K.; Yang, C. *Sens. Actuators A* **2008,** *141*, 530–535.
86. Vasilescu, D.; Cornillon, R.; Mallet, G. *Nature* **1970,** *225*, 635.
87. Lemanov, V. V.; Popov, S. N.; Pankova, G. A. *Phys. Solid State* **2011,** *53*, 1191–1193.

88. Guerin, S.; Tofail, S. A. M.; Thompson, D. *NPG Asia Mater.* **2019,** *11*, 1–5.
89. Sun, Y.; Zeng, K. Y.; Li, T. *Sci. China: Phys., Mech. Astron.* **2020,** *63*, 278701.
90. Denning, D.; Kilpatrick, J. I.; Fukada, E.; Zhang, N.; Habelitz, S.; Fertala, A.; Gilchrist, M. D.; Zhang, Y.; Tofail, S. A. M.; Rodriguez, B. J. *ACS Biomater. Sci. Eng.* **2017,** *3*, 929–935.
91. Liu, Y.; Wang, Y.; Chow, M. J.; Chen, N. Q.; Ma, F.; Zhang, Y.; Li, J. *Phys. Rev. Lett.* **2013,** *110*, 1–5.
92. Stapleton, A.; Ivanov, M. S.; Noor, M. R.; Silien, C.; Gandhi, A. A.; Soulimane, T.; Kholkin, A. L.; Tofail, S. A. M. *Ferroelectrics* **2018,** *525*, 135–145.
93. Skinner, D. P.; Newnham, R. E.; Cross, L. E. Flexible Composite Transducers. *Mater. Res. Bull.* **1978,** *13*, 599–607.
94. Pilgrim, S. M.; Newnham, R. E.; Rohlfing, L. L. An Extension of the Composite Nomenclature Scheme. *Mater. Res. Bull.* **1987,** *22*, 677–684.
95. Akdogan, E. K.; Allahverdi, M.; Safari, A. *IEEE Trans. Ultrason. Ferroelectr. Freq. Control* **2005,** *52*, 746.
96. Kitayama, T.; Sugawarda. *Proc. Study Commun. Electron. Circ. Components Mater.*, 1972. Ref. No. CPM72–17 (1972–0).
97. Uchino, K., *Advanced Piezoelectric Materials Science and Technology*, 2nd Edn.; Elsevier, 2017.
98. Siddiqui, S.; Kim, D. I.; Roh, E.; Duy, L.T.; Trung, T. Q.; Nguyen, M. T.; Lee, N. E. *Nano Energy* **2016,** *30*, 434.
99. Xu, X.; Cao, D.; Yang, H.; He, M. Application of Piezoelectric Transducer in Energy Harvesting in Pavement. *Int. J. Pave. Res. Tech.* **2018,** *11*, 388–395.
100. Sarker, M. R.; Julai, S.; Sabri, M. F. M.; Said, S. M.; Islam, M. M.; Tahir, M. Review of Piezoelectric Energy Harvesting System and Application of Optimization Techniques to Enhance the Performance of the Harvesting System. *Sens. Actuators A* **2019,** 300, 111634.
101. Sapkal, S.; Kandasubramanian, B.; Panda, H. S. A Review of Piezoelectric Materials for Nanogenerator Applications. *J. Mater. Sci.: Mater. Electron.* **2022,** *33*, 26633–26677.
102. Aabid, A.; Raheman, M. A.; Ibrahim, Y. E.; Anjum, A.; Hrairi, M.; Parveez, B.; Parveen, N.; Zayan, J. M. A Systematic Review of Piezoelectric Materials and Energy Harvesters for Industrial Applications. *Sensors* **2021,** *21*, 4145.
103. Shivashankar, P.; Gopalakrishnan, S. Review on the Use of Piezoelectric Materials for Active Vibration, Noise, and Flow Control. *Smart Mater. Struct.* **2020,** *29*, 053001.
104. https://www.worldscientific.com/doi/pdf/10.1142/9789813238947_0001.
105. Gupta, S. Introduction to Ferroelectrics and Related Materials. In: *Ferroelectric Materials for Energy Harvesting and Storage*; Woodhead Publishing Series in Electronic and Optical Materials, 2021; pp 1–41. DOI: 10.1016/b978–0-08–102802–5.00001–7.
106. Sawyer, C. B.; Tower, C. H. Rochelle Salt as a Dielectric. *Phys. Rev.* **1930,** *35* (3), 269–273.
107. Uchino, K. *Ferroelectric Devices*; Marcel Dekker Inc.: New York, 2000.
108. Chawla, A.; Verma, S.; Godara, S.; Bhadu, G. R.; Singh, A.; Singh, M. Understanding Phase Segregation Using Rietveld Analysis and the Dielectric, Ferroelectric Properties of Ba (1_x)CaxTiO_3 Solid Solutions. *J. Elec. Mater.* **2002,** *49* (7).
109. Chawla, A.; Singh, A.; Babu, P. D.; Singh, M. Composition Dependent High M-E Coupling Strength, Multiferroic Behaviour and Phase Analysis of Fe Doped BCT Solid Solutions. *J. Alloy Compos.* **2020,** *857*, 158255.

110. Chawla, A.; Singh, A.; Babu, P. D.; Singh, M. Magnetic Ion Oxidation State Dependent Magnetoelectric Coupling Strength in Fe Doped BCT Ceramics. *RSC Adv.* **2020,** *10,* 21019–21027.
111. Zhang, H.; Wei, T.; Zhang, Q.; Ma, W.; Fan, P.; Salamon, D.; Zhang, S.-T.; Nan, B.; Tan, H.; Ye Haibo, Z.-G. A Review on the Development of Lead-Free Ferroelectric Energy-Storage Ceramics and Multilayer Capacitors. *J. Mater. Chem. C* **2020,** *8,* 16648.
112. Kao, K. C. *Dielectric Phenomena in Solids*; Elsevier, 2004.
113. Govinda Raju, G. *Dielectric Loss and Relaxation—I*; CRC Press, 2016.
114. Cole, K. S.; Cole, R. H. Dispersion and Absorption in Dielectrics I. Alternating Current Characteristics. *Chem. Phys.* **1941,** *9,* 341.
115. Davidson, D. W.; Cole, R. H. Dielectric Relaxation in Glycerin. *J. Chem. Phys.* **1950,** *18,* 1417.
116. Havriliak, S.; Negami, S. A Complex Plane Analysis of α-Dispersions in Some Polymer Systems. *J. Polym. Sci., Pt. C.* **1966,** *14,* 99.
117. Busch, G. Early History of Ferroelectricity. *Ferroelectrics* **1987,** *11* (1), 267–284. DOI: 10.1063/1.1750821.
118. Cohen, R. E. Origin of Ferroelectricity in Perovskite Oxides. *Nature* **1992,** *358* (6382), 136–138.
119. Ederer, C.; Spaldin, N. A New Route to Magnetic Ferroelectrics. *Nat. Mater.* **2004,** *3,* 849–851.
120. Liu, H.; Yang, X. A Brief Review on Perovskite Multiferroics. *Ferroelectrics* **2017,** *507* (1), 69–85.
121. Spaldin, N. A. *Magnetic Materials: Fundamentals and Applications*; Cambridge University Press, 2010.
122. Valasek, J. Piezo-Electric and Allied Phenomena in Rochelle Salt. *Phys. Rev.* **1921,** *17* (4), 475.
123. Kaur, R.; Sharma, V.; Kumar, M.; Singh, M.; Singh, A. Conductivity Relaxation in $Pb_{0.9}Sm_{0.10}Zr_{0.405}Ti_{0.495}Fe_{0.10}O_3$ Solid Solution. *J. Alloys Comp.* **2018,** *735,* 1472–1479.
124. Trainer, M. Ferroelectrics and the Curie-Weiss Law. *Eur. J. Phys.* **2000,** *21,* 459–464.
125. Uchino, K.; Nomura, S. *Ferroelectr. Lett. Sect.* **1982,** *44,* 55–61.
126. Kirillov, V. V.; Isupov, V. A. Relaxation Polarization of $PbMg_{1/3}Nb_{2/3}O_3$ (PMN)—A Ferroelectric with a Diffused Phase Transition. *Ferroelectrics* **1973,** *5,* 3–9.
127. Smolenskii, G. A.; Bokov, V. A. Coexistence of Magnetic and Electric Ordering in Crystal. *J. Appl. Phys.* **1964,** *35,* 915–918.
128. Shukla, A.; Choudhary, R. N. P. High-Temperature Impedance and Modulus Spectroscopy Characterization of La^{3+}/Mn^{4+} Modified $PbTiO_3$ Nanoceramics. *J. Phys. B.: Condens Matter.* **2011,** *406,* 2492–2500.
129. Mohanty, S.; Choudhary, R.; Padhee, R. et al. Dielectricand Impedance Spectroscopy of $BiFeO_3$–$NaTaO_3$ Multi-Ferroics. *Ceram Int.* **2014,** *40,* 9017–9025.
130. Sen, S.; Choudhary, R. N. P.; Tarafdar, A.; Pramanik, P. Impedance Spectroscopy Study of Strontium Modified Lead Zirconate Titanate Ceramics. *J. Appl. Phys.* **2006,** *99,* 124114.
131. Schmidt, R.; Eerenstein, W.; Winiecki, T.; Morrison, F. D.; Midgley, P. A. Impedance Spectroscopy of Epitaxial Multiferroic Thin Films. *Phys Rev B.* **2007,** *75,* 245111.
132. Chawla, A.; Singh, A.; Singh, M.; Malhi, P. S. Small Polaron Hopping-Assisted Electrical Conduction and Relaxation in BCT and Mn-Doped BCT Samples. *J. Asian Ceram. Soc.* **2019,** *7* (4), 558–568.

133. Amine Fersi, M.; Chaabane, I.; Gargouri, M. Experimental and Theoretical Study of AC Electrical Conduction Mechanisms of Bis (4-acetylanilinium) Tetrachloridozincate. *Phys. B: Condens. Matter.* 2014, *444*, 89–93.
134. Mikolajick, T.; Slesazeck, S.; Mulaosmanovic, H.; Park, M. H.; Fichtner, S.; Lomenzo, P. D.; Hoffmann, M. Schroeder, U. Next Generation Ferroelectric Materials for Semiconductor Process Integration and Their Applications. *J. Appl. Phys.* **2021,** *129*, 100901.
135. Koruza, J.; Kodumudi Venkataraman, L.; Malič, B. Magnetic, Ferroelectric, and Multiferroic Metal Oxides. *Metal Oxides* 2018;, pp 51–69. https://doi.org/10.1016/B978–0-12–811180–2.00003–7.
136. Meng, K.; Li, W.; Tang, X.-G.; Liu, Q.-X.; Jiang, Y.-P. A Review of a Good Binary Ferroelectric Ceramic: $BaTiO_3$–$BiFeO_3$, *ACS Appl. Electron. Mater.* **2022,** *4* (5), 2109–2145.
137. Kumari, P.; Rai, R.; Sharma, S.; Shadilya, M.; Tiwari, A. State-of-the-art Of Lead Free Ferroelectrics: A Critical Review. *Adv. Mater. Lett.* **2015,** *6* (6), 453–484.
138. Curie, P. Sur la Symetrie dans les Phenomenes Physiques.' *Journal de physique* **1894,** *Ser. 3*, 393–415.
139. Béa, H.; Gajek, M.; Bibes, M.; Barthélémy, A. Spintronics with Multiferroics. *J. Phys.: Condens. Matter* **2008,** *20* (43), 434221.
140. Lu, C.; Wu, M.; Lin, L.; Liu, J.-M. Single-Phase Multiferroics: New Materials, Phenomena, and Physics. *Nat. Sci. Rev.* **2019,** *6* (4), 653–668.
141. Eerenstein, W.; Mathur, N. D.; Scott, J. F. Multiferroic and Magnetoelectric Materials. *Nature* **2006,** *442* (7104), 759–765.
142. Nan, C.-W.; Liu, J.-M. Multiferroics: A Beautiful But Challenging Multi-Polar World. *Nat. Sci. Rev.* **2019,** *6* (4), 620.
143. Martin, L. W.; Ramesh, R. Multiferroic and Magnetoelectric Heterostructures. *Acta Materialia 60* (6), 2449–2470.
144. Schmid, H. Multi-Ferroic Magnetoelectrics. *Ferroelectrics 162* (1), 317–338.
145. Nan, C.-W. Magnetoelectric Effect in Composites of Piezoelectric and Piezomagnetic Phases. *Phys. Rev. B* **1994,** *50* (9), 6082.
146. Smolenskiĭ, G. A.; Chupis, I. E. Ferroelectromagnets. *Soviet Phys. Uspekhi* **1982,** *25* (7), 475.
147. Dzyaloshinskii, I. On the Magneto-Electrical Effects in Antiferromagnets. *Soviet Physics JETP* **1960,** *10*, 628–629.
148. Astrov, D. N.; *Zh. Éksp. Teor. Fiz.* **1960,** *38*, 984 [*Sov. Phys. JETP* **1960,** *11*, 708].
149. Venevtse, V. Y.u. N.; Zhdanov, G. S.; Solovijev, S. P.; Bezus, Ye. V.; Ivanov, (A. V. V.; Fedulo, V. S. A.; Kapyshe, A. V. G. *Kristallografya* **1960,** *5*, 620.
150. Bertaut, E. F.; Forrat, E. F.; Fang, P. H. *C. R. Acad. Sci.* **1963,** *256*, 1958.
151. Kiselev, S. V.; Ozerov, R. P.; Zhdanov, G. S. Detection of Magnetic Order in Ferroelectric $BiFeO_3$ by Neutron Diffraction. *Sov. Phys. Dokl.* **1963,** *7*, 742–744.
152. Coeuré, P.; Guinet, F.; Peuzin, J. C.; Buisson, G.; Bertaut, E. F. Ferroelectric Properties of Hexagonal Orthomanganites of Yttrium and Rare Earths. *Proc. Int. Meet Ferroelectr. Prag.* **1966.**
153. Ascher, E.; Rieder, H.; Schmid, H.; Stössel, H. Some Properties of Ferromagnetoelectric Nickel-Iodine Boracite, $Ni_3B_7O_{13}I$. *J Appl Phys.* **1966,** *37*, 1404–1405.
154. Aizu, K. Possible Species of Ferromagnetic, Ferroelectric, and Ferroelastic Crystals. *Phys. Rev. B.*, **1970,** *2*, 754–772.
155. Bokov, V. A.; Myl'nikova, I. E.; Smolenskii, G. A. *Zh. Eksp. Teor. Fiz.* **1962,** *42*, 643; *Sov. Phys. JETP (English Transl.)* 1962; *15*, 447.

156. Smolenskii. G. A.; Isupov, V. A.; Agranovskaya, A. I. New Ferroelectrics of Complex Composition of the Type A22+ (BI3+, BII5+)O6. I. *Sov. Phys. Solid State* **1959,** *1*, **1959,** 150–151.
157. Tong, X.; Lin, Y.-H.; Zhang, S.; Wang, Y.; Nan, C.-W. Preparation of Mn-Doped $BaTiO_3$ Nanoparticles and Their Magnetic Properties. *J. Appl. Phys.* **2008,** *104*, 066108.
158. Hur, N.; Park, S.; Sharma, P. A.; Ahn, J. S.; Guha, S.; Cheong, S.-W. Electric Polarization Reversal and Memory in a Multiferroic Material Induced by Magnetic Fields. *Nature* **2004,** *429* (6990), 392–395.
159. Kimura, T.; Goto, T.; Shintani, H.; Ishizaka, K.; Arima, T.; Tokura, Y. Magnetic Control of Ferroelectric Polarization. *Nature*, 426 (6962), 55–58.
160. Fiebig, M.; Lottermoser, T.; Meier, D.; Trassin, M. The Evolution of Multiferroics. *Nat. Rev. Mater.* **2016,** *1* (8), 16046.
161. Cohen, R. E. Origin of Ferroelectricity in Perovskite Oxides. *Nature* **1992,** *358*, 136–138.
162. Cullity, B. D.; Graham, B. D. C. D. *Introduction to Magnetic Materials*; Wiley, 2011.
163. van den Brink, J.; Khomskii, D. I. Multiferroicity Due to Charge Ordering. *J. Phy.: Cond. Matter.* **2008,** *20* (43), 434217.
164. Wright, J. P.; Attfield, J. P.; Radaelli, P. G. Charge Ordered Structure of Magnetite Fe_3O_4 Below the Verwey Transition. *Phys. Rev. B* **2002,** *66* (21), 214422.
165. Gupta, R.; Kotnala, R. K. A Review on Current Status and Mechanisms of Room-Temperature Magnetoelectric Coupling in Multiferroics for Device Applications. *J. Mater. Sci.* **2022,** *57*, 12710–12737.
166. Ebnabbasi, K.; Mohebbi, M.; Vittoria, C. Room Temperature Magnetoelectric Effects in Bulk Poly-Crystalline Materials of M- and Z-Type Hexaferrites. *J. Appl. Phys.* **2013,** *113*, 17C703.
167. Wang, N.; Luo, X.; Han, L.; Zhang, Z.; Zhang, R.; Olin, H.; Wang, Y. Y. Structure, Performance, and Application of $BiFeO_3$ Nanomaterials. *Nano Micro Lett.* **2020,** *12*, 81.
168. Lage, E.; Kirchhof, C.; Hrkac, V.; Kienle, L.; Jahns, R.; Knöchel, R.; Quandt, E.; Meyners, D. Exchange Biasing of Magnetoelectric Composites. *Nat. Mater*. **2012,** *116* (11), 523–529.
169. Hill, N. A.; Rabe, K. M. First-Principles Investigation of Ferromagnetism and Ferroelectricity in Bismuth Manganite. *Phys. Rev. B* **1999,** *59* (13), 8759–8769..
170. Wang, J.; Neaton, J. B.; Zheng, H.; Nagarajan, V.; Ogale, S. B.; Liu, B.; Viehland, D.; Vaithyanathan, V.; Schlom, D. G.; Waghmare, U. V.; Spaldin, N. A.; Rabe, K. M.; Wuttig, M.; Ramesh, R. Epitaxial $BiFeO_3$ Multiferroic Thin Film Heterostructures. *Science* **2003,** *299* (5613), 1719–1722.
171. Sosnowska, I.; Neumaier, T. P.; Steichele, E. Spiral Magnetic Ordering in Bismuth Ferrite. *J. Phys. C: Solid State Phys*. **1982,** *15*, 4835.
172. Fischer, P.; Polomska, M.; Sosnowska, I.; Szymanski, M. Temperature Dependence of the Crystal and Magnetic Structures of $BiFeO_3$. *J. Phys. C: Solid State Phys*. **1980,** *13*, 1931.
173. Albrecht, D.; Lisenkov, S.; Ren, W.; Rahmedov, D.; Kornev, I. A.; Bellaiche, L. Ferromagnetism in Multiferroic $BiFeO_3$ Films: A First-Principles-Based Study. *Phys. Rev. B* **2010,** *81* (14), 140401.
174. Ederer, C; Spaldin, N. A. Weak Ferromagnetism and Magnetoelectric Coupling in Bismuth Ferrite. *Phys. Rev. B* **2005,** *71* (6), 060401.
175. Ramazanoglu, M.; Laver, M.; Ratcliff, II, W.; Watson, S. M.; Chen, W. C.; Jackson, A.; Kothapalli, K.; Lee, S.; Cheong, S.-W..; Kiryukhin, V. Local Weak Ferromagnetism in Single-Crystalline Ferroelectric BiFeO3. *Phys. Rev. Lett.* **2011,** *107* (20), 207206.

176. Wang, Y. P.; Zhou, L.; Zhang, M. F.; Chen, X. Y.; Liu, J.-M.; Liu, Z. G. Room-Temperature Saturated Ferroelectric Polarization in $BiFeO_3$ Ceramics Synthesized by Rapid Liquid Phase Sintering. *Appl. Phys. Lett.* **2004,** *84*, 1731.
177. Cheng, Z. X.; Wang, X. L.; Du, Y.; Dou, S. X. A Way to Enhance the Magnetic Moment of Multiferroic Bismuth Ferrite. *J. Phys. D: Appl. Phys.* **2010,** *43*, 242001.
178. Achary, S. N.; Jayakumar, O. D.; Tyagi, A. K. Multiferroic Materials, Functional Materials. *Preparation, Processing and Applications* **2012,** 155–191. DOI:10.1016/B978–0-12–385142–0.00004–0.
179. Sugawara, F.; Iiida, S.; Syono, Y.; Akimoto, S. Magnetic Properties and Crystal Distortions of $BiMnO_3$ and $BiCrO_3$. *J. Phys. Soc. Jpn.* **1968,** *25*, 1553–1558.
180. Moreira dos Santos, A.; Cheetham, A. K.; Atou, T.; Syono, Y.; Yamaguchi, Y.; Ohoyama, K.; Chiba, H.; Rao, C. N. R. Orbital Ordering as the Determinant for Ferromagnetism in Biferroic $BiMnO_3$. *Phys. Rev. B* **2002,** *66*, 064425.
181. McGuire, T. R.; Shafer, M. W.; Joenk, R. J.; Alperin, H. A.; Pickart, S. J. Magnetic Structure of $EuTiO_3$. *J. Appl. Phys.* **1966,** *37* (3), 981–982.
182. Chien, C.-L.; DeBenedetti, S.; Barros, F. De S. Magnetic Properties of $EuTiO_3$, Eu_2TiO_4, and $Eu_3Ti_2O_7$. *Phys. Rev. B* **1974,** *10* (9), 3913–3922..
183. Bessas, D.; Rushchanskii, K. Z.; Kachlik, M.; Disch, S.; Gourdon, O.; Bednarcik, J.; Maca, K.; Sergueev, I.; Kamba, S.; Ležaić, M.; Hermann, R. P. Lattice Instabilities in Bulk $EuTiO_3$. *Phys. Rev. B* **2013,** *88* (14), 144308.
184. Haeni, J. H.; Irvin, P.; Chang, W.; Uecker, R.; Reiche, P.; Li, Y. L.; Choudhury, S.; Tian, W.; Hawley, M. E.; Craigo, B.; Tagantsev, A. K.; Pan, X. Q.; Streiffer, S. K.; Chen, L. Q.; Kirchoefer, S. W.; Levy, J.; Schlom, D. G. Room-Temperature Ferroelectricity in Strained $SrTiO_3$. *Nature* **2004,** *430*, 758–761,.
185. Choi, K. J.; Biegalski, M.; Li, Y. L.; Sharan, A.; Schubert, J.; Uecker, R.; Reiche, P.; Chen, Y. B.; Pan, X. Q.; Gopalan, V.; Chen, L.-Q.; Schlom, D. G.; Eom, C. B. Enhancement of Ferroelectricity in Strained $BaTiO_3$ Thin Films. *Science* **2004,** *306*, 1005–1009.
186. Rabe, K. M.; Ahn, C. H.; Triscone, J.-M. (Eds.). *Physics of Ferroelectrics: A Modern Perspective*; Springer Science & Business Media, 2007.
187. Rushchanskii, K. Z.; Spaldin, N. A.; Ležaić, M. First-Principles Prediction of Oxygen Octahedral Rotations in Perovskite-Structure $EuTiO_3$. *Phys. Rev. B* **2012,** *85* (10), 104109.
188. Hemberger, J.; Schrettle, F.; Pimenov, A.; Lunkenheimer, P.; Ivanov, V. Yu, Mukhin, A. A.; Balbashov, A. M.; Loidl, A. Multiferroic Phases of $Eu_{1-x}Y_xMnO_3$. *Phys. Rev. B* **2007,** *75* (3), 035118.
189. Morozovska, A. N.; Gu, Y.; Khist, V. V.; Glinchuk, M. D.; Chen, L.-Q.; Gopalan, V.; Elisee, E. A. Low-Symmetry Monoclinic Ferroelectric Phase Stabilized by Oxygen Octahedra Rotations in Strained $Eu_xSr_{1-x}TiO_3$ Thin Films. *Phys. Rev. B* **2013,** *87* (13), 134102.
190. Eliseev, E. A.; Glinchuk, M. D.; Khist, V. V.; Lee, C.-W.; Deo, C. S.; Behera, R. K.; Morozovska, A. N. New Multiferroics Based on EuxSr1−xTiO3 Nanotubes and Nanowires. *J. Appl. Phys.* **2013,** *113* (2), 024107.
191. Katsufuji, T.; Takagi, H. Coupling Between Magnetism and Dielectric Properties in Quantum Paraelectric $EuTiO_3$. *Phys. Rev. B* **2001,** *64* (5), 054415.
192. Howard, C. J.; Kennedy, B. J.; Woodward, P. M. Ordered Double Perovskites—A Group-Theoretical Analysis. *Acta Crystallogr. Sect. B* **2003,** *59* (4), 463–471.
193. Yáñez-Vilar, S.; Mun, E. D.; Zapf, V. S.; Ueland, B. G.; Gardner, J. S.; Thompson, J. D.; Singleton, J.; Sánchez-Andújar, M.; Mira, J.; Biskup, N.; Señarís-Rodríguez, M. A.;

Batista, C. D. Multiferroic Behavior in the Double-Perovskite Lu_2MnCoO_6. *Phys. Rev. B* **2011,** *84* (13), 134427.
194. Azuma, M.; Takata, K.; Saito, T.; Ishiwata, S.; Shimakawa, Y.; Takano, M. Designed Ferromagnetic, Ferroelectric $Bi_{(2)}NiMnO_{(6)}$. *J. Am. Chem. Soc.* **2005,** *127* (24), 8889–8892.
195. Su, J.; Yang, Z. Z; Lu, X. M.; Zhang, J. T.; Gu, L.; Lu, C. J.; Li, Q. C.; Liu, J.-M.; Zhu, J. S. Magnetism-Driven Ferroelectricity in Double Perovskite Y_2NiMnO_6. *ACS Appl. Mater. Interf.* **2015,** *7* (24), 13260–13265.
196. Nechach, R.; Harnagea, C.; Pignolel, A. Multiferroic Properties—Structure Relationships in Epitaxial Bi_2FeCrO_6 Thin Films: Recent Developments. *J. Phys.: Condens. Matter* **2012,** *24* (9), 096001.
197. Jeong, I.-K.; Ahn, J. S.; Kim, B. G.; Yoon, S.; Singh, S. P.; Pandey, D. Short- and Medium-Range Structure of Multiferroic Pb $(Fe_{1/2}Nb_{1/2})O_3$ Studied Using Neutron Total Scattering Analysis. *Phys. Rev. B* **2011,** *83* (6), 064108.
198. Blinc, R.; Cevc, P.; Zorko, A.; Holc, J.; Kosec, M.; Trontelj, Z.; Pirnat, J.; Dalal, N.; Ramachandran, V.; Krzystek, J. Electron Paramagnetic Resonance of Magnetoelectric Pb $(Fe_{1/2}Nb_{1/2})O_3$. *J. Appl. Phys.* **2007,** *101* (3), 033901.
199. R. Martinez, R. Palai, H. Huhtinen, J. Liu, J. F. Scott, and R. S. Katiyar, Nanoscale ordering and multiferroic behavior in Pb $(Fe_{1/2}Ta_{1/2})O_3$. *Phys. Rev. B* **2010,** *82* (13), 134104.
200. Ye, Z.-G.; Toda, K.; Sato, M.; Kita, E.; Schmid, H. Synthesis, Structure and Properties of the Magnetic Relaxor Ferroelectric Pb $(Fe_{2/3}W_{1/3})O_3$ [PFW]. *J. Korean Phys. Soc.*, **1998,** *32* (93), S1028–S1031.
201. Yang, Y.; Liu, J.-M.; Huang, H. B.; Zou, W. Q.; Bao, P.; Liu, Z. G. Magnetoelectric Coupling in Ferroelectromagnet Pb $(Fe_{1/2}Nb_{1/2})O_3$ Single Crystals. *Phys. Rev. B* **2004,** *70* (13), 132101.
202. Kumar, A.; Rivera, I.; Katiyar, R. S.; Scott, J. F. Multiferroic Pb $(Fe_{0.66}W_{0.33})_{0.80}Ti_{0.20}O_3$ Thin Films: A Room-Temperature Relaxor Ferroelectric and Weak Ferromagnetic. *Appl. Phys. Lett.* **2008,** *92* (13), 132913.
203. Fraygola, B. M.; Coelho, A. A.; Garcia, D.; Eiras, J. A. Magnetic and Dielectric Proprieties of Multiferroic (1-x) Pb (Fe2/3W1/3) O3-xPbTiO3 Ceramics Prepared via a Modified Two-Stage Solid-State Reaction. *Mater. Res.* **2011,** *14*, 434–441.
204. Fraygola, B.; Coelho, A. A.; Garcia, D.; Eiras, A. J. Magnetic and Ferroelectric Phase Coexistence in Multiferroic PFW-PT Ceramics. *Process. Appl. Ceram.* *6* (1), 65–75.
205. Lottermoser, T.; Lonkai, T.; Amann, U.; Hohlwein, D.; Ihringer, J.; Fiebig, M. Magnetic Phase Control by an Electric Field. *Nature* **2004,** *430*, 541–544.
206. Zhang, G.; Dong, S.; Yan, Z; Guo, Y.; Zhang, Q.; Yunoki, S.; Dagotto, E.; Liu, J.-M. Multiferroic Properties of $CaMn_7O_{12}$. *Phys. Rev. B* **2011,** *84* (17), 174413.
207. Vaz, C. A. F.; Hoffman, J.; Ahn, C. H.; Ramesh, R. Magnetoelectric Coupling Effects in Multiferroic Complex Oxide Composite Structures, Hoffman et al. *Adv. Mater.* **2010,** *22*, 2900.
208. Zheng, H.; Wang, J.; Lofland, S. E.; Ma, Z.; Mohaddes-Ardabili, L.; Zhao, T.; Salamanca-Riba, L.; Shinde, S. R.; Ogale, S. B.; Bai, F.; Viehland, D.; Jia, Y.; Schlom, D. G.; Wuttig, M.; Roytburd, A.; Ramesh, R. Multiferroic $BaTiO_3$- $CoFe_2O_4$ Nanostructures. *Science* **2004,** *303*, 661.
209. Ramesh, R.; Spaldin, N. A. Multiferroics: Progress and Prospects in Thin Films. *Nat. Mater*. **2007,** *6*, 21.

CHAPTER 3

Perovskites in the Womb of Science: Roadmap to Global Market

BHUMIKA SHARMA

Department of Physical & Chemical Sciences, Lovely Professional University, Jalandhar, India

ABSTRACT

This chapter aims to provide a clear roadmap of emerging perovskite family from science to technology and from technology to global market. Nano-perovskites exhibit many applications in the field of optoelectronics, transparent conducting oxides, spintronics, thermoelectric energy devices, superconductors, supercapacitors, hydroelectric energy devices, electrode modifiers, hydrophobic films, cleaning agents, fuel cells, 3-D printing, photocatalyst, and many more. This chapter highlights exceptional properties and potential applications of perovskites. In addition, fast and cheaper synthesis routes are also discussed and how to tune and tailor structural, electronic, magnetic, optical, and thermal behavior is mentioned in detail. Perovskites are the future of upcoming technical era and how they will be a gamechanger to the world economy is explained beautifully in this chapter. The world is seeking to replace Indium and Lead halides in the future. This chapter gives a brief analysis of many potential, sustainable, affordable, and ecofriendly perovskites which can replace Indium and Lead halides.

3.1 INTRODUCTION

"Any sufficiently advanced technology is equivalent to magic." said by Arthur C. Clarke. If we see this in the context of bright shade of nano-perovskites

Contemporary Advancements in Materials Technology: Novel Solutions to Complex Problems.
Shrikaant Kulkarni, Vipul Srivastava, & P. William (Eds.)

which are transparent and conducting, then it is nothing short of magic. Science, technology, and innovation can be alone until they are just an idea but once the idea is implemented, innovation and market are the uplifting tools to drive the world of tomorrow.[1] This is the high time to work on and understand hybrid perovskites for application purposes. 3D and 2D hybrid perovskites are the ingredient to replace present technologies in a more efficient way. From the band theory of solids, it is well known that electrical conductivity and optical transparency are mutually exclusive properties for optoelectronic devices. However, due to advancement in science and technology, controlled doping has allowed merging these properties in few materials such as thin films and bulk oxides. Such materials are collectively referred to as transparent conducting oxides (TCO) and transparent conducting films, and include thin films, bulk oxides, conducting polymers, and carbon-based materials (TCF). This remarkable quality has recently been discovered in doped nano-perovskites, which can further address the varied demand in the fields of photovoltaics, LCDs, FETs, diodes, supercapacitors, and many other applications. Doped perovskite TCOs have grown significantly in importance and cost effectiveness for these applications in the future global markets. The status and development of various TCOs, notably those with a perovskite structure, will be outlined in this chapter. The technological era's problems and doped perovskites' potential future are also examined. A description of the ferroic, magnetic, electrical, and optical properties of several TCO-perovskites is included. Also, a thorough discussion is given of the worries regarding the thermal and chemical stability of TCO-doped perovskites.[1] According to global reports, the market for transparent electrodes is expected to increase significantly by 2028. It has been discovered that nano-perovskites have enormous characteristics, making them a potential replacement for lead halides in the future. High optical transmission (>80%), tunable direct band gap between 2 and 5 eV, and strong electrical conductivity are all characteristics of transparent conducting oxides. The transparent conducting oxide characteristics of perovskites are known to decrease when oxygen octahedra in the material are tilted.[2] Tin oxide with indium doping is utilized commercially, but it is both costly and hazardous. Researchers have discovered that perovskites can function in a variety of photovoltaic applications in place of ITOs. Therefore, doping is the only method to achieve property of TCOs in perovskite structured material. Synthesis of $SrVO_3$ TCO films solves the purpose and can be utilized as an electrode for building a transparent and self-powered perovskite photodetector.[3] Al-substituted $SrSnO_3$ perovskite for TCO applications gives hope that opting chemical substitutions rather than doping can frame environment friendly p-type materials and could compete with current n-type materials for TCO applications.[3]

3.2 HISTORY OF PEROVSKITES

Individual atoms and molecules can be controlled as well as visualized with the help of nanoscience and nanotechnology. In 1959, Richard Feynman said that "There is plenty of room at the bottom."[4] Nano-Physics was in thought process in early 20th century and now in 21st century; the technology has a new synonym as Nanotechnology. Researchers are working on nanoscale to take advantage of highly advanced essence of nanotechnology. The project described in this chapter lines up to understand the effect of doping on nano-perovskites. Impact of doping on dielectric properties, optical, electrical, magnetic, structural, and thermal properties of nano-perovskites is studied by millions of researchers till date. The structural formula of a perovskite crystal is ABX_3, where A and B are two cations and X is an anion, typically oxygen. Gustav Rose made the discovery of the perovskite crystal in the Ural Mountains in 1839 and L.A. Perovski was honored as the crystal's name sake. The journey of perovskite materials seems to be aimed to achieve technical goals in laboratory but globally acceptable and can be best fit to achieve sustainable development. Nano-perovskites which are highly electrically conductive with comparatively low absorption of light, can be put into TCOs family. In-depth study of nano-perovskites and their detailed segmentation leads to enormous applications in the field of memory devices, multilayer ceramics, opto-electronics, supercapacitive components, photovoltaics, and many more applications. Perovskites are the leading advocates which are replacing silicon material in semiconductor applications. Cost effectiveness of perovskite materials makes them cheaper than silicon and germanium. Perovskites which are transparent and conducting are the most prominent and demanding materials in the digital world. LCDs, LEDs, capacitors, sensors, FETs, diodes, and mermistors are the key applications of transparent and conducting nano-perovskites.

Ha et al.[6] stated that nanomaterials in the form of quantum dots, nanowires, nanoparticles, and nanofilms exhibit significant anisotropic behavior at nanoscale (i.e., 100 nm). It usually happens due to quantum confinement and variable structural arrangements boost thermal, electrical, optical, and magnetic behavior which is so different from their bulk counterparts. In Photovoltaic applications, many researchers are focusing on metal halide perovskites to manufacture many photonic and optoelectronic devices. Enhanced inverse photoelectric effect gives rise to remarkable performances in applications like single photon emitters and LEDs. The future of upcoming generations will be brighter and shiner under the limelight of photovoltaics and optoelectronic applications. That is why researchers must need to better understand the mechanism, timeline, and need of the hour for the production of nanomaterial form of perovskites. These mechanisms include higher stability under ambient circumstances and toxicity.[6] They emphasized the most recent developments in low-dimensional metal-halide perovskites production, their special characteristics, and their innovative optoelectronic and photonic applications.[7]

Ouyang et al.[8] reported that the preparation of metal oxide nanomaterials with well-defined size and morphology is an important challenge for various applications, especially in photovoltaic applications. When the material dimension is reduced to the nanoscale, they gain a new identity in all the aspects. Such features make nanomaterial attractive for unique applications and at the same time cause complication in their synthesis. Even though varieties of techniques have been applied for the synthesis of metal oxide nanocrystals, it still needs to exploit new synthesis methods according to its application.[8] In addition, nonlinear optical interactions and charge carrier dynamics in (pseudo-) 2D perovskites and photonic structures are also incorporated in further research.[9]

Ma et al.[10] proved that for energy conversion and storage, electrochemical activities of perovskite materials should be examined and understood from the reaction at the active B site. This research showed the impact of heteroatom doping at site A to prevent multiphase electrochemical reaction. Major outline of this research can be beneficial for those who are willing to pursue their research in the area of ion extraction.[10]

Dhiman and Singhal[11] fabricated different rare earth substituted nanoperovskites using the bottom up approach, that is, sol–gel process with different dopants in unequal proportions. They found low band gaps with orthorhombic structure. A successful catalysis cum oxidative destruction of dye molecules was observed by all substituted perovskites. There was an excellent improvement in the rate constant values of photocatalytic processes of perovskites.[11]

Ravichandran et al.[12] used the solution combustion method to prepare undoped and doped metal ferrite nanoparticles which are perovskite in structure. To get a pure phase, they have kept in mind the significant role of annealing temperature. For characterization, XRD, FTIR, FESEM, EDX,

TEM, UV-Vis, PL, and Raman spectroscopy were used for the samples. Nano powders synthesized were found to exhibit rhombohedral structure. As proportion of doping increases in base material, average particle size decreases and strain increases. For surface morphology, shape, and size at nanoscale, FE-SEM images were studied. The TEM micrograph technique showed agglomeration and nonuniformity in grain distribution which indicates the presence of magnetic interaction of particles. EDX mapping was in good agreement and revealed that there is no impurity in the samples. Blue emission was confirmed due to the presence of intensity peak at 464 nm. If doping concentration of Al is increased in metal ferrite nano-perovskite, then these nanoparticles can be a promising material in the field of antibacterial activity.[12]

Srivastav and Gajbhiye[13] synthesized phase pure nanosized $BiFeO_3$ (BFO) powder by sol–gel method based on the propylene glycol-gel reaction. The formation mechanism, crystal structure, magnetic, and optical properties were investigated by thermogravimetry, differential thermal analysis, infrared spectroscopy, X-ray diffractometry, Raman spectroscopy, transmission electron microscopy, and ultraviolet-visible spectroscopy. X-ray diffraction and Raman studies revealed the formation of pure and well-crystallized BFO nanocrystallites at a low temperature of 400°C. The BFO nanoparticles showed weak ferromagnetic property at room temperature due to size-confinement effects. The UV-visible spectra showed strong band-gap absorption at ~590 nm (2.1 eV) for BFO nanoparticles suggesting their potential application as visible-light-response photocatalyst.[13]

In the current study, a straightforward low-temperature synthesis method employing iron and bismuth nitrates and soft chemicals was disclosed for producing bismuth ferrite nanopowders (Ghosh et al.[14]). Nitric acid is utilized as an oxidizing agent and tartaric acid as a template material. X-ray diffractometry, thermogravimetry and differential thermal analysis, infrared spectroscopy, and scanning electron microscopy are used to characterize the synthesized powders. The powder's particles range in size from 3 to 16 nm. In contrast to the coprecipitation approach, which requires 550°C, the process allows for the production of phase-pure bismuth ferrite at as low as 400°C. On the other hand, we discover that Pechini's autocombustion technique of synthesis produces several impurity phases in addition to bismuth ferrite, much as the solid-state reaction approach.[14]

Khan et al.[15] reported enhanced anomalous photovoltaic effects and switchable photovoltage generation in pure and Pr–Cr codoped $BiFeO_3$ (BFO) nanotubes (NTs). Influence of metal doping on short-circuit current, open-circuit voltage, power conversion efficiency, and fill factor are investigated. The power conversion efficiency of pure BFO NTs (~0.207%) is found to be enhanced by several orders of magnitude in comparison with the reported bulk effect. Pr-doped NTs provide highest values of power conversion efficiency (~0.5%).[15]

Akram et al.[16] showed improvement in the dielectric properties and high T_{max} of the optimum specimen makes the system viable for applications that require high-temperature dielectric stability.[16]

Yongshang et al. (2019) have prepared $Ba_{(1-x)}Ca_xTi_{0.92}Zr_{0.08}O_3$ lead-free ceramics at a low temperature of 1210°C. They studied structure, permittivity, ferroelectricity, piezoelectricity, and thermal expansion characteristics of the ceramics that are influenced by varying calcium contents. X-ray powder diffraction and Raman results suggest that calcium does not induce the rhombohedral phase transition, and impurity phases are detectable with excessive X.[17]

Ma et al.[17] synthesized nanocomposite which is lead free and ecofriendly. A regular structure change from tetragonal to pseudo cubic phase was noticed in Raman spectra and XRD studies. There was a transition from typical ferroelectric behavior to diffusive and dispersive relaxor-like properties. As doping concentration increased, temperature stability of permittivity got enhanced and high relative permittivity, low loss were also seen over a broad temperature range from 30°C to 150°C at 1 kHz. This research showed thermal stability of composite and can have enormous stable dielectric applications for ceramics.[18]

Sareecha et al.[18] emphasized on electrical characteristics of ceramics of rare earth elements. They examined these nanomaterials in a wide range of temperatures (40–700°C) at 1 kHz. The aim of the study was to study resistive behavior and conduction process in samples. The synthesized ceramic exhibits cubic and tetragonal crystal formations at room temperature. As concentration at A o B site decreases, Curie temperature increased by 10°C and relaxor behavior gives rise to changes in dielectric irregularities. The negative temperature coefficient of resistance decreases with increase in temperature. Ionic conduction and conductivity were also studied to notice the impact of single- and double-ionized mechanism of oxygen vacancies.[19]

Yang et al[20] stated that codoping of elements of similar atomic radii reduces the band structure of $BaTiO_3$.[15] Band structure of pure $BaTiO_3$ shows an indirect band gap of 1.56 eV.[20]

Johnsson and Lemmens et al.[21] stated that the distortion of the crystal structure can be assigned to Jahn–Teller active ions at the B position in some perovskites. In the present study, the main focus will be on codoping of zinc at B position. As zinc has direct band gap and it would show tremendous electronic effects in perovskite morphology.[21]

Butler et al.[22] stated that by controlling the cation ratio and distribution in the double perovskite Bi_2FeCrO_6 (Fe and Cr ions are distributed over the perovskite B site), the optical band gap could be tuned by several eV. It is noted that in the synthesis of perovskite structure, ABO_3, Coordination number of cation A must be 12 and coordination number of cation B must be 6, only then, one can have a stable perovskite structure.[22]

Jia et al.[23] concluded that high-purity $BiFeO_3$ R-phase powders have been successfully synthesized by a simple sol–gel process at a temperature as low as 450°C. Dielectric behavior of the $BiFeO_3$ powders is considered for the coupling of electric and magnetic order parameters. In the present study, we will synthesize and analyze dielectric behavior of doped perovskite nanomaterials with different concentrations of Zn at site B in high-purity $BiFeO_3$.[23]

Ha et al.[6] concluded, due to the vulnerability of perovskite to surrounding environments such as moisture and oxygen, the performance of the devices may dramatically drop within only tens to hundreds of exposed hours, which casts a shadow toward their practical applications. In order to bring light to this dark side, we will perform analysis of perovskites in ambience atmosphere.[6]

Kharbanda et al.[24] used a solid-state technique to create polycrystalline $Mg_{1-x}Li_xFe_2O_4$ ferrite samples. They identified the purity of the phases using XRD analysis. There is a reduction in grain size and increase in saturation magnetization with doping of Lithium. In $Mg_{1-x}Li_xFe_2O_4$ samples the study of voltage, current, and power generation in hydroelectric cells reveals maximum power response in $Mg_{0.8}Li_{0.2}Fe_2O_4$. The resistivity response with frequency of Ferrites:$Mg_{1-x}Li_xFe_2O_4$ reveals that ferrite $Mg_{1-x}Li_xFe_2O_4$ materials are what cause the dissociation of the water molecules into ions which indicates magnetic materials can be a ecofriendly source of electrical energy.[24]

Han et al.[25] acquired a vast knowledge about environment friendly and easily affordable high-energy storage devices. Using DFT, they studied bimetallic oxides of strontium and bismuth. XRD, SEM, TEM, bandgap studies, and electrochemical measurements (high specific capacity and capacitive retention rate) show that perovskite $SrBiO_3$ will be a promising candidate for high storage battery-type supercapacitors.[25]

Bougerol-Chaillout et al.[26] synthesized $SrBiO_3$ with doping of potassium to enhance superconducting behavior under high pressure. They have also studied variation of lattice parameters of $Sr_{1-x}K_xBiO_3$ with temperature. From this study, one can have idea that doped Bismuthates can have great electrical and thermal properties at nanoscale.[26]

Ahmed et al.[27] implemented the co-precipitation method to synthesize La-doped BFO on A site. Rare earth element doping in Bismuth ferrite in different proportions enhanced leakage current in BFO. They have characterized samples of $Bi_{1-x}La_xFeO_3$ using XRD, SEM, EDX, TEM, FTIR, UV, BET, LCR, and VSM. Electrical and magnetic properties showed that La-doped BFO can be a promising material for high-energy storage devices.[27]

Royen et al.[28] framed a mechanism for the synthesis of Bismuth Ferrite. They have studied crystallography at different temperatures and elaborated phase group and grain behavior of Bismuth Ferrite. Royen et al.[28] were the first ones to synthesize Bismuth Ferrite in 1957.

Dar et al.[29] worked on double perovskites materials with rich diversity of properties and applications. Using DFT, they concluded that electronic profile of Ba_2InTaO_6 reveals its semiconductor behavior. Elastic properties and thermodynamic studies indicate its mechanical stability and longer life and this material can be used in thermoelectric and optical devices.[29]

Khadtare et al.[30] studied piezoelectric behavior of strontium-doped BFO on A-site with different amounts of dopant. These nanomaterials exhibit longer life and high mechanical strength. It has the features of a good capacitor for market circuit components.[30]

Hussain et al.[31] have synthesized $Bi_{1-x}Sr_xFeO_3$ using sol-gel process and emphasized on frequency-dependent dielectric behavior at 300K for every sample. They notice that with small quantity of dopant, there is no significant change in dielectric properties of BFO. Also there is a noticeable decrease in the value of dielectric constant with increased amount of doping. As strontium component increases, there is a remarkable change in magnetic behavior of BFO.[31]

Wang et al.[32] used the self-combustion method for harmonize nanoparticles of Sr-doped Bismuth ferrite. After XRD, SEM, TEM, and UV-Vis spectroscopy, they concluded that particle size and dopant concentration are responsible for magnetic and photocatalytic behavior of Sr-doped BFO nanomaterials. Due to noticeable photocatalytic activity, Sr-doped BFO can be a good photocatalyst.[32]

Bhushan et al.[33] framed their research work on Ca, Ba, and Sr-doped BFO. After doing structural studies, they have diverted themselves toward magnetoelectric coupling. There is a consistent change in band gap values and this narrow band gap material shows ferromagnetic character which makes it useful for practical applications in the field of technology.[33]

Song et al.[34] invented lead-free thin-film capacitor made up of Sr-doped $BiFeO_3$ which are capable of high electrostatic storage and thermally stable (up to 115°C) as well. Thin-film composite of doped Barium Titanate and doped Bismuth ferrite exhibits a storage density of 47.6 Jcm^{-1} and an efficiency of ~65.68%. Due to the presence of oxygen vacancies, Schottkey and F-N tunneling calculations show reduction in leakage current. The 2% Mn-doped $0.7Sr_{0.7}Bi_{0.2}TiO_3$-$0.^{3BiFeO3}$ is a promising dielectric material suitable for many applications.[34]

Hegab et al.[35] found that there are some point defects in Sr-doped BFO due to the presence of strontium ion and oxygen vacancies. Structural analysis was done by XRD data. Conductivity, magnetic, and dielectric studies have been showcased to explain the transport behavior of samples. The presence of Sr induces more magnetization in BFO.[35]

3.3 SYNTHESIS ROUTE

Nanomaterials have many excellent properties than bulk material like high mechanical strength, thermal stability, electrical conductivity, catalytic activity, magnetic, and optical properties. The journey of a material from bulk to nano scale can be achieved by chemical route methods. The major challenges are:

- Size and shape control
- Reproducibility
- Scale up
- Building complex nanostructures
- Means to achieve mono dispersity

The good quality of nanomaterials using sol-gel process is variation of PH, temperature, time, concentration of precursors and catalyzer, phase transition from sol to gel, and drying process. Sol-gel process is a bottom-up approach to synthesize nanoparticles using a wet chemical route. The key advantages of sol-gel processes are:

- Able to get uniform and small-sized powder
- High uniformity, multicomponent systems
- Can get nanoparticles with special porosity
- Very cheap and user friendly
- Allows the fine control on the product's chemical composition

Generally speaking, the sol-gel process entails changing a solution system's liquid "sol" phase into a solid "gel" phase. Initially, metal salts are used to prepare solution and further, a precursor or chelating agent is added to form gel. The final step is drying and annealing at several hundred degree temperature to get a cluster of fine nanoparticles. Its disadvantage is sensitivity to moisture which reduces the lifetime of synthesized material. But neglecting this, we can inherent many astonishing facts about a material at nanoscale using the sol-gel process. Annealing temperature plays a great role in phase formation and tuning and tailoring of band structure. Also the sol-gel method is very time consuming, but yield is sure. There is one more method to synthesize nanomaterials within hours, that is, solution combustion method. In this method, yield is not in our control.

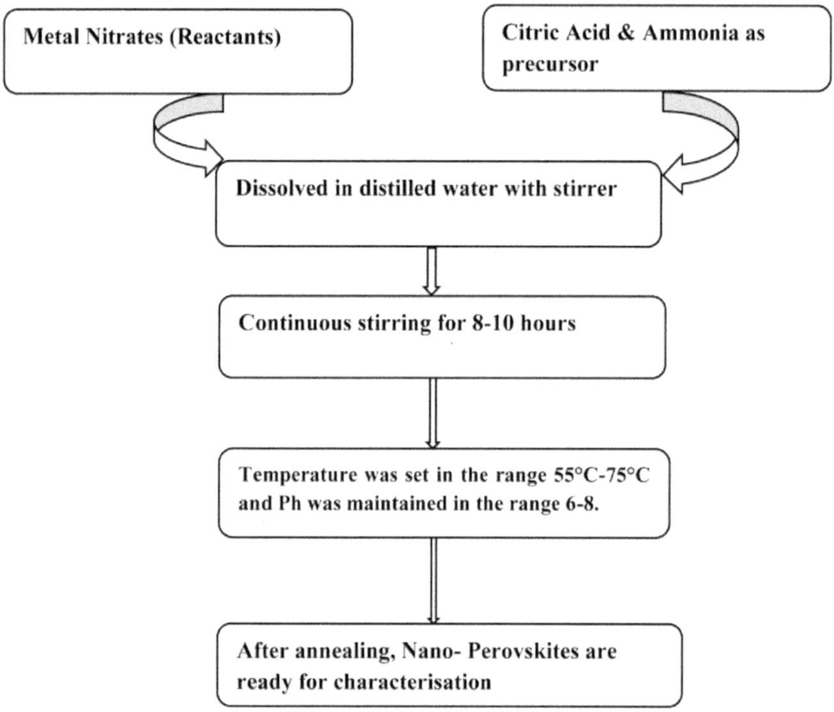

FIGURE 3.1 Flow chart of sol-gel process.

3.4 IMPACT OF STRUCTURAL TAILORING ON VARIOUS PROPERTIES OF PEROVSKITES

Structural tailoring leads to introducing lattice strain, variable particle size, grain shape and size, dislocation density and to overcoming toxicity issues at nanoscale. The question arises: How can we tailor structure of a particular material. Firstly, in case of perovskites, follow the fundamental rule before the synthesis route. Every synthesis route follows some fundamentals for their phase formation. In a similar way, perovskite structure: ABO_3 requires value of Goldschmidt tolerance factor nearly equal to unity. The size geometry factor in the form of effective ionic radii can be expressed by the Goldschmidt tolerance factor (t).

$$t = [(r_A + r_X)/ \sqrt{2}\,(r_B + r_X)] \tag{3.1}$$

where r_A, r_B, and r_X are the ionic radii of A, B, and X, respectively.[36,37]

If t>1, then structure formation is hexagonal and tetragonal as A ion is too big or B ion is too small.[37] If t ranges in 0.9–1 then the structure is cubic as A and B have ideal size.[2] Similarly, if t is 0.7–0.9 then the structure is either orthorhombic or rhombohedral as A ions are too small to fit into B ion interstices.[37] Goldschmidt tolerance factor is the tool to determine the compatibility of an ion in crystal structure.

It is noted that in the synthesis of Perovskite structure, ABO_3, Coordination number of cation A must be 12 and coordination number of cation B must be 6, only then, one can have a stable Perovskite structure..[42] The correct formula for perovskite structure is $A^{xii}B^{vi}O_3$.

Tailoring of crystal structure gives rise to phenomenon of tunable bandgap, super-capacitive behavior, photocatalysis, and many more fascinating properties which will be discussed in the next section.

3.5 FASCINATING PROPERTIES OF NANO-PEROVSKITES

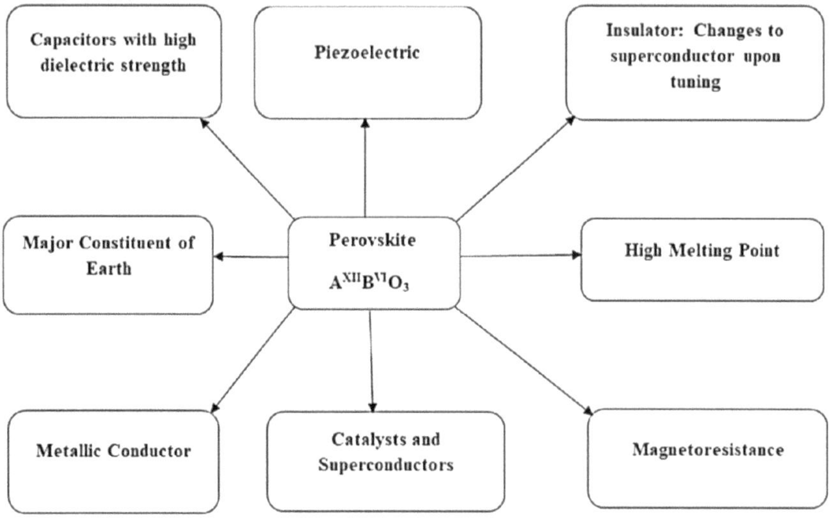

FIGURE 3.2 Applications and properties of perovskite materials in the ceramic industry.

3.6 APPLICATIONS OF NANO-PEROVSKITES

- IT Industry
- Automobile Industry
- Aerospace Technology
- Aircraft Industry
- Defence Technology
- Materials Industry
- Medical Sciences
- Entertainment Industry
- Healthcare Tools and Medical Industry
- Infrastructural Development
- Smart roads and transport Industry
- Smart cities and what not.

3.7 RESEARCH GAP IN PEROVSKITE FAMILY

Yang et al.[20] stated that codoping of elements of similar atomic radii reduces the band structure of $BaTiO_3$. The band structure of pure $BaTiO_3$ shows an indirect band gap of 1.56 eV.

1. So, one can wish to study band structure after codoping in $BiFeO_3$.
2. The tailoring of band gap may lead to significant and worthful electrical properties.

Johnsson and Lemmens[21] stated that the distortion of the crystal structure can be assigned to Jahn–Teller active ions at the B position in some perovskites. Doping of any direct bandgap material would show tremendous electronic effects in perovskite morphology.[21]

Affiliated To

Stockholm University, Sweden

TU Braunschweig, Germany

Butler et al.[22] stated that by controlling the cation ratio and distribution in the double perovskite Bi_2FeCrO_6 (Fe and Cr ions are distributed over the perovskite B site), the optical band gap could be tuned by several eV. By controlling the cation ratio and distribution of Fe and Zn ions on B site ($BiZn_xFe_{1-x}O_3$), band gap and analysis of induced electrical properties should be studied.[42]

Affiliated To

University of Bath, UK

Jia et al.[23] concluded that high-purity $BiFeO_3$ R-phase powders have been successfully synthesized by a simple sol–gel process at a temperature as low as 450°C. Dielectric behavior of the $BiFeO_3$ powders is considered for the coupling of electric and magnetic order parameters. In the present study, we will synthesize and analyze dielectric behavior of doped perovskite nanomaterials with different concentrations of Zn at site B in high-purity $BiFeO_3$.[23]

Affiliated To

Harbin Institute of Technology, China

Ha et al.[6] (2016) raised the voice that metal halide perovskites have received substantial attention from the research community due to their applications in solar cells, lasers, light-emitting diodes (LED), water splitting, laser cooling etc. The material is now considered the most promising material for the future of opto-electronics because of its high performance, low cost, and abundance. The major challenge for this type of material is its stability. Due to the vulnerability and issue caused by oxygen vacancies of perovskite to surrounding environments such as moisture, the device performance may dramatically drop within only tens to hundreds of exposed hours, which casts a shadow toward their practical applications. In order to bring light to this dark side, we will perform analysis of perovskites in ambience atmosphere.[6]

Affiliated To

Nanyang Technological University, Singapore

Peking University, Beijing

3.8 PEROVSKITES: GAMECHANGER TO WORLD ECONOMY

Researchers of the 21st century wish to pursue their research to boost economy by fulfilling the daily needs of diverse lifestyles. Recently,

IITRDF and General Insurance Corporation of India Limited signed an MoU for the development of semi-transparent perovskite solar cells for building integrated photovoltaics—a road to net zero carbon economy.[38] In the same way, a circular economy is also required for waste management system. Many researchers are working to synthesize photocatalytic hybrid perovskite which can help in the recycle of plastic, e-waste, garbage, and any unwanted polluted materials. Nowadays, automation is replacing humans in many work profiles. There is huge market for sensors, LEDs, optoelectronic devices, photovoltaics, FETs, memristors, etc. There is one stressing point to leading economies that China is uprising faster and faster due to their research and development in every field. Indeed, Ural Mountain perovskites are now gamechanger to world economy in many ways.

3.9 DISADVANTAGES OF LEAD HALIDES AND INDIUM

Perovskites based on lead halides possess high efficiency and potential candidacy for low manufacturing costs. However, the use of toxic lead creates worries and negative impacts on environment and public health persist. Also, patents and research which are not ecofriendly are rejected immediately by industrialists. If research could not make bridge between science and market, then it is of no use.

Secondly, indium is very expensive and only few countries own it naturally. If research based on it gets commercialized, then it would be injustice for those who are not in creamy layer. Good research aims at the fact that it should be achievable and can be everyone's cup of tea.

3.10 WHY PEROVSKITES?

United Nations are attracting the world toward sustainable development goals for the betterment of this planet Earth. As we are aware of the fact that we need more and more inventions to have routes to consume renewable energy sources at cheap rates so that one who is at poverty line, can also afford the system. Perovskites are inorganic materials and are the best fit for solar panels to meet out the electricity demands of the society. There are many novel functional cum versatile materials which are applicable in many efficient devices.

3.11 JOURNEY OF PEROVSKITES FROM URAL MOUNTAINS TO OUR DAILY NEEDS

Today's industrial heritage is being created through nanotechnology. The improvement of doped perovskite-structured materials has benefited from the accumulation and application of knowledge thanks to their numerous special features. According to a 2015 report from Nano Market, the market for metal mesh transparent conductors is expected to reach $560 million (USD) in 2020 and $1 billion by 2022. Two popular and developing market segments for TCOs are touchscreen technology and transparent photovoltaics. In order to find doped perovskites, research is being done in a transparent manner. Let us concentrate on various TCOs with perovskite structures. Furthermore, light absorption and photoelectric conversion are fundamental functions of the perovskite materials. The field of photovoltaics has advanced due to the low cost and excellent power conversion efficiency of devices made of perovskite materials.[1]

KEYWORDS

- **nano-perovskites**
- **sol-gel process**
- **optical properties**
- **spintronics**
- **3-D printing**
- **hydrophobic films**
- **TCO**

REFERENCES

1. Sharma, B.; Rana, P. S. The Bright and Dark Shades of Transparent Conducting Perovskites: From Science to Global Market. *J Phys. Conf. Ser.* **2022,** *2267* (1), p012030.
2. Kumar, Y.; Kumar, R.; Choudhary, R. J.; Thakur, A.; Singh, A. P. Reduction in the Tilting of Oxygen Octahedron and its Effect on bandgap with La Doping in SrSnO3. *Ceramics Int.* **2020,** *46* (11), 17569–17576.
3. Amor, L. B.; Belgacem, B.; Filhol, J. S.; Doublet, M. L.; Yahia, M. B.; Hassen, R. B. New p-type Al-substituted SrSnO3 Perovskites for TCO Applications. *Chem. Commun.* **2020,** *56* (17), 2566–2569.

4. Feynman, R. P. There's Plenty of Room at the Bottom: An Invitation to Enter a New Field of Physics. In *Miniaturization*; Reinhold, 1961.
5. Xu, R.; Min, L.; Qi, Z.; Zhang, X.; Jian, J.; Ji, Y.; Yang, H. Perovskite Transparent Conducting Oxide for the Design of a Transparent, Flexible, and Self-Powered Perovskite Photodetector. *ACS Appl. Mater. Interfaces* **2020,** *12* (14), 16462–16468.
6. Ha, S. T.; Su, R.; Xing, J.; Zhang, Q.; Xiong, Q. Metal Halide Perovskite Nanomaterials: Synthesis and Applications. *Chem. Sci.* **2017,** *8* (4), 2522–2536.
7. Zhang, Y.; Liu, J.; Wang, Z.; Xue, Y.; Ou, Q.; Polavarapu, L.; ... Bao, Q. Synthesis, Properties, and Optical Applications of Low-Dimensional Perovskites. *Chem. Commun.* **2016,** *52* (94), 13637–13655.
8. Ouyang, D.; Huang, Z.; Choy, W. C. Solution-Processed Metal Oxide Nanocrystals as Carrier Transport Layers in Organic and Perovskite Solar Cells. *Adv. Funct. Mater.* **2019,** *29* (1), 1804660.
9. Zhang, Y.; Lim, C. K.; Dai, Z.; Yu, G.; Haus, J. W.; Zhang, H.; Prasad, P. N. Photonics and Optoelectronics using Nano-Structured Hybrid Perovskite Media and their Optical Cavities. *Phys. Rep.* **2019.**
10. Ma, C. J.; Chen, Y.; Zhu, C.; Chen, Q.; Song, W. L.; Jiao, S.; ... Fang, D. Bismuth Ferrite: An Abnormal Perovskite with Electrochemical Extraction of Ions from A Site. *J. Mater. Chem. A* **2019,** *7* (19), 12176–12190.
11. Dhiman, M.; Singhal, S. Effect of Doping of Different Rare Earth (Europium, Gadolinium, Dysprosium and Neodymium) Metal Ions on Structural, Optical and Photocatalytic Properties of LaFeO3 Perovskites. *J. Rare Earths* 2019.
12. Ravichandran, A. T.; Srinivas, J.; Karthick, R.; Manikandan, A.; Baykal, A. Facile Combustion Synthesis, Structural, Morphological, Optical and Antibacterial Studies of Bi1− xAlxFeO3 (0.0≤ x≤ 0.15) Nanoparticles. *Ceram. Int.* **2018,** *44* (11), 13247–13252.
13. Srivastav, S. K.; S. Gajbhiye, S. N. Low Temperature Synthesis, Structural, Optical and Magnetic Properties of Bismuth Ferrite Nanoparticles. *J. Am. Ceram. Soc.* **2012,** *95* (11), 3678–3682.
14. Ghosh, S.; Dasgupta, S.; Sen, A.; Maiti, H. S. Low-Temperature Synthesis of Nanosized Bismuth Ferrite by Soft Chemical Route. *J. Am. Ceram. Soc.* **2005,** *88* (5), 1349–1352.
15. Khan, G. G.; Das, R.; Mukherjee, N.; Mandal, K. Effect of Metal Doping on Highly Efficient Photovoltaics and Switchable Photovoltage in Bismuth Ferrite Nanotubes. *Phys. Status Solidi (RRL)–Rapid Res. Lett.* **2012,** *6* (7), 312–314.
16. Akram, F.; Malik, R. A.; Song, T. K.; Lee, S.; Kim, M. H. Thermally-Stable High Dielectric Properties of (1–x)(0.65 Bi1. 05FeO3–0.35 BaTiO3)–xBiGaO3 Piezoceramics. *J. Eur. Ceram. Soc.* **2019,** *39* (7), 2304–2309.
17. Ma, D.; Chen, X.; Huang, G.; Chen, J.; Zhou, H.; Fang, L. Temperature Stability, Structural Evolution and Dielectric Properties of BaTiO3–Bi (Mg2/3Ta1/3) O3 Perovskite Ceramics. *Ceram. Int.* **2015,** *41* (5), 7157–7161.
18. Sareecha, N.; Shah, W. A.; Anis-ur-Rehman, M.; Mirza, M. L.; Awan, M. S. Electrical Investigations of BaTiO3 Ceramics with Ba/Ti Contents under Influence of Temperature. *Solid State Ion.* **2017,** *303*, 16–23.
19. Eerenstein, W.; Morrison, F. D.; Dho, J.; Blamire, M. G.; Scott, J. F.; Mathur, N. D. Comment on Epitaxial BiFeO3 Multiferroic Thin Film Heterostructures. *Science* **2005,** *307* (5713), 1203–1203.
20. Yang, F.; Yang, L.; Ai, C.; Xie, P.; Lin, S.; Wang, C. Z.; Lu, X. Tailoring Bandgap of Perovskite BaTiO3 by Transition Metals Co-Doping for Visible-Light Photoelectrical Applications: A First-Principles Study. *Nanomaterials* **2018,** *8* (7), 455.

21. Johnsson, M.; Lemmens, P. Crystallography and Chemistry of Perovskites. In *Handbook of Magnetism and Advanced Magnetic Materials*, 2007.
22. Butler, K. T.; Frost, J. M.; Walsh, A. Ferroelectric Materials for Solar Energy Conversion: Photoferroics Revisited. *Energy Environ. Sci.* **2015,** *8* (3), 838–848.
23. Jia, D. C.; Xu, J. H.; Ke, H.; Wang, W.; Zhou, Y. Structure and Multiferroic Properties of BiFeO3 Powders. *J. Euro. Ceram. Soc.* **2009,** *29* (14), 3099–3103.
24. Kharbanda, P.; Madaan, T.; Sharma, I.; Vashishtha, S.; Kumar, P.; Chauhan, A.; ... Verma, V. Ferrites: Magnetic Materials as an Alternate Source of Green Electrical Energy. *Heliyon* **2019,** *5* (1), e01151.
25. Han, Y.; Jiao, Y.; Lv, G.; Pang, Y.; Zhou, J.; Xue, Z.; ... Liu, Y. High-Performance Strontium and Bismuth Bimetallic Oxides Electrode: Combine First-Principles Calculations with Electrochemical Tests. *Mater. Today Commun.* **2020,** *24*, 100927.
26. Bougerol-Chaillout, C.; Bordet, P.; Kazakov, S.; Pshirkov, J.; Putilin, S. N.; Antipov, E. V.; Nunez-Regueiro, M. Structural Studies of New Superconducting Bismuthates (Sr, K) BiO3. *Phys. C Supercond.* **2000,** *341*, 1813–1816.
27. Ahmed, I.; Mustafa, G.; Subhani, M. U.; Hussain, G.; Ismail, A. G.; Anwar, H. A Detailed Investigation of Lanthanum Substituted Bismuth Ferrite for Enhanced Structural, Pptical, Dielectric, Magnetic and Ferroelectric Properties. *Results Phys.* **2022,** *38*, 105584.
28. Royen, P.; Swars, K. Das system Wismutoxyd-Eisenoxyd im Bereich von 0 bis 55 mol% Eisenoxyd. *Angew. Chem.* **1957,** *69* (24), 779–779.
29. Dar, S. A.; Sharma, R.; Srivastava, V.; Sakalle, U. K. Investigation on the Electronic Structure, Optical, Elastic, Mechanical, Thermodynamic and Thermoelectric Properties of Wide Band Gap Semiconductor Double Perovskite Ba 2 InTaO 6. *RSC Adv.* **2019,** *9* (17), 9522–9532.
30. Khadtare, S.; Kumar, A.; Jung, U.; Sang, C.; Park, J. Piezoelectric Nanogenerator Based on Lead-Free BiFeO3: Sr perovskite. *J. Alloys Compd.* **2022,** *926*, 166738.
31. Hussain, T.; Siddiqi, S. A.; Atiq, S.; Awan, M. S. Induced Modifications in the Properties of Sr Doped BiFeO3 Multiferroics. *Prog. Natl. Sci. Mater. Int.* **2013,** *23* (5), 487–492.
32. Wang, B.; Wang, S.; Gong, L.; Zhou, Z. Structural, Magnetic and Photocatalytic Properties of Sr2+-doped BiFeO3 Nanoparticles Based on an Ultrasonic Irradiation Assisted Self-Combustion Method. *Ceram. Int.* **2012,** *38* (8), 6643–6649.
33. Bhushan, B.; Basumallick, A.; Bandopadhyay, S. K.; Vasanthacharya, N. Y.; Das, D. Effect of Alkaline Earth Metal Doping on Thermal, Optical, Magnetic and Dielectric Properties of BiFeO3 Nanoparticles. *J. Phys. D Appl. Phys.* **2009,** *42* (6), 065004.
34. Song, B.; Wu, S.; Yan, H.; Zhu, K.; Xu, L.; Shen, B.; Zhai, J. Fatigue-Less Relaxor Ferroelectric Thin Films with High Energy Storage Density via Defect Engineer. *J. Mater. Sci. Technol.* **2021,** *77*, 178–186.
35. Hegab, A. F.; Ahmed Farag, I. S.; El-Shabiny, A. M.; Nassaar, A. M.; Ramadan, A. A.; Moustafa, A. M. Role of Sr Doping on Transport and Magnetic Properties of Bismuth Ferrite in Correlation with Defect Formation. *J. Mater. Sci. Mater. Electron.* **2017,** *28*, 14460–14470.
36. Goldschmidt, V. M. *Naturwissenschaften* **1926,** *14*, 477.
37. https://handwiki.org/wiki/Physics:Goldschmidt_tolerance_factor
38. https://indiaeducationdiary.in/iitrdf-and-general-insurance-corporation-of-india-limited-signed-an-mou-for-the-development-of-semi-transparent-perovskite-solar-cells-for-building-integrated-photovoltaics-a-road-to-net-zero-carbon-e/
39. Zhang, K. H.; Du, Y.; Papadogianni, A.; Bierwagen, O.; Sallis, S.; Piper, L. F.; Chambers, S. A. Perovskite Sr-doped LaCrO3 as a New p-type Transparent Conducting Oxide. *Adv. Mater.* **2015,** *27* (35), 5191–5195.

40. Kumar, Y.; Kumar, R.; Asokan, K.; Choudhary, R. J.; Phase, D. M.; Singh, A. P. High Mobility Transparent and Conducting Oxide Films of La-doped SrSnO $$_3 $$3. *J. Mater. Sci. Mater. Electron.* **2021,** *32* (9), 11835–11844.
41. Minami, T. New n-Type Transparent Conducting Oxides. *MRS Bull.* **2000,** *25* (8), 38–44.
42. Smirnov, Y.; Holovsky, J.; Rijnders, G.; Morales-Masis, M. Origins of Infrared Transparency in Highly Conductive Perovskite Stannate BaSnO3. *APL Mater.* **2020,** *8* (6), 061108.
43. Kumar, Y.; Kumar, R.; Asokan, K.; Singh, A. P. In *Effect of Swift Heavy Ion Irradiation on Structural, Electrical and Optical Properties of Zinc-Stannate Thin Films*, AIP Conference Proceedings; AIP Publishing LLC, 2019; vol *2142* (1), p 080004.
44. Rhodes, C.; Franzen, S.; Maria, J. P.; Losego, M.; Leonard, D. N.; Laughlin, B.; Weibel, S. Surface Plasmon Resonance in Conducting Metal Oxides. *J. Appl. Phys.* **2006,** *100* (5), 054905.
45. Kim, U.; Park, C.; Ha, T.; Kim, Y. M..; Kim, N.; Ju, C.; Char, K. All-perovskite Transparent High Mobility Field Effect using Epitaxial BaSnO3 and LaInO3. *APL Mater.* **2015,** *3* (3), 036101.
46. Park, J.; Paik, H.; Nomoto, K.; Lee, K.; Park, B. E.; Grisafe, B.; Schlom, D. G. Fully Transparent Field-Effect Transistor with High Drain Current and on-off Ratio. *APL Mater.* **2020,** *8* (1), 011110.
47. Liu, Q.; Jin, F.; Dai, J.; Li, B.; Geng, L.; Liu, J. Effect of Thickness on the Electrical and Optical Properties of Epitaxial (La0. 07Ba0. 93) SnO3 Thin Films. *Superlattices Microstruct.* **2016,** *96*, 205–211.
48. Niedermeier, C. A.; Rhode, S.; Fearn, S.; Ide, K.; Moram, M. A.; Hiramatsu, H.; Kamiya, T. Solid Phase Epitaxial Growth of High Mobility La: BaSnO3 Thin Films Co-doped with Interstitial Hydrogen. *Appl. Phys. Lett.* **2016,** *108* (17), 172101.
49. Grätzel, M. The Light and Shade of Perovskite Solar Cells. *Nat. Mater.* **2014,** *13* (9), 838–842.
50. Sharma, B.; Sharma, R.; Kour, S.; Sharma, M.D.; Amin, O.; Maity, A. R.; Mukherjee, R. Fractional Exponents of Electrical and Thermal Conductivity of Vanadium Intercalated Layered 2H-NbS 2 Bulk Crystal. *Indian J. Phys.* **2021,** 1–5.
51. Sarkar, D.; Xie, X.; Liu, W.; Cao, W.; Kang, J.; Gong, Y.; Banerjee, K. A Subthermionic Tunnel Field-Effect Transistor with an Atomically Thin Channel. *Nature* **2015,** *526* (7571), 91–95.
52. Pan, Q.; Zuo, P.; Mu, T.; Du, C.; Cheng, X.; Ma, Y.; Yin, G. Improved Electrochemical Performance of Micro-Sized SiO-Based Composite Anode by Prelithiation of Stabilized Lithium Metal Powder. *J. Power Sources* **2017,** *347*, 170–177.

CHAPTER 4

Rietveld Analysis: An Essential Tool for Structural Analysis

M. NATH[1] and A. BANDYOPADHYAY[2]

[1]*Department of Physics, Deshabandhu Mahavidyalaya, Chittaranjan, West Bengal, India*

[2]*Department of Physics, University of Gour Banga, Malda, West Bengal, India*

ABSTRACT

The structural analysis of crystalline materials got a vital kick when Hugo M. Rietveld developed a code for pattern fitting technique from the data of its X-ray or neutron diffraction pattern. It allows the investigation of size-strain analysis, structural changes during phase transition, quantitative phase analysis, etc. even when the peaks in the diffraction pattern are not well-separated. The basic principle behind the analysis is to compare the observed diffraction pattern with a calculated pattern and determine the most probable values for the crystal structure's parameters. This technique can extract information about the microstructure of a material and handle complex crystal structures, including the materials with multiple phases or disordered regions. This method is indispensable for verifying the phase purity of spintronic material and can shed light on the site-specific selective response of dopants. In this chapter, we focus on the details of this technique in analyzing the structure of the materials.

Contemporary Advancements in Materials Technology: Novel Solutions to Complex Problems.
Shrikaant Kulkarni, Vipul Srivastava, & P. William (Eds.)

4.1 INTRODUCTION TO RIETVELD METHOD

Powder diffraction started its journey when Walter Friedrich and Paul Knipping conducted their famous X-ray diffraction experiment in 1912.[1] They investigated the diffraction of X-rays from grounded copper sulfate crystals (which are in powdered form with a large number of randomly oriented crystallites instead of a single crystal). Their investigation put forward a significant milestone in the field of X-ray crystallography and paved a new way for the understanding of the atomic and molecular structure of matter. The diffraction pattern contains information about the crystal structure of the sample, including the position and intensity of the diffraction peaks.

The Rietveld analysis is a powerful refinement technique used in materials science for analyzing the crystal structures of crystalline materials. The method is named after the Dutch scientist Hugo Rietveld, who developed it in the 1960s,[2] and since then, it has become a widely used method for the quantitative analysis of X-ray and neutron powder diffraction data. A variety of crystallographic information about a material may be obtained from it, including the crystal structure, lattice parameters, atomic positions, thermal vibrations, the degree of preferred orientation of the crystals, defects, impurities, and disorder. It is particularly useful for analyzing complex crystal structures and for quantifying the amount of different phases present in a material that give us an insight into its properties, such as its mechanical, electrical, and magnetic behavior. In this chapter, we have discussed the mathematical aspect of Rietveld analysis, including various important parameters like the R-factor, profile shape function, orientation model, reliability factor, background modeling, etc.

4.2 MATHEMATICAL ASPECT OF RIETVELD ANALYSIS

In 1969, the Journal of Applied Crystallography presented the world with a pathbreaking paper titled "A Profile Refinement Method for Nuclear and Magnetic Structures" by Hugo Rietveld.[3] He had written a 162 kilobyte program, which was the beginning of today's Rietveld refinement simulation program. The simplified algorithm of the refinement procedure is shown in Figure 4.1.

There are very few resources available in the literature explaining the mathematical aspect with the basic physics of the Rietveld analysis with which a researcher can start the refinement of his data. However, without understanding the underlying theory of the refinement, it is unsafe to play with the refinable parameters. Sometimes, good fitting occurs without yielding a valid structural model or physical value of the parameters. Therefore, in this section, we have tried to give the basic idea in a nutshell with the hope that it will be beneficial for beginners in this field.

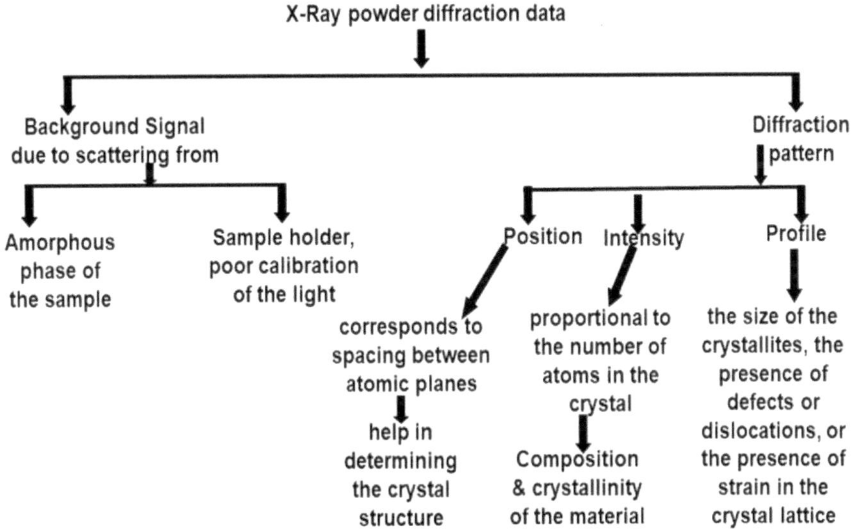

FIGURE 4.1 Simplified algorithm of the Rietveld refinement procedure.

Source: Adapted from Ref. [4]

Rietveld refinement requires prior knowledge of the crystal structure of all phases present in the sample.[5] It is the fitting of a calculated diffraction pattern with the experimental data by adjusting each of the variables that build the diffraction pattern. The basic equation[6] on which Rietveld refinement stands is as follows:

$$I_i^{cal} = I_F \sum_{j=1}^{Nphase} \frac{f_j}{V_j^2} \sum_{k=1}^{Npeaks} L_k \left|F_{k,j}\right|^2 S_j\left(2\theta_i - 2\theta_{k,j}\right) P_{k,j} A_j + bkg_i \quad (4.1)$$

Here, the symbols have the following meanings:

I_F = beam intensity

S_j = The profile shape function

f_j, V_j = volume fraction $\wedge$ volume of each phase

L_K = Lorentz polarization factor

$$F_k = \sum_{k=1}^{N} f_k N_k exp\left[2\pi i\left(hx_k + ky_k + lz_k\right)\right]$$

$P_{k,j}$ = Preferred orientation function
A_j = Absorption factor

$$bkg_i\left(2\theta\right) = background\ function = \sum_{n=0}^{n(degree\ of\ polynomial)} a_n$$

4.2.1 KNOWLEDGE ABOUT DIFFRACTOMETERS

Prior knowledge of the instrument is one of the most important criteria before starting the Rietveld refinement. Most diffractometers adhere to Bragg–Brentano parafocusing geometry, where the diffraction vector is perpendicular to the sample. The Bragg–Brentano focusing arrangement in the powder diffractometer is shown in Figure 4.2.

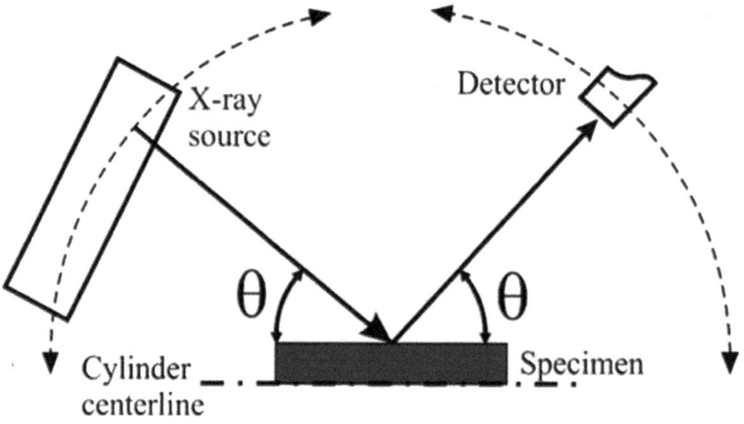

FIGURE 4.2 Bragg–Brentano focusing arrangement in a powder diffractometer.
Source: Reprinted with permission from Ref. [7]. © 2013 AIP Publishing.

The diffraction vector bisects the angle between the incident and the scattered beam. Depending upon the rotation of the X-ray tube, sample, and detector, the instrument may be classified into two types:

- θ:2θ instrument: In this instrument type, the tube is held fixed and the sample and detector rotate θ° and 2θ°/min, respectively. Example: Rigaku H3R
- θ:θ instrument: In this instrument type, the sample is held fixed and the tube and detector rotate –θ° and θ°/min, respectively. Example: PANalytical X'Pert pro

In addition, goniometer radius, slit type (fixed width/programmable), and different types of X-ray tubes (Cu/Co/Mo Kalpha) are required in some Rietveld programs.

4.2.2 SCATTERING FACTOR

The scattering factor F_k is proportional to the intensity of the scattered beam from the unit cell, which contains the term f called the atomic scattering factor (the ratio of the wave scattered by an atom to that of an electron). It seems to be $f = z$, where z is the total number of electrons within the atom. However, it is not so simple because as θ increases, the waves scattered by electrons are more out of phase, thereby decreasing f. f is therefore written as[6]

$$f = \int_0^{\infty} 4\pi r^2 \rho(r) \frac{sin\left(\frac{4\pi r sin\theta}{\lambda}\right)}{\left(\frac{4\pi r sin\theta}{\lambda}\right)} dr \qquad (4.2)$$

where $\rho(r)$ is the charge density at point "r", which can be determined from the prior knowledge of the "Thomas–Fermi" model. In this calculation, it is assumed that atoms in a unit cell are vibrating about their mean position, which is an oversimplified assumption. The temperature at which the measurement is carried out can affect the vibration. The degree of atomic displacement around their average positions in the crystal lattice is quantified as the Debye–Waller temperature factor. Thus, this factor plays an important role in accurately modeling the thermal vibrations of atoms in a crystal lattice and is a key parameter in the Rietveld analysis of crystal structures. The Debye–Waller factor is defined as

$$B = 8\pi^2 U^2 \qquad (4.3)$$

where U^2 is the mean-square amplitude of the vibration. Therefore, considering only the temperature factor, the correct expression for the atomic scattering factor becomes

$$\left[f_k\right]^2 = \left| \int_0^{\infty} 4\pi r^2 \rho(r) \frac{sin\left(\frac{4\pi r sin\theta}{\lambda}\right)}{\left(\frac{4\pi r sin\theta}{\lambda}\right)} exp\left(\frac{-B sin^2\theta}{\lambda^2}\right) dr \right|^2 \qquad (4.4)$$

It is to be noted that B can have six components (B_{11}, B_{22}, B_{33}, B_{12}, B_{23}, and B_{31}) if the vibrations are not similar in all directions. Due to this temperature factor, the intensities of the diffraction line as well as the background line decrease.

4.2.3 LORENTZ POLARIZATION FACTOR

It is a well-known fact that if the reciprocal lattice points touch the Ewald sphere, diffraction occurs. The intensity of the related reflection will be correspondingly higher if a reciprocal lattice point is in the diffracting location for a longer period. The periods needed for various lattice points to cross the Ewald sphere vary depending on the technique used to measure Bragg's intensities and the location of the reciprocal lattice points. Therefore, a correction factor is required to account for this, known as the Lorentz correction factor. Mathematically, the factor is $\frac{1}{sin2\theta}$, where θ is the diffraction angle. This factor accounts for the geometric factors that affect the intensity of diffracted X-rays.

If the incident beam is not polarized, then the radiation scattered by the incident beam has angular dependence. This is known as the polarization factor and can be calculated as $\frac{\left(1+cos^2 2\theta\right)}{2}$. Thus, this factor accounts for the anisotropic scattering of X-rays by the crystal lattice. According to this factor, radiation scattering is highest in the direction of the incident beam and least in the direction perpendicular to the main beam. The Lorentz polarization factor is the product of the Lorentz correction factor and the polarization factor and is therefore mathematically equal to $\frac{\left(1+cos^2 2\theta\right)}{2sin2\theta}$.

4.2.4 PROFILE SHAPE FUNCTION

The success of Rietveld refinement depends on the perfect matching of peak shapes in the diffraction pattern. Different factors like crystallite size, lattice strain, crystal defects/vacancies, the geometry of the X-ray source, slit-width, etc. have their own contribution to the shape of the diffraction peak. In the Rietveld refinement procedure, it is assumed that the peak profiles in the diffraction pattern vary systematically as a function of 2θ. For peak profile analysis of powder diffraction data, the pseudo-Voigt peak function is widely used. The pseudo-Voigt function (*pV*) is essentially a linear combination of the Lorentzian and Gaussian peak functions (eq. 4.5)[8]

$$pV(x) = \eta L(x) + (1-\eta)\,G(x) \tag{4.5}$$

where η, $L(x)$ and $G(x)$ are the mixing parameters, Gaussian, and Lorentzian functions, respectively. It is to be noted that in some cases, pseudo-Voigt with FJC asymmetry correction is also needed to fit the peak. Thompson, Cox, and Hastings modified the pV function to give a relation between the full width at half maximum (FWHM) parameters and the instrument resolution and specimen broadening effects. This one was relatively easy compared to η.[9]

Another function that is used to model the peak shapes of X-ray or neutron diffraction patterns in Rietveld analysis is the "Pearson VII peak-shape function". In general, the Gaussian function is broad at the peak and narrow at the tail, while the Cauchy function has just the opposite character of the Gaussian one. A mathematical function that can be varied from Gaussian to Cauchy and, in the intermediate step, pass through the Lorentzian function is called the Pearson VII type function.[10] The function can be written as

$$y = y_0\left[1+\frac{(x-x_0)^2}{ma^2}\right]^{-m} \tag{4.6}$$

The type of the function is Cauchy, modified Lorentzian, and Gaussian when $m = 1, 2$, and ∞, respectively.[11]

After matching the peak positions, the additional profile parameters (such as the peak width, peak asymmetry, etc.) can be refined. Caglioti et al.[12] identified a way to state that the FWHM of the Gaussian component of the peak shape varies with 2θ (eq. 4.7)

$$(FWHM)^2 = \mathrm{U}\,tan^2\,\theta + V tan\theta + W \tag{4.7}$$

and that of the Lorentzian part is described in eq. 4.8.

$$FWHM = X tan\theta + \frac{Y}{cos\theta} \tag{4.8}$$

In Bragg–Brentano geometry, U, V, and W are the positions of the incident, receiving, and diffracted slits, respectively, relative to the sample. The signs of U, W, and V depend on the orientation of the slits and the direction of the diffracted beam. In this geometry, U and W are positive, and V is negative.

The X and Y parameters in Bragg–Brentano geometry describe the orientation of the sample surface relative to the incident and diffracted beams. X and Y are the angles the sample surface makes with the incident and the diffracted beams, respectively. The values of X and Y can be adjusted by tilting the sample or rotating the diffractometer.

4.2.5 INTENSITY EXTRACTION METHOD

To tackle the issue of estimating reflection intensities in the absence of structural knowledge, two separate methods have been commonly used. In Pawley's approach, the reflection intensities are considered refinable variables rather than modifying structural parameters, akin to the Rietveld fit.[13] This necessitates specific handling for reflections that perfectly and very nearly overlap. In order to speed up computations for iterative ab initio structure solving techniques, the covariance matrix from a Pawley fit can be utilized to quantify the uncertainty in intensity apportionment for overlapped reflections. On the other hand, Le Bail approach[14] demands prior knowledge of the unit cell and the space group of the crystal. The adjustments of the unit cell parameter, the profile parameter, and the peak intensities are required to match the measured diffraction data. Therefore, the structural factor and its related structural parameters are not important. Le Bail fitting is advantageous for getting started (it can be used to generate an initial estimate of the crystal structure from powder diffraction data when no prior knowledge of the structure is available), solving structures (very useful for complex crystal structures), and treating impurities (refine the crystal structure separately from the impurities). Almost all software, viz., FULLPROOF, RIETAN, TOPAZ, and GSAS, use the Le Bail method for intensity extraction, whereas the MAUD software provides the option to choose any of the Pawley and Le Bail methods for the user.

4.2.6 ABSORPTION FACTOR

The absorption factor (A) arises due to the finite absorption of the incident beam by the specimen itself, which in turn affects the intensity of the diffracted beam. The geometry of the diffraction method may influence this factor as the geometry of the diffraction experiment can affect the path of the X-rays. There are two main geometries in X-ray diffraction experiments that can affect the absorption factor in Rietveld analysis.

a) Hull/Debye–Scherer: In this method, the powdered sample in the form of a thin cylinder is placed on the camera axis. Qualitatively, the absorption is a function of the angle and position of the detector, but the exact calculation of the correction factor for a cylindrical specimen is a tedious job.
b) Diffractometer: In this case, the specimens are flat-shaped, and thereby, an equal angle is formed with the incident and diffracted beams. Here 'A' is equal to $1/(2\mu)$, where μ is the linear absorption coefficient of the sample.

4.2.7 PREFERRED ORIENTATION

Sometimes, it has been observed that one of the peaks is comparatively more intense and the intensity ratio with the other is larger than the standard JCPDS (Joint Committee on Powder Diffraction Standards) pattern. Most of the thin film systems and nanorods exhibit preferred orientations in the diffraction pattern. A preferred orientation model in X-ray diffraction analysis is the March–Dollase function, which describes the effects of preferred orientation on the intensity of diffraction peaks.[15] Compared to other preferred orientation models, such as the FWHM model or the modified Lorentzian model, the March–Dollase function has several advantages in terms of physical relevance, flexibility, computational efficiency, etc. The equation is

$$T_{hkl} = \frac{1}{N}\sum_{i=1}^{N}(\tau^2 \cos^2 \Phi^i_{hkl} + \frac{1}{\tau}\sin^2(\tau^2 \cos^2 \Phi^i_{hkl})^{\frac{-3}{2}} \tag{4.9}$$

where T_{hkl} is the proposed orientation factor that can be refined, Φ is the angle made by the preferred orientation vector with the normal to the planes, and N is the symmetrically equivalent reciprocal lattice point.[16] The above expression is widely used as it applies to both disc and rod grain shapes with a single variable parameter. The value of $\tau = 0$ means perfect orientation, while $\tau = 1$ means no orientation/random orientation. In addition to this model, harmonic and exponential harmonic M-Tex models are also available.

4.2.8 R-FACTOR IN RIETVELD REFINEMENT

The Rietveld method uses least-square refinement, where the difference between the observed intensity (Y_o) and the calculated intensity (Y_i) is minimized at each 2θ point by refining the parameters. The weighted profile of R factor is defined as follows:[17]

$$R_{wp} = [\frac{\sum_{i=1}^{N} w_i [Y_{io} - Y_{ic}]^2}{\sum_{i=1}^{N} w_i Y_{io}^2}]^{\frac{1}{2}} \tag{4.10}$$

The expected R factor is

$$R_{\exp} = [\frac{N - P}{\sum_{i=1}^{N} w_i Y_{io}^2}]^{\frac{1}{2}} \tag{4.11}$$

The goodness of fit (GOF) is defined as

$$GOF = (\frac{R_{wp}}{R_{\exp}})^{\frac{1}{2}} \tag{4.12}$$

In the above eq. 4.11, N and P denote the number of data points and parameters, respectively. It is to be noted that there should not be any peak without fitting. A small expected profile R-factor bears no relationship to the quality of a profile fit; rather, it means that the average number of counts in a profile is large.[18] In general, $R_{wp} < 10\%$ and GOF between 1 and 2% are usually suggested to be satisfactory. Lutterotti et al.[19] suggested R_{wp} values for the acceptability of the refinement results. If the value is less than 0.1, then it corresponds to the medium complex phase, whereas a value of less than 0.15 corresponds to the complex phase. For cubic symmetry with few peaks, the acceptable value is less than 0.08.

However, these are "just numbers", and there is no thumb rule. R_{wp} values actually depend on the complexity of the crystal structure, the quality of the data, and the degree of agreement between the refined model and the experimental data. Toby et al.[20] demonstrated that when the background attributes account for a large fraction of the total intensity in a pattern, it is possible to get small values of χ^2 or R_{wp} by fitting the background alone, even without a lawful model of the structure of the crystal. Again, we can always improve χ^2 if we have lower-quality data.

4.3 BACKGROUND MODELING

Background modeling is the process of accurately characterizing and subtracting the unwanted signals from the observed diffraction pattern to obtain the true diffraction pattern of the sample. The accuracy of background modeling is critical in obtaining a reliable refinement of the sample structure. Basically, the background contributions can be dealt with in two different ways. The background can be modeled by a polynomial function like Legendre, Chebyshev I and II, or shifted Chebyshev I and II. Background subtraction can be an inelegant process when the background signal is complex or cannot be described well by a simple polynomial function. Many diffraction experiments contain additional noncrystalline scattering components that may arise from a variety of sources, such as amorphous phases, small crystallites, or sample preparation artifacts. Generally, the oscillations of these additional components overlap on the sharp Bragg pattern. Conventional background functions may not be able to fit the background and lead to inaccuracies in the refinement result. In this regard, it is required to use more sophisticated methods to account for the noncrystalline scattering.[20]

4.4 CONSTRAINTS IN RIETVELD REFINEMENT

In X-ray crystallography, sometimes materials cannot be easily crystallized in single crystals, and the information on the atomic arrangement of materials remains unknown. In these cases, determination of the atomic structure of crystalline solids using Rietveld refinement becomes difficult as it uses the local optimization of the atomic positions, which depends on the prior knowledge of the structural model.[21] Thus, in this situation, a good starting model approximation of the crystal structure becomes difficult, thereby making the use of Rietveld analysis a challenging task. If in the sample there are multiple phases present, the random orientation of the crystallites leads to complex diffraction patterns that are difficult to model accurately using Rietveld analysis. Rietveld refinement can have difficulties in determining the structure of disordered materials, such as glasses or liquids, as their structure may be difficult to model with traditional crystallographic methods. The refinement process can struggle when there are multiple phases present in the data. Overlapping peaks and varying peak shapes can make it difficult to accurately assign peak positions to the correct phase. In a multiphase sample, a sequential Rietveld refinement can be used, where each phase is refined individually, and then the individual results are combined to obtain the overall structural information.[22] This complex analysis may reduce the accuracy of the result. In fact, Rietveld analysis gives accurate results for the single-phase samples. As the refinement process is based on a pre-existing model, the iteration process sometimes becomes long and time-consuming if the initial approximation is not good. In some cases, the refinement may not converge at all, and additional information or adjustments to the model may be necessary to obtain an accurate solution. These constraints put limitations on the use of Rietveld refinement. In such a scenario, the ab initio structure solution method might be useful for solving crystal structures from powder diffraction data, especially for complex or disordered structures.

4.5 AB INITIO STRUCTURE SOLUTION WITH POWDER DIFFRACTION DATA

Powder diffraction data plays a very significant role in predicting accurate cell parameters. Condensing the 3D reciprocal lattice information onto a 1D photon intensity profile with respect to detector angle 2θ in powder diffraction is a very challenging task. A strong overlap of diffraction peaks happens during the extraction of structure factor amplitudes from the powder diffraction data. Significant improvement of this method has been achieved through various means such as synchrotron-based powder diffractometers,[23] fast iterative Patterson squaring,[24] the Bayesian method,[25] etc. These are being done using various software programs. Ab initio structure solution methods with powder diffraction data involve solving the phase problem directly from the diffraction data, without the need for any prior knowledge of the crystal structure. The raw data that is obtained from the powder diffraction method is first processed to reduce the noise and correct any systematic errors. Based on these data, a preliminary model of the molecule or material is constructed. This model can be based on experimental data or obtained from first principles calculations. The preliminary model is refined using various computational techniques until it accurately matches the experimental data. A few of these techniques are molecular dynamics (MD) simulations, Monte Carlo (MC) simulations, and density functional theory (DFT) calculations.

To perform a MD simulation of powder diffraction data, one would typically start by configuring a model considering the positions of the atoms in the lattice and by calculating the total energy of the system.[26] Next, the system is typically equilibrated at a particular temperature and pressure to ensure that it is in a stable state. Once the system is equilibrated, the MD simulation is run for a sufficient amount of time to allow the atoms to move and rearrange themselves. During the simulation, the positions of the atoms are recorded at regular intervals. From these positions, the diffraction pattern can be calculated using the principles of X-ray diffraction. This simulated diffraction pattern can then be compared to experimental data to refine the crystal structure model. This process of refinement involves adjusting the parameters of the model, such as the positions of the atoms and their thermal vibrations, to better match the experimental data.

To perform a MC simulation, the unit cell of the preliminary model is configured by a collection of atoms that are under random movement, and each configuration is accepted or rejected through a sampling process that is based on the Metropolis algorithm.[27] The algorithm proposes a new configuration by randomly selecting an atom and moving it to a new position. The acceptance or rejection of this new configuration depends on the energy of the system and the temperature at which the simulation is performed. By repeating this process many times, the system eventually reaches equilibrium. During the simulation, the positions of the atoms are recorded at regular intervals. From these positions, the diffraction pattern can be calculated using the principles of X-ray diffraction. The goodness of fit is compared for the experimental powder diffraction data with the diffraction data obtained from every trial model of the crystal structure.

To perform a DFT analysis of powder diffraction data, one would typically start by constructing a model of the crystal structure of the material, including the positions and types of the atoms in the lattice.[28] The electronic structure of the material is then calculated using DFT, which involves solving the Schrödinger equation for the electrons in the crystal lattice. From the calculated electronic structure, it is possible to obtain the X-ray diffraction pattern of the material. This diffraction pattern can be compared to experimental data to refine the crystal structure model. By adjusting the parameters of the model, such as the positions of the atoms or the lattice parameters, it is possible to obtain a better match of diffraction patterns between the simulated and experimental ones and understand the material and its properties.

In all the above refinement processes, the final crystal structure is validated using a variety of methods, including calculating the R-factor, comparing the calculated and experimental diffraction patterns, and analyzing the geometry of the crystal structure.

Thus, from the above discussion, it can be concluded that the ab initio structure solution with powder diffraction data is a complex process and requires specialized software and expertise in crystallography and materials science. This method has several advantages over traditional single-crystal X-ray diffraction methods for studying materials under nonambient conditions, such as high temperature or high pressure.

4.6 CONCLUSIONS

In conclusion, Rietveld analysis is a powerful method for analyzing the crystal structure of a material using X-ray or neutron powder diffraction data. The technique involves refining a crystal structure model against diffraction data using a least-squares refinement approach. The R-factor, GOF parameter, reliability factor, profile parameters, Lorentz polarization factor, scattering factor, and background parameters are the important parameters for assessing the quality of the refinement result and ensuring that the crystal structure model accurately represents the observed diffraction pattern. Although Rietveld analysis is a powerful method for determining the crystal structure of materials from powder diffraction data, one important limitation of Rietveld analysis is that it requires a good starting model of its structure in order to converge to the correct solution. If the starting model is incorrect, the refined structure may be incorrect or incomplete. In such cases, ab initio methods are useful as they use first principles to predict the crystal structure. This makes these methods particularly useful for studying novel or poorly characterized materials or for predicting the properties of materials under nonambient conditions.

KEYWORDS

- **Rietveld refinement**
- **least-square refinements**
- **powder diffraction**
- **crystal structure**
- **powder pattern fitting**

REFERENCES

1. Walther, F.; Knipping, P.; Laue, M. Bayerische Akademie der Wissenschaften. In *Mathematisch-Physikalische Klasse Sitzungsberichte*; 1912; pp 303–322.
2. Rietveld, H. M. Line Profiles of Neutron Powder-diffraction Peaks for Structure Refinement. *Acta Crystallogr.* **1967,** *22* (1), 151–152.
3. Rietveld, H. M. A Profile Refinement Method for Nuclear and Magnetic Structures. *J. Appl. Crystallogr.* **1969,** *2* (2), 65–71.
4. Para, T. A.; Sarkar, S. K. Challenges in Rietveld Refinement and Structure Visualization in Ceramics. *Adv. Ceram. Mater.* **2021**.
5. Takeshi, E.; Billinge, S. J. Crystallographic Analysis of Complex Materials. *Pergamon Mater. Ser.* **2012,** *16,* 27–54.
6. Cullity, B. D.; Stock, S. R. Elements of X-Ray Diffraction, 3rd edition, Pearson 2001.
7. Zhigachev, A. O. The Effect of Specimen Surface Curvature on X-ray Diffraction Peak Profiles. *Rev. Sci. Instrum.* **2013,** *84* (9), 095105.
8. Karmakar, A.; Arora, H.; Nath, M.; Sutradhar, S.; Sarkar, B. J.; Mandal, G.; Samanta, A.; Bandyopadhyay, A. Determination of Exchange Integrals and Effect of Cationic Site Occupancy (8b/24d) on the Structural and Magnetic Properties of Nanocrystalline Mn-doped Gd_2O_3. *J. Alloys Compd.* **2023,** *931,* 167475.
9. Kaduk, J. A.; Reid, J. Typical values of Rietveld Instrument Profile Coefficients. *Powder Diffr.* **2011,** *26* (1), 88–93.
10. Hall, M. M.; Veeraraghavan, V. G.; Rubin, H.; Winchell, P. G. The Approximation of Symmetric X-ray Peaks by Pearson Type VII distributions. *J. Appl. Crystallogr.* **1977,** *10* (1), 66–68.
11. Elderton, W. P.; Johnson, N. L. Systems of Frequency Curves. Cambridge University Press: New York, 1969; pp 77–78.

12. Caglioti, G.; Paoletti, A. T.; Ricci, F. P. Choice of Collimators for a Crystal Spectrometer for Neutron Diffraction. *Nucl. Instrum.* **1958,** *3* (4), 223–228.
13. Pawley, G. S. Unit-cell Refinement from Powder Diffraction Scans. *J. Appl. Crystallogr.* **1981,** *14* (6), 357–361.
14. Le Bail, A.; Duroy, H.; Fourquet, J. L. Ab-initio Structure Determination of LiSbWO6 by X-ray Powder Diffraction. *Mater. Res. Bull.* **1988,** *23* (3), 447–452.
15. Dollase, W. A. Correction of Intensities for Preferred Orientation in Powder Diffractometry: Application of the March Model. *J. Appl. Crystallogr.* 1986, *19* (4), 267–272.
16. Kundu, S.; Sain, S.; Satpati, B.; Bhattacharyya, S. R.; Pradhan, S. K. Structural Interpretation, Growth Mechanism and Optical Properties of ZnO Nanorods Synthesized by a Simple Wet Chemical Route. *RSc Adv.* **2015,** *5* (29), 23101–23113.
17. Lutterotti, L.; Scardi, P. Simultaneous Structure and Size–strain Refinement by the Rietveld Method. *J. Appl. Crystallogr.* **1990,** *23* (4), 246–252.
18. Toby, B. H. R Factors in Rietveld Analysis: How Good is Good Enough? *Powder Diff.* **2006,** *21* (1), 67–70.
19. Lutterotti, L. Quantitative Rietveld Analysis in Batch Mode with Maud. *IUCR CPD Newsletter*. **2005,** *32,* 53–55.
20. Richardson Jr., J. W. Background Modelling in Rietveld Analysis. Argonne National Lab.: IL (United States), 1989.
21. Putz, H.; Schön, J. C.; Jansen, M. Combined Method for Ab Initio Structure Solution from Powder Diffraction Data. *J. Appl. Crystallogr.* **1999,** *32* (5), 864–870.
22. Reid, J. W.; Hendry, J. A. Rapid, Accurate Phase Quantification of Multiphase Calcium Phosphate Materials using Rietveld Refinement. *J. Appl. Crystallogr.* **2006,** *39* (4), 536–543.
23. Sedigh Rahimabadi, P.; Khodaei, M.; Koswattage, K. R. Review on Applications of Synchrotron-based X-ray Techniques in Materials Characterization. *X-Ray Spectrom.* **2020,** *49* (3), 348–373.
24. Estermann, M. A.; Gramlich, V. Improved Treatment of Severely or Exactly Overlapping Bragg Reflections for the Application of Direct Methods to Powder Data. *J. Appl. Crystallogr.* **1993,** *26* (3), 396–404.
25. Sivia, D. S.; David, W. I. A Bayesian Approach to Extracting Structure-factor Amplitudes from Powder Diffraction Data. *Acta Crystallogr. A Found. Crystallog.* **1994,** *50* (6), 703–714.
26. Szczerba, M.; Kalinichev, A. G. Intercalation of Ethylene Glycol in Smectites: Several Molecular Simulation Models Verified by X-ray Diffraction Data. *Clays Clay Miner.* **2016,** *64* (4), 488–502.
27. Harris, K. D.; Tremayne, M.; Lightfoot, P.; Bruce, P. G. Crystal Structure Determination from Powder Diffraction data by Monte Carlo Methods. *J. Am. Chem. Soc.* **1994,** *116* (8), 3543–3547.
28. Rammohan, A.; Kaduk, J. A. Crystal Structure of Pentasodium Hydrogen Dicitrate from Synchrotron X-ray Powder Diffraction Data and DFT Comparison. *Acta Crystallogr. E: Crystallogr. Commun.* **2017,** *73* (2), 286–290.

PART II

Materials Characteristics for Applications

CHAPTER 5

Mechanical Characteristics of Metals, Ceramics, and Polymers and Processing Role in Controlling Different Properties

RIYADH A. AL-SAMARAI[1] and Y. AL-DOURI[2,3,4]

[1]Electromechanical Engineering Department, Engineering College, University of Samarra, Samarra, Iraq

[2]College of Health and Medical Techniques, Al-Bayan University, Baghdad, Iraq

[3]Department of Mechanical Engineering, Faculty of Engineering, Piri Reis University, EflatunSk, No: 8, 34940 Tuzla, Istanbul, Turkey

[4]Nanotechnology and Catalysis Research Centre, University of Malaya, 50603 Kuala Lumpur, Malaysia

ABSTRACT

The themes are based on the chemistry and physics of molecular structure as well as the engineering principles of thermodynamics, kinetics, and transport phenomena. In this chapter, we will look at mechanical properties, which are often used as the main criterion for choosing materials. Our major emphasis will be on the properties of materials, while we will briefly touch on the mechanics that govern these traits in both the liquid and solid states. We explain how solids (mostly) respond to these pressures as we look at further types of stresses, such as tensile and compressive stresses. Elastic, inelastic, viscoelastic, ductile, fracture, or time-dependent creep are the usual properties. We will see that Newton's law and other theories can help explain all of these topics using the concepts of stress (applied force per unit area)

Contemporary Advancements in Materials Technology: Novel Solutions to Complex Problems.
Shrikaant Kulkarni, Vipul Srivastava, & P. William (Eds.)

and strain (dimensional change). We will differentiate between stress–strain diagrams for several kinds of materials qualitatively using the stress–strain diagram and determine the strength and modulus. We will describe the differences between the viscoelasticity models and appreciate the impact of strain rate on mechanical properties and explain how temperature, molecular weight, and strain rate affect the mechanical properties of polymers.

5.1 INTRODUCTION

The topics are built on the engineering principles of thermodynamics, kinetics, and transport properties, as well as the chemistry and physics of molecular structure. It will examine the mechanical qualities, which, for many applications, serve as the primary selection criterion for materials. Although the mechanics control both liquid and solid stages, our main focus will be on the mentioned properties.[1–3] We will examine more forms of stresses in this chapter, such as tensile and compressive stresses, and discuss how solids (mainly) react to these pressures. This response typically takes on one of the following characteristics: elastic, inelastic, viscoelastic, ductile, fractured, or time-dependent creep.

We will see that some of these reactions may be explained by Newton's law, which states that the ideas of stress (applied force per unit area) and strain (dimensional change) are applicable to all of these subjects.

This chapter enables you to conclude the following:

- Qualitatively discriminate between stress–strain diagrams for several classes of materials based on the strength and modulus from the stress–strain diagram.
- Outline the variations between models of viscoelasticity.
- Recognize the bending, volume, and tension components.
- Differentiate between yield and tensile strength on one side and brittle and ductile materials on the other side.
- Recognize how strain rate affects mechanical characteristics.
- Application of time–temperature superposition concepts to polymer moduli and determination of shear factors.
- Different types of mechanical creep and creep rates.
- Describe how the mechanical properties of polymers are affected by temperature, molecular weight, and strain rate.
- Multilayer composites with unidirectional, continuous, and discontinuous fiber reinforcement have their axial and transverse moduli, as well as their strength.

- Recognize how the physical makeup of soft and hard biologics affects particular mechanical characteristics.
- To choose the best candidate materials for a certain application, considering mechanical qualities.

5.1.1 MECHANICS OF METALS AND ALLOYS

In this chapter, we explain a material's atomic structure-based response to an applied voltage. Prior to discussing topics pertaining to polycrystalline materials, we shall first concentrate on the single crystal of a metallic element.[4]

5.1.1.1 ELASTICITY

5.1.1.1.1 The Molecular Originsof Elasticity

Interatomic bonds bind atoms together. Recall those equations, which connect the interatomic force F to the potential energy function U between atoms and the distance r between them:

$$F = \frac{-dU}{dr} \tag{5.1}$$

It should be obvious by this point that, at least in theory, we can connect how relatively difficult or simple it is to pull the atoms apart or bring them closer together under an applied voltage to the potential energy function. If we separate atoms at minor deformations, where the deformation is fully recoverable, that is, the atoms can return to their original undeformed positions, without being irrevocably displaced from one another, we can create a quantitative description of this process. Elastic reaction is the term for this. Similar to its more common usage, the word "elastic" has no particular connotations for polymers in this context. It has the same application as in physics, a totally recoverable deformation, and chemistry. Placing an imagined spring between the atoms makes it easy to visualize an elastic link between two atoms. The atoms return to the undeformed equilibrium separation distance r_0, where the forces are zero, when they are slightly deformed, either away from each other with the so-called stretching force or toward each other with the so-called compression force. Tensile forces are opposed by attractive forces between atoms, and compressive forces are opposed by repulsive forces.[5]

5.1.1.1.2 Generalized Hooke's Law

The deformation of an elastic item or substance is proportional to stress, according to the current theory of elasticity, which generalizes Hooke's law. In practice, a force applied to a volume will produce shear stresses in the plane of the surface in addition to normal forces or forces perpendicular to the surface. With one normal and two shear components in each of the three directions, the applied voltage has a total of nine components. Remember at the beginning that the direction of the applied force (outward normal to the surface) is indicated by the first subscript, and the direction of the resulting stress is indicated by the second subscript.[6] Thus, xy indicates the tangential stress caused by the force acting in the y direction along the x-axis since the x-component in this notation for normal forces is slightly redundant.

5.1.2 DUCTILITY

Since all ensuing deformations are fully recoverable, we have thus far restricted the discussion to minor deformations, i.e., modest departures from the equilibrium bond distance. Nearly all materials have an elastic reaction in this area (see Figure 5.1). However, a few factors determine what occurs at huge deformations based on the kind of material being thought of. We now enter the region of permanent deformation, which is the boundary of the elastic region. Permanent deformation mostly occurs in two ways: viscous flow and plastic flow.

Since this is where viscous flow primarily occurs, we delay consideration of it to the pertinent section on the mechanical properties of polymers. Fracture and creep are two more inelastic responses to applied forces that will be postponed to later locations. We shall now concentrate on a kind of irrecoverable deformation known as plastic flow or more frequently known as plasticity in the context of metals and alloys.[7] Figure 5.1 illustrates how the plastic region should ideally come right after the elastic reaction region and cause elongation (deformation) without the requirement for increased stress, which is why segment BC in Figure 5.1a is a flat region. Practically speaking, microstructural processes, specifically the increase in stress with percent strain in the region of plastic deformation, cause the behavior depicted in Figure 5.1b. We examine the molecular underpinnings of these observations in the following section.

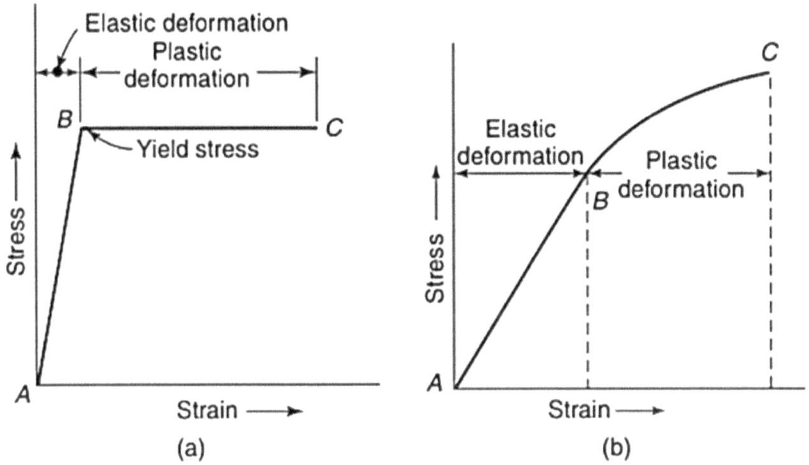

FIGURE 5.1 Illustrations of (a) ideal elastic deformation followed by (b) ideal plastic deformation in solids.

5.1.3 ANNEALING AND RECRYSTALLIZATION

In addition to the strain hardening and increased dislocation density mentioned above, deformation in polycrystalline materials can also result in a change in the grain structure. In some circumstances, a heat treatment known as annealing can be used to reverse all of these processes. An example of a thermal treatment called annealing is prolonged exposure to high temperatures followed by delayed cooling. The three steps of the annealing process are typically as follows: (1) heating to a predetermined temperature; (2) holding or "soaking" at that temperature; and (3) slow cooling, typically to room temperature.[8] At every point, time is a key factor.

There are a number of processes that can take place during annealing that lead to "recovery" or the restoration of properties and structures that are more like those present in the states before hardening. In this section, we will concentrate on the annealing reduction processes that are unique to metals and alloys (see Figure 5.2).

We will return to the explanation of the annealing of other materials, particularly inorganic glasses, in later parts. Metal annealing might first cause stress relaxation, which is a process in which a thermally stimulated dislocation motion releases the internal strain energy caused by plastic deformation (see Figure 5.2). The dislocation density may drop during the reduction process because higher temperatures are associated with enhanced atomic mobility. Recrystallization is a procedure that can occur at even greater temperatures when a new set of the resulting grains are deformation-free and equiaxed (have the same size in all directions). The grains in this recrystallized material resemble those in prewound metal and have a low dislocation density. Similar to the crystallization processes detailed, the recrystallization process is a diffusion phenomenon. Figure 5.2 illustrates how recrystallization results in a drop in tensile strength and an increase in ductility. It is important to reiterate that this phenomenon is also time-dependent. How long the deformed material is held at a specific temperature determines the degree of recrystallization and recovery. The recrystallization temperature is sometimes described as the temperature at which recrystallization is finished in one hour because complete recrystallization typically takes many hours. The temperature of recrystallization is about between one-third and fifty percent of the absolute melting point, Tm.[9–11]

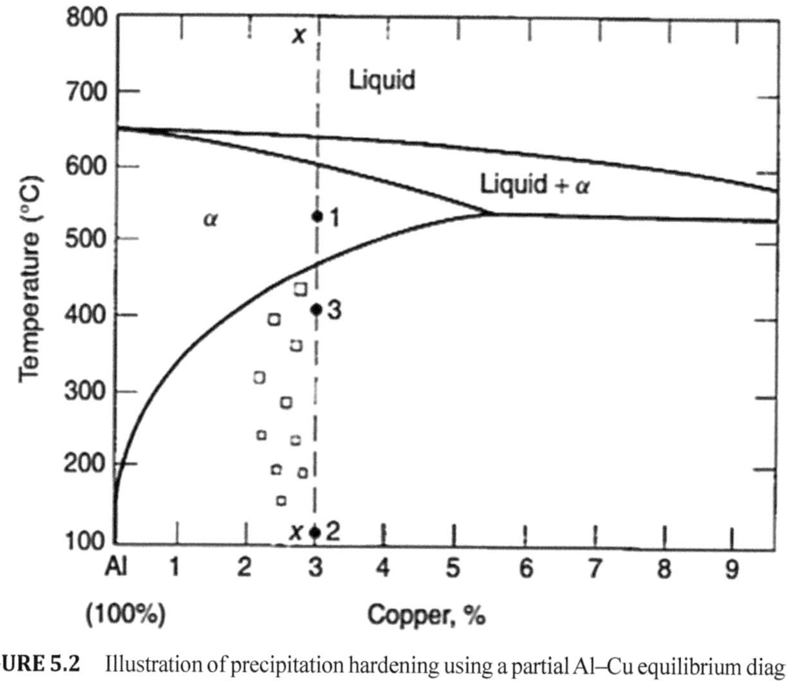

FIGURE 5.2 Illustration of precipitation hardening using a partial Al–Cu equilibrium diagram.

The recrystallization temperature for pure metals is roughly 0.3 Melt, and it can exceed 0.7 Melt for industrial alloys. Table 5.1 displays recrystallization temperatures for a few popular metals and alloys.

TABLE 5.1 Melting Points and Recrystallization of a Few Popular Metals and Alloys.

	Recrystallization temperature		Melting temperature	
Metal	°C	F	°C	F
Lead	−4	25	327	620
Tin	−4	25	232	450
Zinc	10	50	420	788
Aluminum (99.999 wt%)	80	176	660	1220
Copper (99.999 wt%)	120	250	1085	1985
Brass (60 Cu–40 Zn)	475	887	900	1652
Nickel (99.99 wt%)	370	700	1455	2651
Iron	450	840	1538	2800
Tungsten	1200	2200	3410	6170

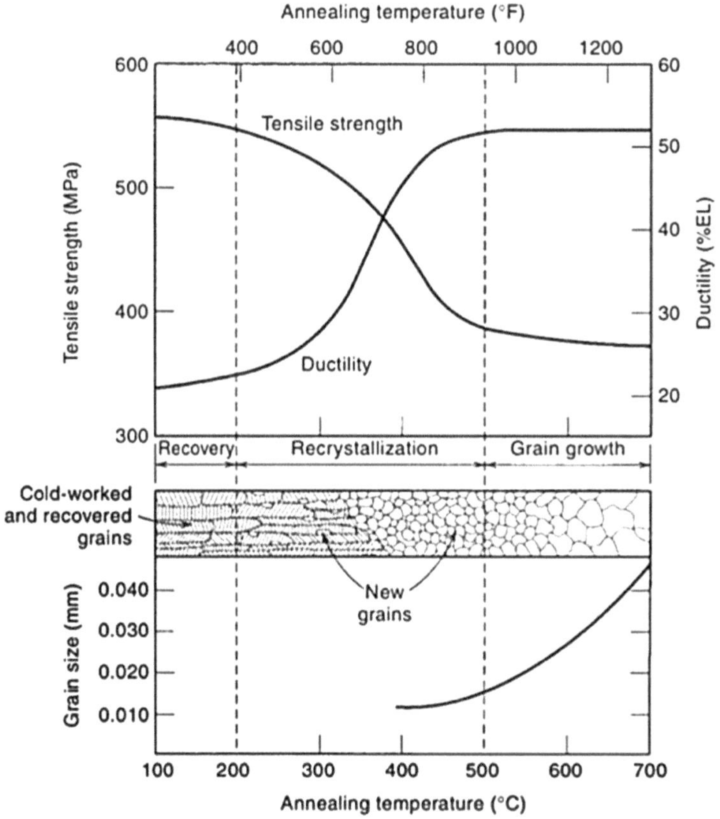

FIGURE 5.3 Effect of annealing temperature on brass alloy's tensile strength and ductility. Schematic representations of grain structures during recovery, recrystallization, and grain development are provided.

5.1.4 HARDNESS AND STRENGTH

A material's resistance to local plastic deformation, such as a scratch or minor dent, is measured by its hardness. This section includes an explanation of hardness since it shares similarities with strength and modulus in its relationship to molecular bond forces. As metal is cold-worked, its tensile strength increases, and this process is referred to as "hardening." Hardness is, therefore, a recoupable quality, similar to yield strength and ductility upon annealing. Dispersion-strengthened alloys and metals frequently exhibit greater hardness than pure metals due to the principles of solid solution strengthening and precipitation strengthening, respectively. It is problematic to quantify hardness in an absolute sense, which is one of its annoying characteristics. There are several different hardness scales, most of which are relative and largely qualitative. In Figure 5.4, a few of the most typical scales are depicted. In a typical hardness test, a standard-shaped hard indenter is pressed into a material's surface while being subjected to a specific load, causing the substance to first deform elastically and then plastically. A measurement and numerical value are assigned to the resulting indentation depth or indentation area. The price is determined by using scales and a tool. The Mohs scale is one of the most basic scales, where a material's ranking is determined solely by what it will scratch and what will scratch it, in comparison to the hardest (diamond) and softest (talc) materials known. One would anticipate that hardness and strength would be related to one another as both are relative indices of resistance to plastic deformation. This is the case, but it is challenging to give this relationship a general equation because it is challenging to estimate the absolute values of hardness. Figure 5.5 depicts the correlation between tensile strength and Brinell hardness for steel, brass, and cast iron.[12–14]

5.1.5 THERMOMECHANICAL EFFECTS

There are two fundamental topics that are frequently discussed in the context of heat transfer properties or material thermodynamics, such as thermal conductivity or specific heat, but are more related to mechanical properties because they are related to dimensional changes. These two fundamental topics are temperature dependence and properties such as strength and modulus, which we will discuss separately for each class of material.[15–17] Despite their close relationship, thermoelasticity and thermal expansion will be discussed separately (see Figure 5.6).

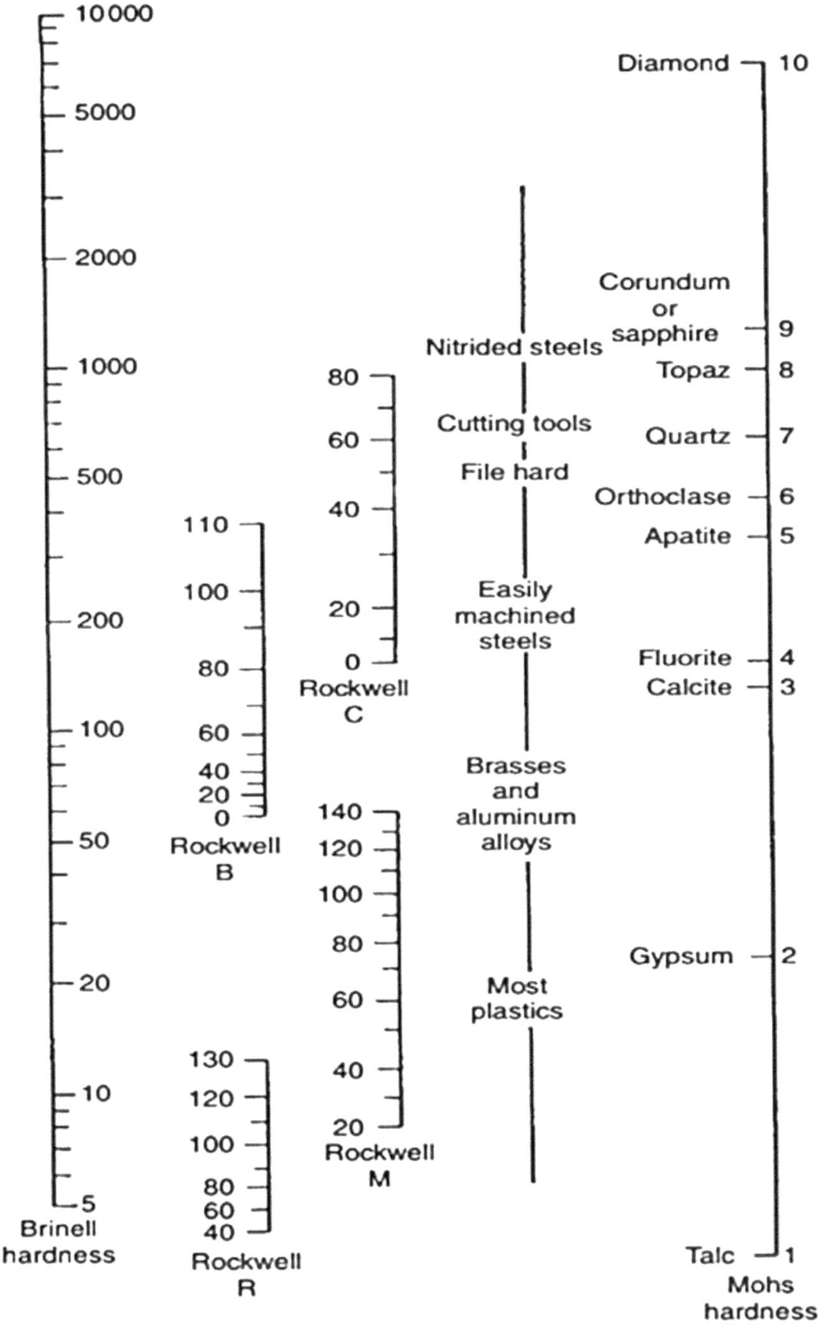

FIGURE 5.4 A comparison of some widely used hardness measures.

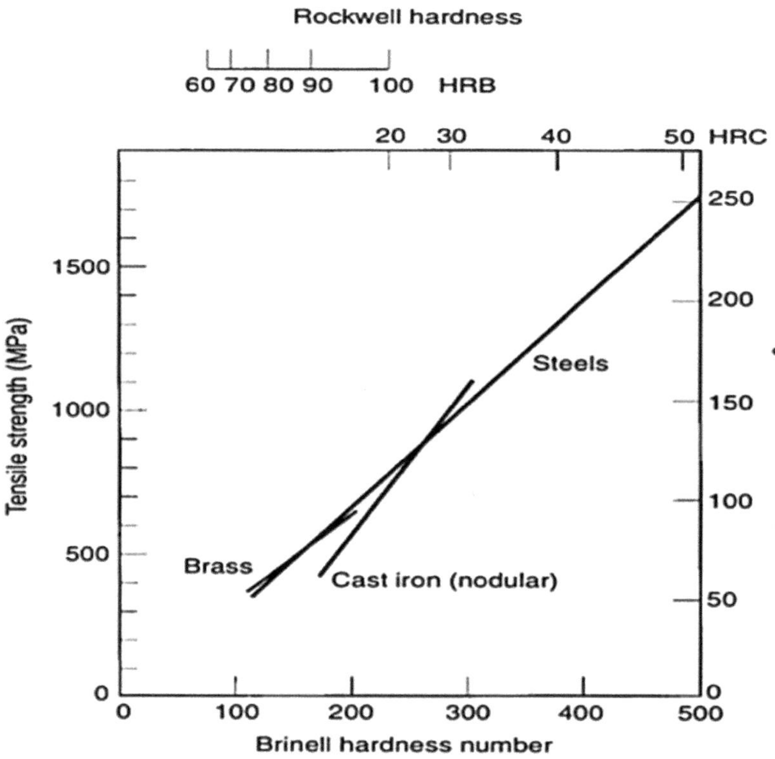

FIGURE 5.5 Steel, brass, and cast iron have a relationship between hardness and tensile strength.

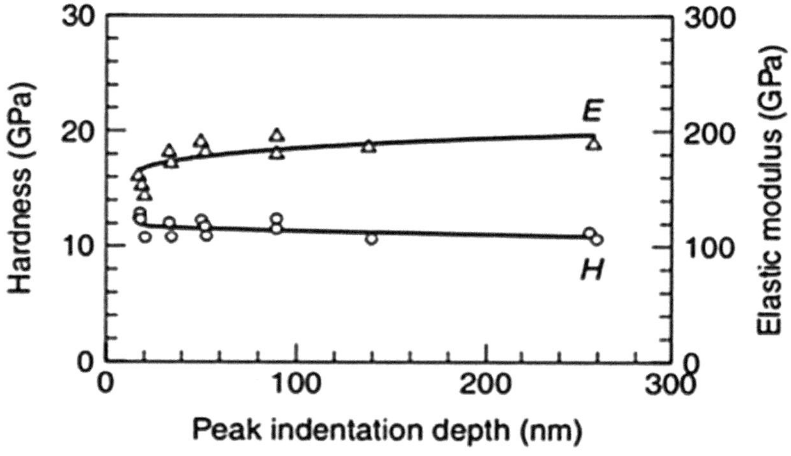

FIGURE 5.6 For Si, hardness and elastic modulus are related to indentation depth at different peak loads (100).

5.1.5.1 DIAGRAMS OF STRESS–STRAIN

So, let us discuss some useful facets of identifying mechanical qualities. On a stress–strain diagram, significant properties like modulus, strength, and ductility are typically shown graphically. For a few typical types of applied forces, the procedure for conducting the experiment and producing the stress–strain diagram is described below in detail.[18]

5.1.5.2 THE TENSILE TEST

A so-called universal test rig, which is schematically depicted in Figure 5.7, is used to conduct a conventional tensile test. In order to make insertion into the specimen grips easier, test specimens are typically designed like "dog bones," with the extremities being wider than the center. Any geometry for the sample's cross-section is acceptable as long as the space between the grips is maintained, which is identical and referred to as the calibration length. The most popular geometries are those with round and rectangular cross-sections.[19–21]

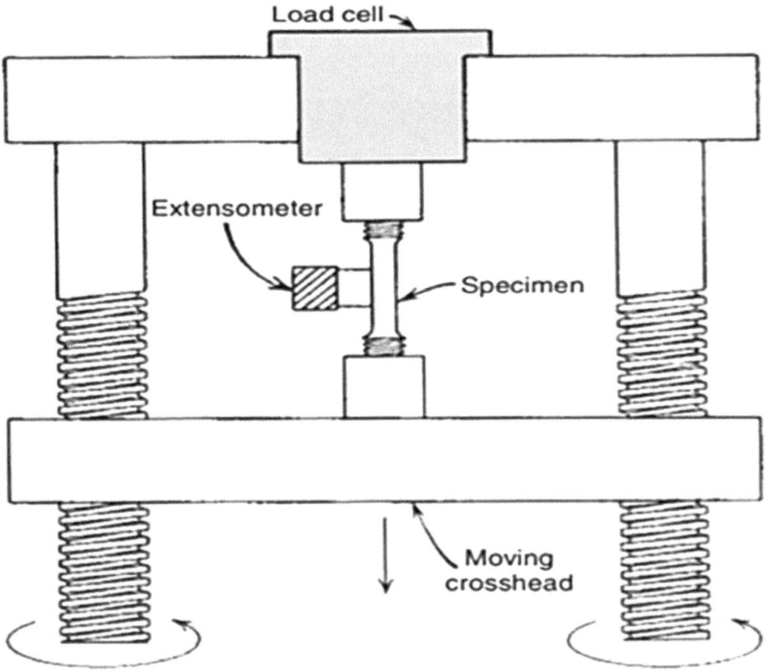

FIGURE 5.7 Tensile test plan using examples.

Although 1 inch (25 mm) and 2 inch (50 mm) measuring lengths are frequently employed, the measured length might be anything and is not defined. A fixed load cell mounted on the test frame is connected to one handle, typically the top handle. The force exerted on the sample is measured by the load cell. The other handle is fastened to the crosshead, a moving component of the frame. The specimen has a known stretching rate since the traverse descends at a given pace, known as the traverse speed. Typically, these rates are quite slow, on the order of a few thousandths of an inch per minute. The greatest span stretching can be inferred from the rate and duration of stretching, but extensometers are typically used to measure it considerably more precisely. Tensile tests typically result in the sample being stretched to the point of plastic distortion or rupture, though this is not essential if simply the elasticity modulus needs to be determined. As was said in the previous section, once ductile materials approach the yield point, they are subject to plastic deformation, which reduces their strength. The necking phenomenon's cross-sectional area and the load versus elongation measurements obtained by the load cell and extensometer, respectively, constitute the "raw data" of the tensile test. The elongation is calculated without taking into account the specimen geometry. The dimensionless strain, which is sometimes multiplied by 100 and represented as % elongation for the specimens, is obtained by dividing by the original length.[22]

5.1.6 METAL AND ALLOY MECHANICAL PROPERTIES

We eventually get to the characteristics of the mechanical properties of metals and alloys after a rather extensive introduction to the theoretical and experimental aspects of mechanical properties, which, as we shall see, apply to all classes of materials. Strength depends on temperature. Young's modulus, also called the modulus of elasticity, determines stiffness. Young's modulus is constant for any given metal and is unaffected by external stresses like heat treatment, processing, or cold work, whereas strength can differ between grades. Nonetheless, there are a few trends and generalizations that can be made about the mechanical characteristics of metals and alloys that should be noted.[23]

5.1.6.1 TEMPERATURE DEPENDENCE OF STRENGTH

Tempering is a heat treatment method used on ferrous alloys, like cast iron or steel, to reduce the alloy's hardness and increase its toughness. The metal becomes less brittle due to an increase in flexibility that typically coincides with a decrease in hardness. The latter result results in a reduction of the gap between the yield strength and the tensile strength until they almost equal each other at suitably high temperatures. The deformation of crystalline solids is clearly influenced by temperature at lower temperatures. The number of active slip systems might vary depending on temperature in some metals, especially HCP metals, and thus affect polycrystalline metals like Zn and Mg's ductility. Dislocations also move less readily as the temperature drops. As a result, polycrystalline materials' yield strength and tensile strength rise, and when temperature falls, their ductility, expressed as a percentage reduction in area, or %PA, declines, as illustrated in Figure 5.8. Metals can act brittle below a specific temperature, known as the brittle–ductile transition temperature, which means they can break before undergoing significant plastic deformation. For BCC metals below the brittle–ductile transition temperature, this is especially true. In contrast, FCC metals rarely experience this transition, and when temperature drops, their ductility only marginally declines. This occurrence is connected to the critical permissible shear stress' temperature dependence, as was previously depicted.[24]

5.1.6.2 BOTH EXTREME FLEXIBILITY AND PLASTICITY

Superplastic solids are polycrystalline substances that can withstand significant, homogeneous strains before breaking. Although some materials can attain an elongation of more than 1000% under uniaxial tensile strain, superplasticity is often indicated by an elongation to failure of more than 200%. Microgranular or microstructural superplasticity and environmental transformation or superplasticity are the two basic subtypes of superplastic behavior. Materials having a tiny grain size, typically less than 10 m, display microgranular superplasticity when they are deformed at strain rates between 10^{-5} and 10^{-1}/s at temperatures over 0.5 Melt, where Melt is the melting point in degrees Kelvin. Low flow stresses characterize superplastic deformation; superplastic molding of components has attracted a lot of commercial interest due to this and the great homogeneity of plastic flow. As stated previously in the equation, the most crucial mechanical property of a superplastic material is its great sensitivity to the strain rate of the flow stress, which must be greater than or equal to 0.5 in order for superplastic behavior to occur; for the majority of superplastic materials, this falls between 0.4 and 0.8 m. When a material is stretched, the existence of a neck causes a localized increase in strain rate and, at large values of m, a sudden increase in flow stress in the neck region. As a result, the neck experiences strain hardening, which hinders continued development. As a result, superplastic materials have strong necking resistance and high tensile elongation due to their great sensitivity to strain rate. The two primary varieties of superplastic alloys are microduplex and pseudo-single-phase.[25]

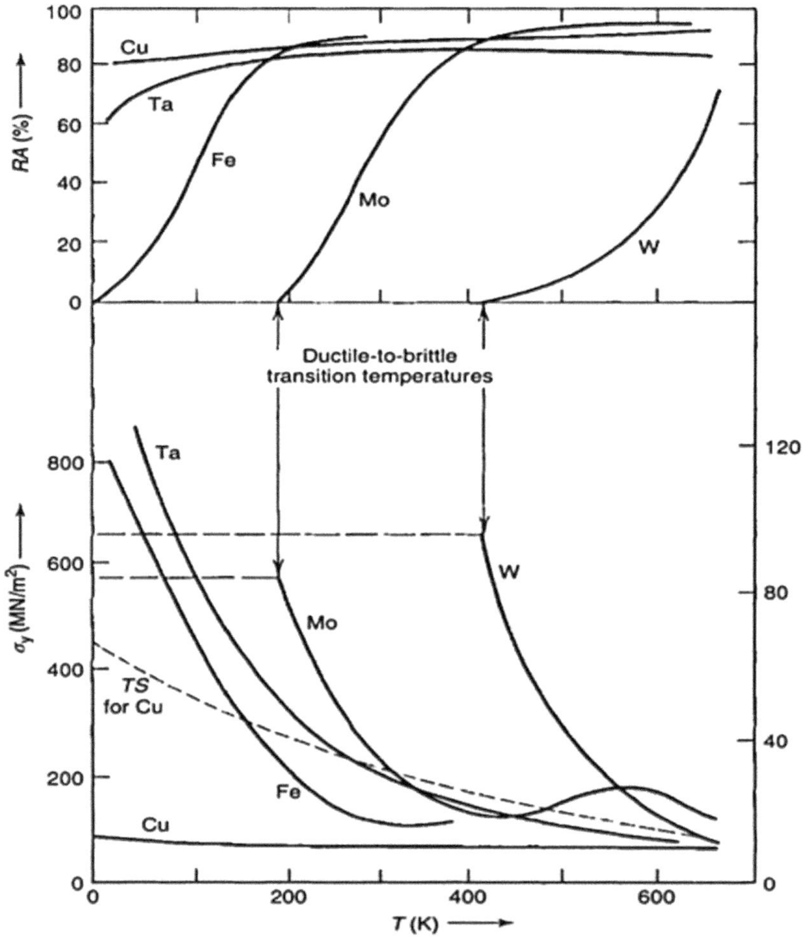

FIGURE 5.8 Yield strength (lower graph) and percentage reduction in area (%RA, upper graph) vs temperature for a few metals.

The first-class materials combine hot and cold working with heat treatment to achieve a fine-scale distribution of dispersoids, resulting in an alloy with a recrystallization grain size of less than 5 m. Particle dispersion should ideally also stop grain growth during superplastic deformation, but at superplastic molding temperatures, the majority of second-phase particles often dissolve back into solution. Dispersion-strengthened aluminum alloys based on Al–Cu, Al–Mg, and Al–Zn–Mg, as well as Al–Li and Al–Cu–Li alloys, are included in this category. Table 5.2 lists a few typical characteristics of superplastic alloys based on aluminum. Additional materials include copper alloys that have been precipitation-hardened.[25]

TABLE 5.2 Characteristics of a Few Alloys with Superplasticity.

Alloy composition	Temperature of applicability (°C)	Superplasticity (% elongation)	m
	Al-Based Alloys (Balance Al)		
6%Cu–0.4%Zr–0.3%Mg	400–480	1800	0.45–0.7
5.5% Zn–2.0%Mg–1.5% Cu–0.2%Cr	510–530	1400	0.5–0.8
2.7%Cu–2.2%Li–0.7% Mg–0.12%Zr	510–530	800	0.4–0.6
4.8%Cu–1.3%Li–0.4% Mg–0.4% Ag–0.14%Zr	470–530	1000	0.45
5%Ca–5%Zn	450–550	600	0.4–0.5
4.7%Mg–0.7%Mn–0.15%Cr	480–550	670	0.4–0.65
2.5%Li–1.2%Cu–0.6% Mg–0.1%Zr	500–540	1000	0.4–0.6
	Ti-Based Alloys (Balance Ti)		
6%Al–4%V	790–940	1400	0.6–0.8
5.8%Al–4%Sn–3.5%Zr–0.5% Mo–0.3% Si–0.05%C	950–990	400	0.35–0.65
4%Al–4%Mo–2%Sn–0.5%Si	810–930	1600	0.48–0.65
4.5%Al–3%V–2%Fe–2%Mo	750–830	700	0.5–0.55
14%Al–20%Nb–3%V–2%Mo	940–980	1350	0.4–0.6
6%Al–2%Sn–4%Zr–2%Mo	880–970	900	0.5–0.7

Atoms of aluminum occupy a face-centered cubic configuration when in the solid state. Regions of solid aluminum, called "crystallites," begin to emerge during the solidification process when all of the atoms are oriented in the same direction. The most significant pseudo-single-phase material in terms of commerce is aluminum alloys, which can be further separated into those that recrystallize prior to superplastic forming and those that only acquire their fine grain structure after a very small amount of deformation at the temperature for molding. Thermomechanical processing is used to reduce microduplex materials to a small grain or phase size. The presence of a microstructure made up of roughly equal amounts of two or more chemically and physically distinct phases restricts the formation of grains. The last set of materials consists of eutectics, copper alloys, stainless steels, titanium alloys, and steels.[26]

5.1.7 CERAMICS AND GLASS MECHANICS

As we saw in earlier sections, materials can exhibit both elastic and plastic behavior. Plastic flow after yield can be viewed as a type of failure brought on by permanent deformation. This section will demonstrate a few different titanium-based superplastic alloys. Superelasticity is a characteristic of superplasticity or pseudoelasticity. Several of the same alloy types that are superplastic are also superelastic, particularly those based on titanium. These alloys were also referred to as form-memory alloys. Remember that below their transformation temperature (see Figure 5.9), these materials are in the austenitic–martensitic state, and above it, they are in the austenitic state. When a material is tested just above the temperature at which it transforms from austenite to martensite, the applied stress causes martensite to form, and the material then experiences increasing strain at constant applied stress, meaning that significant deformation happens at low applied stress. When the stress is released, the material transforms from martensite to austenite and takes on its former form. Due to this result, the alloy is very elastic. In exceptional circumstances, deformations of up to 10% can be fully repaired (see Figure 5.9). Almost indestructible eyeglass frames are made of superelastic shape memory alloys, which also make great springs. Superelastic wires and tubing are used in numerous biomedical applications, including as catheters, orthodontic guidewires, and catheters that direct other catheters. We shall pay particular attention to two failure types known as creep and brittle failure. Although they are not exclusive to brittle materials like glass and ceramics, these kinds of failures are frequent among them.[27]

Lack of plastic deformation is a characteristic of brittle materials, as depicted in Figure 5.10. Metals and other ductile materials have a particular yield strength at which plastic flow starts necking, or a reduction in cross-sectional area. Figures 5.10a and 5.10b depict the necking phenomenon. At this point, either severe ductility failure (also known as rupture) or moderate ductility failure can lead to failure. Failure in brittle materials is preceded by little or no plastic deformation, as seen in Figure 5.10c. Brittle materials show a largely elastic response on the stress–strain curve, with little or no deformation beyond the elastic limit of failure (Figure 5.11). The modulus

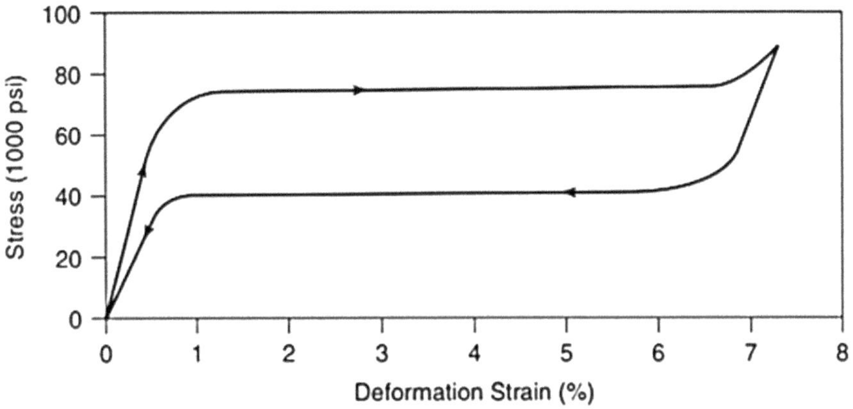

FIGURE 5.9 Superelasticity in a Ni–Ti alloy as an example. The entire 7% distortion is recovered. ShapeMemory applications are also included.

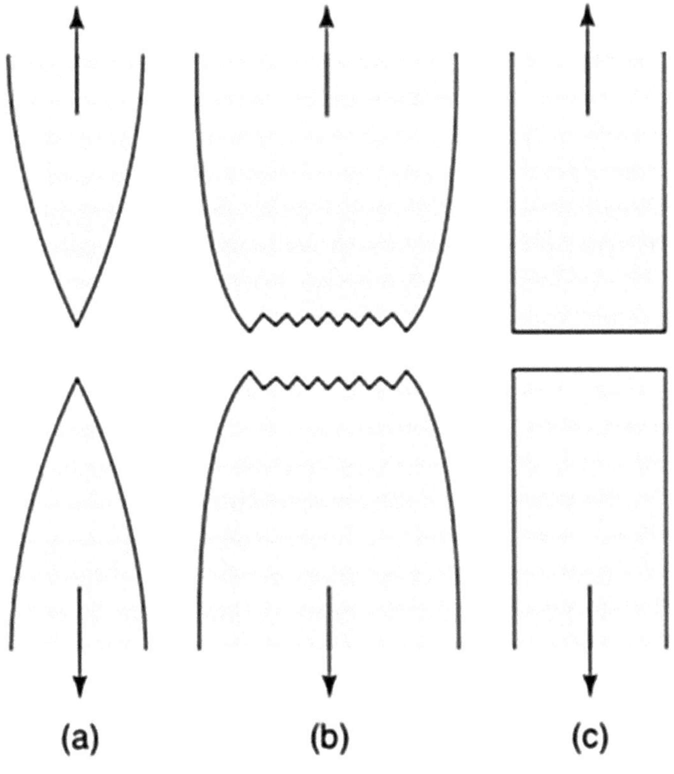

FIGURE 5.10 Illustration of the fractures on materials that are: (a) brittle, (b) ductile, and (c) brittle with ductile mechanism.

can differ from the plastic material's modulus in both directions. The stress at failure is also referred to as the ultimate tensile strength, or f, as earlier demonstrated by point R in Figure 5.11. The development of theoretical formulations for fracture resistance will be the focus of the following section. The mechanical characteristics of glass and ceramics will next be discussed once a description of creep is given.[28]

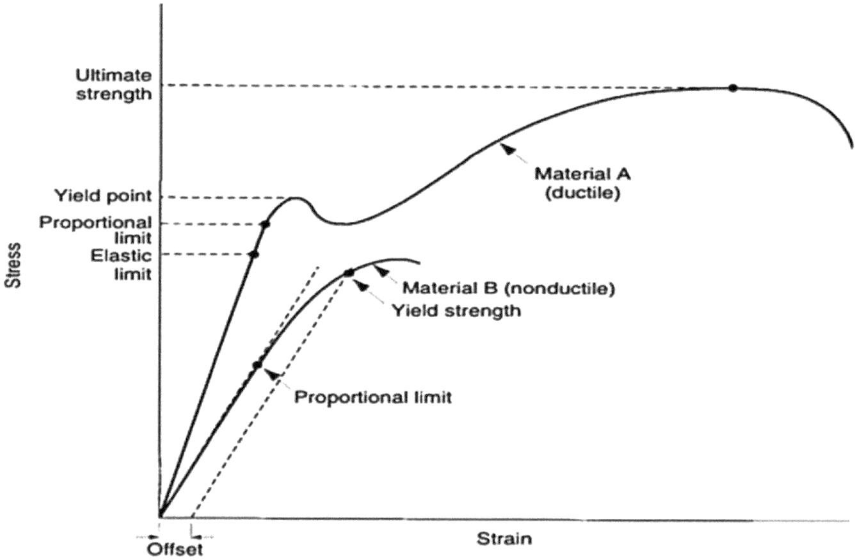

FIGURE 5.11 Typical strain diagrams for brittle (lower curve) and ductile (upper curve) materials are compared.

5.1.7.1 FRACTURE MECHANICS

5.1.7.1.1 Molecule-Based Reasons for Failure

First, we determine the theoretical material's fracture strength or the energy needed to split the atomic solid's planes. Consider extruding a rod with a single cross-section in the same manner as we did schematically, but this time, we apply a tensile force to try to pull the atomic planes apart rather than a shear stress. The interplanar distance x, where x = a0, affects the cohesive stresses between atomic planes. Thus, a0 is the interplanar spacing for any applied voltage and is the initial (equilibrium) interplanar distance at which the associated stress is zero.[29]

5.1.7.2 FATIGUE

The fracture phenomenon, which is closely connected to brittle fracture, is brought on by cyclic forces being applied repeatedly. For brittle or ductile materials, this form of failure, known as fatigue failure, causes failure below the tensile strength f or below the tensile strength u. Fatigue causes brittle fractures even in ductile materials because there is little to no plastic deformation prior to fracture. Similar to destruction, this process is brought on by the beginning and growth of cracks. Although polymers and ceramics are equally susceptible to similar wear, fatigue is thought to be the cause of about 90% of all metal failures. Because slight, nearly undetectable voltage variations over extended periods of time can cause even the largest structures to collapse catastrophically, fatigue is particularly significant over time. Despite the fact that pressures, particularly environmental stresses, can take on a variety of forms, four types of cyclic stress should be taken into account. The first is referred to as the reverse stress cycle (Figure 5.12a), in which compressive and tensile loads are equally alternated in terms of stress.[30]

When the reverse voltage cycle's maximum and lowest voltages are uneven around the zero voltage level, as shown in Figure 5.12b, a repeating voltage cycle is created. A random stress cycle, in which the stresses are neither periodic nor uniform in amplitude, is another type of cyclic stress. It has a maximum and minimum tensile or compressive load (but not both). The stress cycle is often defined by an average stress of (max + min)/2, a stress range of (max + min), and an amplitude of stress S = (max –min)/2. The average voltage for a reverse voltage cycle is zero; hence, the voltage amplitude is typically a more important quantity. One of the stress cycles mentioned above is typically applied during a fatigue test until the test item fails (N_f = Number of cycles to failure).

A specific stress magnitude. S, also known as fatigue life, describes the properties of the substance. The fatigue limit, which is defined as the value S below which failure does not occur regardless of the number of cycles N, is reached for some materials (see Figure 5.12). The voltage level at which failure will occur for a specific numerical value of N_f, such as 107 cycles, is defined as fatigue strength, often known as the fatigue limit. As was already mentioned, crack initiation and propagation lead to fatigue.[30]

5.1.7.3 CREEP

An illustration of how time affects a material's mechanical properties is fatigue. Creep is another instance of a mechanical characteristic that changes with time. The term creep, also known as viscoplasticity, refers to the time-dependent deformation of materials constantly under stress, typically in hot temperatures. Because creep often only becomes significant at $T_{mp}/2$, where T_{mp} is the material's absolute melting point, higher temperatures are required. In a typical creep experiment, the amount of deformation, or creep strain, is measured over a lengthy period of time, typically thousands of hours, while being subjected to constant tensile loads and temperatures. The resulting creep rate, which is the slope of the curve, is displayed in the ensuing plot of creep strain vs. time (Figure 5.13).

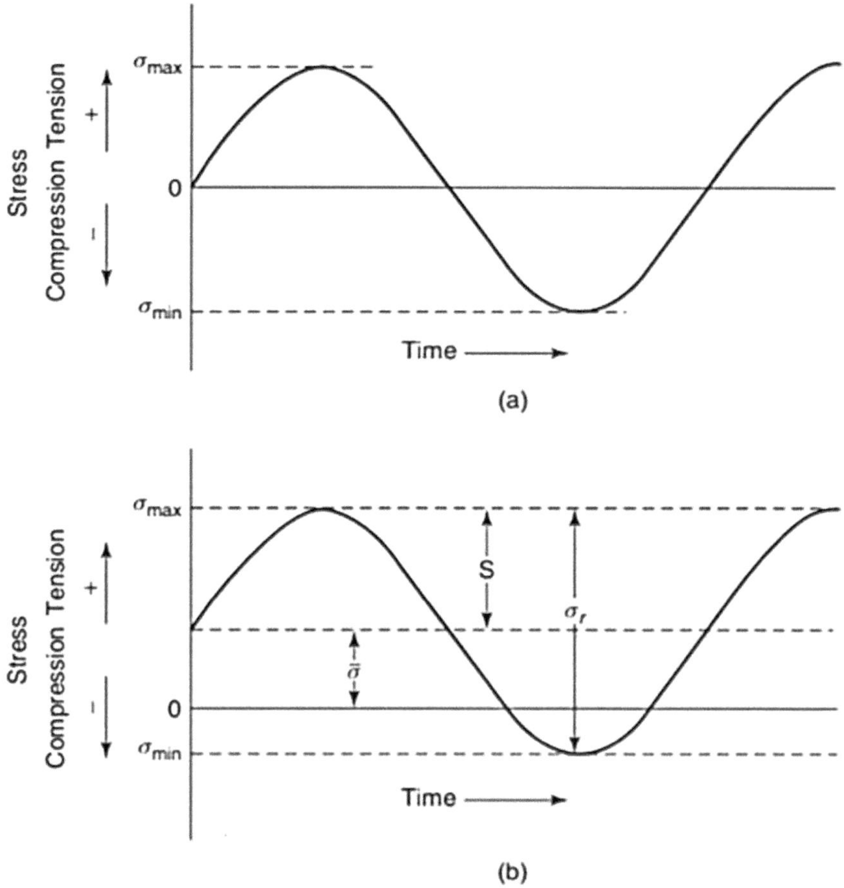

FIGURE 5.12 The reverse stress cycle and repeating stress cycle are the changes in stress over time, underlying fatigue failure.

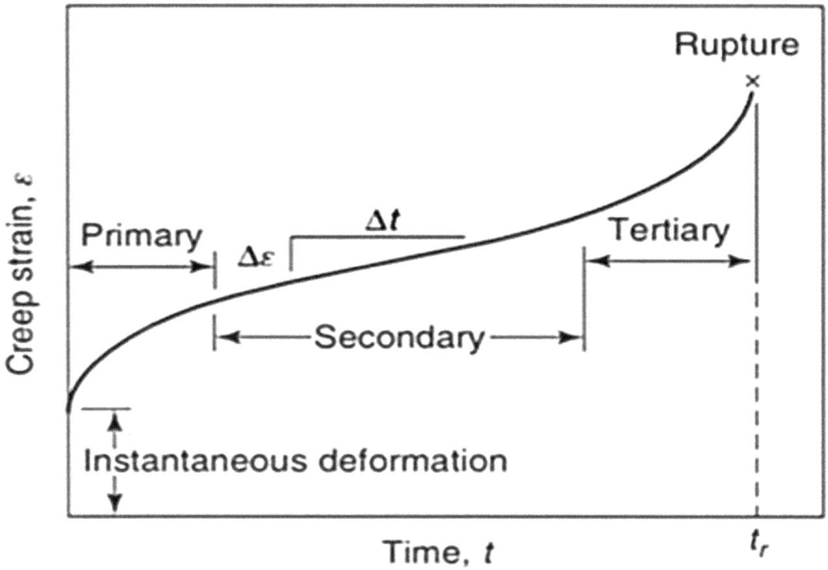

FIGURE 5.13 Primary, secondary, and tertiary creep are illustrated by a representative creep curve.

The maximum capacity of structural elements that are mostly subjected to bending moments and, as a result, have a bending failure mode is determined by moment-bending diagrams. The three segments of a typical creep curve each have a unique functional dependency on the deformation over time. The primary creep rate p continually declines over time in the first region, which is also known as transitory creep.[27] You can use force dependence in this area.

5.1.8 GLASS AND CERAMIC MATERIALS' MECHANICAL CHARACTERISTICS

The elastic, ductile, fracture, fatigue, and creep qualities unique to particular classes of ceramic and glass materials are discussed in this section, with an emphasis on the mechanical characteristics that make the material acceptable for particular applications from a structural standpoint.

Generally speaking, the modules of ceramic materials are higher than those of metals and range from 70 to 500 GPa. In Figure 5.14, for instance, the moduli for three common metals are displayed alongside the moduli for a common technical ceramic, polycrystalline Al_2O_3, as a function of temperature. It should be noted that the modulus is mostly temperature-independent, which is typical of polycrystalline ceramics. Depending on the crystal orientation in relation to the applied voltage, modules for single crystals may even be higher than for their polycrystalline counterparts.[26]

For instance, the modulus of MgO has a maximum value in the [111] direction, where E[111] = 336 GPa, and a minimum value in the [100] direction, where E[100] = 245 GPa. The modulus of a polycrystalline MgO sample lies in the same range as that of most polycrystalline materials, between these two numbers, which is roughly 310 GPa.

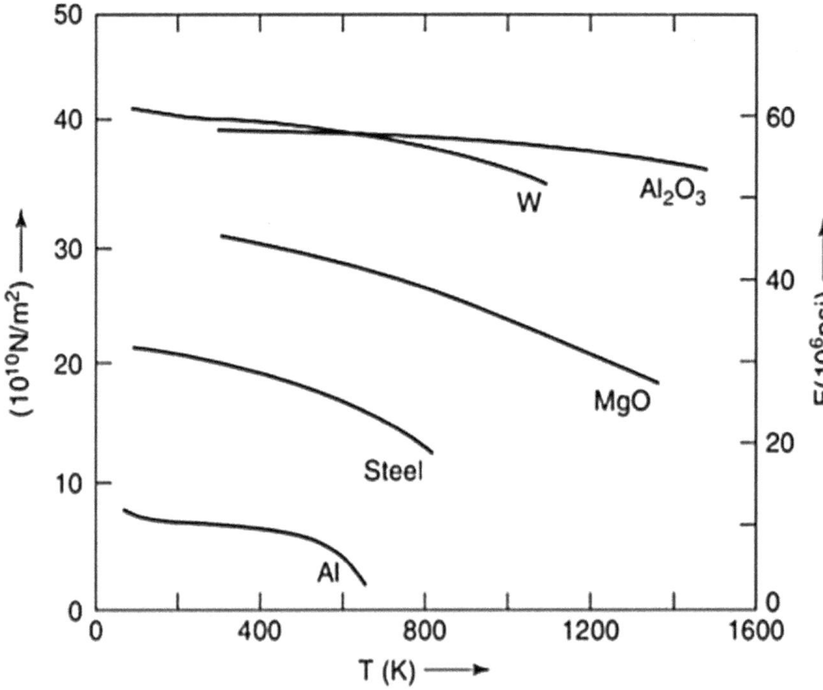

FIGURE 5.14 A comparison of the temperature-dependent elasticity of polycrystalline Al_2O_3 and various common metals.

Ceramics are usually brittle materials that normally fail close to their yield strength; hence, they cannot generally be considered to have a considerable plastic reaction. Nonetheless, slippage is very conceivable, especially in single-crystal ceramics. Crystalline ceramics experience plastic deformation similar to how metals do, thanks to the slip-dislocation mechanism.

Table 5.3 displays the slip mechanisms of a few typical ceramic crystals. Be aware that there are not as many independent sliding systems as there are metal structures in Table 5.3, particularly those made of BCC metals. Temperature and strain rate have a significant impact on the yield strength and flexibility of single crystals. For an Al_2O_3 single crystal, these effects are depicted in Figure 5.15. Be aware that there is As with metals, the material that has both an upper yield strength and a lower yield strength, and both weaken exponentially as temperature and strain rate rise. It is possible to explain this drop in yield strength figures.[25]

TABLE 5.3 Certain Ceramic Crystals Have Slip Systems.

Crystal	Slip system	Number of independent systems	Comments
C (diamond), Si, Ge	$(111)[1\bar{1}0]$	5	At $T > 0.5\ Tm$
NaCl, LiF, MgO, NaF	$(110)[1\bar{1}0]$	2	At low temperatures
NaCl, LiF, MgO, NaF	$(110)[1\bar{1}0]$		At high temperatures
	$(001)[1\bar{1}0]$	5	
	$(111)[1\bar{1}0]$		
TiC, UC	$(111)[1\bar{1}0]$	5	At high temperatures
PbS, PbTe	$(001)[1\bar{1}0]$	3	
	(110)[001]		
CaF_2, UO_2	$(001)[1\bar{1}0]$	3	
CaF_2, UO_2	$(001)[1\bar{1}0]$		At high temperatures
	(110)[111]	5	
	$(0001)[11\bar{2}0]$		
C (graphite), Al_2O_3, BeO	$(101)[10\bar{1}]$	2	
TiO_2	(110)[001]	4	
	$(111)[1\bar{1}0]$		
$MgAl_2O_4$	(110)	5	

5.1.8.1 *SPECIFIC PROPERTIES*

The low density of ceramics in comparison to most structural metals, especially for high-temperature applications, is one feature that makes them particularly appealing for structural applications. Reporting mechanical characteristics, such as modulus and strength per unit weight, which result in amounts referred to as specified qualities, is one way to demonstrate this. Tensile strength divided by density is hence known as specific strength, while modulus of elasticity divided by density is known as specific modulus. Ceramics do not expand significantly when heated, can tolerate high temperatures, and are good thermal insulators. Because of this, they are excellent thermal barriers that can be used for a variety of purposes, such as lining industrial furnaces or shielding the space shuttle from extremely high temperatures during reentry.[27]

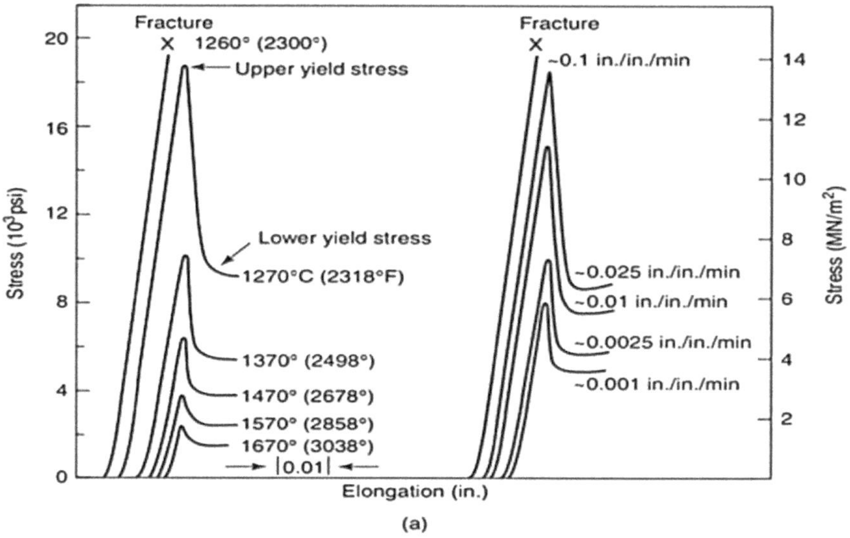

FIGURE 5.15 Influence of temperature and strain rate on single-crystal Al_2O_3 strain behavior.

Metals are used to make appliances, wires, cookware, and cookware because of their thermal and electrical conductivity. For example, the most common metals used for these purposes are copper and aluminum.[27] Although the usage of Ashby design diagrams is covered in length, we shall explain it here with the aid of two crucial diagrams. Figure 5.16 shows a graph showing Young's modulus as a function of strength (yield strength, compressive strength, tensile strength, or tensile strength, as appropriate). When feasible, individual materials are included along with groupings of materials (technical ceramics, technical alloys, etc.). The equivalent graph was created by dividing the strength and modulus values in Figure 5.16 by density. The materials on the top right of both diagrams have the ideal balance of strength and modulus. Here, we pick an extreme case to highlight the significance of certain characteristic ties. Let us say we are hunting for a refractory substance.

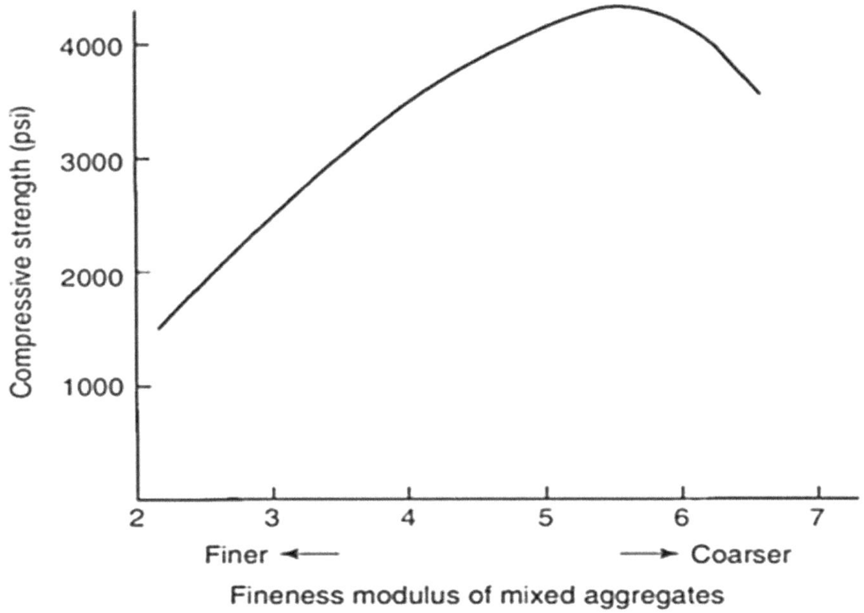

FIGURE 5.16 Concrete compressive strength and aggregate size.

A well-known refractory metal with a melting point of 3387 °C is tungsten, symbol W. In Figure 5.16, its modulus range and strength values are emphasized. Al_2O_3, which is similarly depicted, will be a comparable ceramic material in terms of modulus and strength, with a melting point.

5.1.9 POLYMER MECHANICS

In this section, we go through a group of materials' mechanical characteristics that are increasingly being used in structural applications. As energy use and concerns about global warming drive up demand for lightweight, recyclable materials, research and development in this field will continue to be dominated by the creation of novel polymers and the characterization of recycled polymers. Stress–strain curves for polymers can display many of the same properties as brittle materials (Figure 5.17, curve A) and ductile materials (Figure 5.17, curve B) in terms of mechanical behavior. Very crystalline polymers typically display brittle behavior (curve A), but amorphous polymers can show plastic deformation, as in the case of bend B. As we will see, these phenomena are quite temperature-dependent, more so for polymers than for metals or ceramics. We will also show that some cross-linked polymers' distinctive structural characteristics allow them to exhibit recoverable deformations at high stresses, even to the point of rupture (Figure 5.17, curve C).

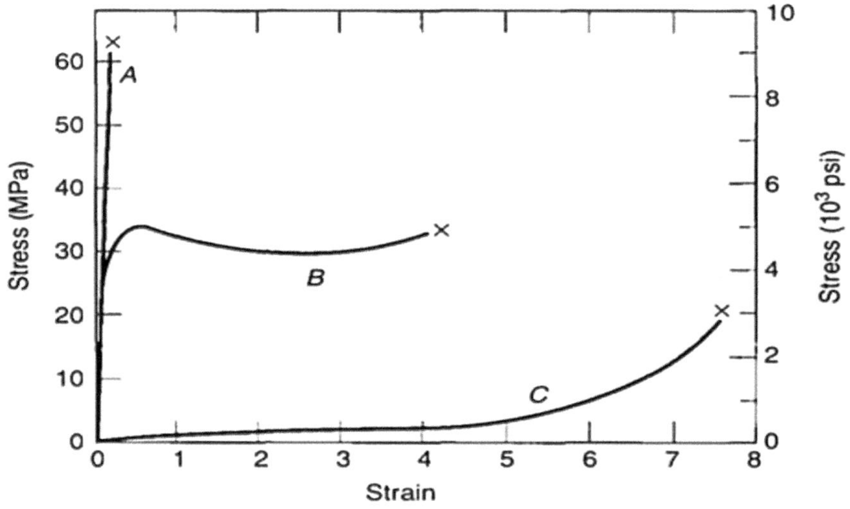

FIGURE 5.17 A brittle polymer (curve A), a plastic polymer (curve B), and a highly elastic polymer's stress–strain behavior (curve C).

Although polymers exhibit brittle and ductile behaviors that are similar to those of metals and ceramics, respectively, the mechanisms underlying elastic and permanent deformation in polymers are entirely different as a result of variations in the structure and scale of the objects that are subjected to motion. While dislocations and slip planes in metals and ceramics can be used to characterize plastic deformation (or lack thereof), the polymer chains that must be deformed are on a far larger scale.However, before going into detail about the mechanical characteristics of polymers in this context, we must first describe a phenomenon that is peculiar to polymers and endows these materials with some remarkable characteristics. This characteristic, called viscoelasticity, can be explained in terms of the fundamental mechanisms that we have just discussed.[28]

5.1.10 VISCOELASTICITY

Viscoelasticity, as the name suggests, is a material's response to applied stress that combines viscous and elastic properties. As was the case with metals and some glasses, as previously mentioned, polymers can experience permanent deformation at high stresses in addition to a recoverable elastic response to an applied force. Yet, as we showed, the mechanism of irreversible deformation in polymers can resemble liquid or viscous flow (Chapter 4). Before describing how some polymers display this elasticity, let us first create two crucial theoretical models that define viscoelasticity, an essential quality.

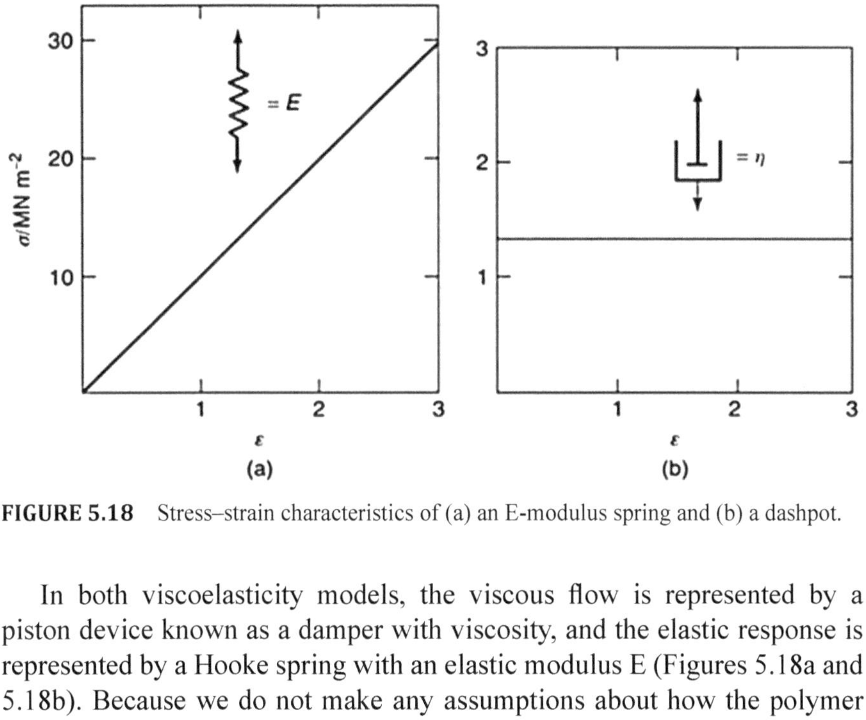

FIGURE 5.18 Stress–strain characteristics of (a) an E-modulus spring and (b) a dashpot.

In both viscoelasticity models, the viscous flow is represented by a piston device known as a damper with viscosity, and the elastic response is represented by a Hooke spring with an elastic modulus E (Figures 5.18a and 5.18b). Because we do not make any assumptions about how the polymer will behave viscously, we utilize the non-Newtonian viscosity rather than the Newtonian viscosity. Using either the elastic modulus E or the shear modulus G, we may additionally extend these models to include both tension and shear. Compared to earlier, we will actually use these moduli more interchangeably. Just with the "frictionless pulley" from beginning physics, keep in mind that the spring and the damper are wholly fictitious objects that are solely employed to create two viscoelasticity models.[29]

5.1.10.1 POLYMER MECHANICAL PROPERTIES

The mechanical properties of polymers exhibit a considerable dependency on a number of factors, including temperature, strain rate, and morphology, to a far greater extent than those of metals or ceramics. Moreover, elements that are not present in other kinds of materials, like molecular weight and glass transition temperature, are significant. It goes without saying that it is impossible to discuss all of these impacts, even briefly. Here, we will concentrate on the factors that have the most potential to affect the choice of polymers for mechanical design. They are the effect of strain rate and temperature. Brittle materials are defined as polymers with a high modulus of elasticity at room temperature, such as polystyrene and polymethyl methacrylate, or PMMA. They typically disintegrate before becoming liquid. Cellulose acetate is a common illustration of "rigid" polymers or those polymers that experience some deformation prior to breaking. Figure 5.19 displays these two types of behavior along with the corresponding temperature dependencies.

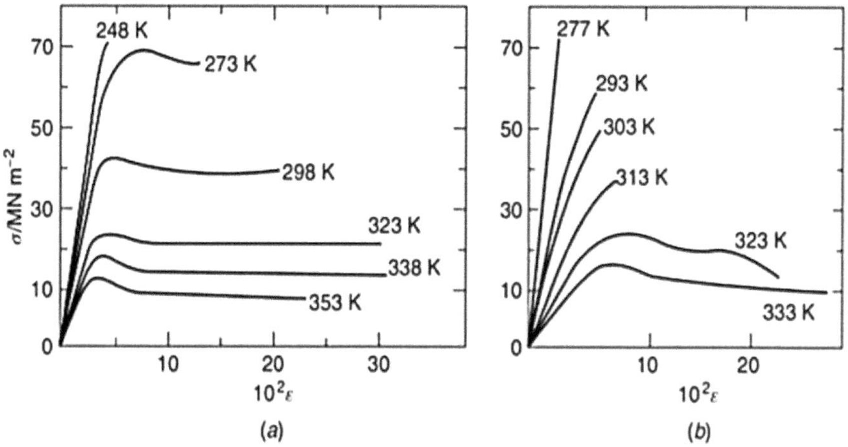

FIGURE 5.19 Temperature effects on the stress–strain responses of (a) polyethylene and (b) cellulose acetate (methyl methacrylate).

The major impact of temperature on the curves' distinctive shape may be seen in Figure 5.19. Elongation normally grows as temperature rises, but stiffness and yield strength fall. A transition from ductility to brittleness occurs in the cellulose acetate polymer at a temperature of around 273 K, after which the polymer becomes softer and stiffer than the hard, brittle polymer at temperatures below 273 K. PMMA maintains its hardness and brittleness at even greater temperatures. But ultimately, at roughly 320 K, it enters a soft and strong condition. Thus, temperature is crucial if high requirements for stiffness and toughness are to be reached.[31–33]

5.2 CONCLUSIONS

- The effect of mechanical properties was effective on four major classes of materials: metals, ceramics, high polymers (plastics), and composites.
- Depending on their mechanical properties, metals, ceramics, and polymers differ from each other in terms of bending, size, and tensile components.
- Heat treatments, in particular, significantly improve mechanical properties.
- Valence electrons form an electronic cloud in a nondirectional metallic bond, and atoms can easily exchange electrons.
- The effects of mechanical properties on the stress rate were significant.
- We noticed that ceramics are inorganic materials with ionic and covalent atomic bonds. Ceramics also have extremely high melting temperatures and exceptional chemical stability thanks to their extremely stable atomic bonding.
- The mechanical properties of polymers are affected by temperature, molecular weight, and strain rate. Primary (covalent) and secondary (van der Waals and hydrogen) bonds are what hold polymers together. The s and p shells, which make up an octet (a set of eight electrons), are shared by the atoms in a covalent connection.

KEYWORDS

- **mechanical characteristics**
- **metals**
- **ceramics**
- **polymers**
- **properties**

REFERENCES

1. Li, X.; Bhushan, B. A Review of Nanoindentation Continuous Stiffness Measurement Technique and its Applications, *Mater. Char.* **2002,** *48,* 11–36.
2. Jenkins, C. H. M. The Strength of Cd–Zn and Sn–Pb Alloy Solder, *J. Inst. Metals* **1928,** *40,* 21–39.

3. Pearson, C. E. The Viscous Properties of Extruded Eutectic Alloys of Lead-tin and Bismuth-tin, *J. Ins. Metals* **1934,** *54,* 111–124.
4. Bochavar, A. A.; Sviderskaya, Z. A. Superplasticity in Zn–Al Alloys. *Bull. Acad. Sci. (USSR)* **1945,** *9,* 821–824.
5. Underwood, E. E. A Review of Superplasticity, *JOM* **1962,** 914–919.
6. Holt, D. L.; Backofen, W. A. Superplasticity in the Al–Cu Eutectic Alloy. *Trans. ASM* **1966,** *59,* 755–767.
7. Ashby, M. F. *Materials Selection in Mechanical Design,* 2nd ed.; Butterworth Heinemann: Oxford, 1992.
8. *Concise Encyclopedia of Composite Materials,* Kelly, A., Ed.; Pergamon: Elmsford, NY, 1994; p 100.
9. *Concise Encyclopedia of Composite Materials,* Kelly, A., Ed.; Pergamon: Elmsfords NY, 1994; p 161.
10. Powell, P. C.; Ingen Housz, A. J. *Engineering with Polymers,* Stanley Thornes: Cheltenham, UK, 1998.
11. Suchanek, W.; Yoshimura, M. Processing and Properties of Hydroxyapatite-based Bio-materials for use as Hard Tissue Replacement Implants, *J. Mater. Res.* **1998,** *13* (1), 94.
12. Knott, J.; Withey, P. *Fracture Mechanics: Worked Examples,* Institute of Materials: London, 1993.
13. *Metals Handbook, Desk Edition,* American Society for Metals, 1985.
14. Wachtman, J. B. *Mechanical Properties of Ceramics,* John & Sons, Wiley: New York, 1996.
15. McCrum, N. G.; Buckley, C. P. Bucknall, C. B. *Principles of Polymer Engineering,* Oxford Science: Oxford, 1997.
16. Kelly, A. *Composite Materials,* American Elsevier: New York, 1966.
17. *Composite Materials Technology,* Mallick, P. K., Newman, S., Eds.; Hanser Publishers: Munich, 1990.
18. *Mechanical Testing of Engineering Ceramics at High Temperatures,* Dyson, B. F., Lohr, R. D., Morrell, R., Ed.;, Elsevier Applied Science: London, 1989.
19. Lee, W. E.; Rainforth, W. M. *Ceramic Microstructures: Property Control by Processing,* Chapman & Hall: London, 1994.
20. *Composite Applications: The Role of Matrix, Fiber and Interface,* Vigo, T. L., Kinzig, B. J., Eds.; VCH Publishing: New York, 1992.
21. The Mechanical Properties of Biological Materials. *Symposium, Society of Experimental Biolo-gists,* Cambridge University Press: New York, 1980.
22. Biomaterials Properties Database, University of Michigan, Dr. William J. O'Brien, Director *http://www.lib.umich.edu/dentlib/Dental tables/toc.html*
23. Evans, F. E. *Mechanical Properties of Bone,* C. Charles: Springfield, IL, 1973.
24. Muccio, E. A. *Plastics Processing Technology,* ASM International: Materials Park, OH, 1994.
25. Powell, P. C.; Housz, A. J. *Engineering with Polymers,* 2nd ed.; Nelson Thornes: Cheltenham, UK, 1998.
26. Rosen, S. L. *Fundamental Principles of Polymeric Materials,* 2nd ed.;, Wiley: New York, 1993.
27. Rudin, A. *The Elements of Polymer Science and Engineering,* 2nd ed.; Academic Press: San Diego, 1998.

28. Strong, A. B. *Plastics: Materials and Processing,* 3rd ed.; Prentice Hall PTR: Paramus, IL, 2006.
29. Tobolsky, A. V. *Properties and Structures of Polymers,* Wiley: New York, 1960. Advanced Treatment.
30. Ward, I. M.; Sweeney, J. *An Introduction tothe Mechanical Properties of Solid Polymers,* 2nd ed.; John Wiley & Sons: Hoboken, NJ, 2004.
31. Riyadh A. Al-Samarai, Y. Al-Douri, *Friction and Wear in Metals*, Springer, Germany. 2024
32. Maria S. da Dunla, Exp. Theo. NANOTECHNOLOGY8(2024) 23–26
33. Ziyad Khalf Salih, Angham AyadKamall-Eldeen, Exp. Theo. NANOTECHNOLOGY 8 (2024) 27–32

CHAPTER 6

Physical Properties of Heusler Alloys

L. BOUMIA[1], F. SEMARI[2], F. DAHMANE[2,3], V. SRIVASTAVA[4], and R. KHENATA[2]

[1]*Département Sciences de la Matière, Faculté des Sciences et Technologies, Tissemsilt University, Algeria*

[2]*Laboratoire de Physique Quantique et de Modélisation Mathématique (LPQ3M), Département de Technologie, Université de Mascara, Mascara, Algeria*

[3]*Laboratoire de Simulation et de Modélisation des Propriétés Magnétique des Hétéro-Structures (LPMH), Département des Sciences de la Matière, Tissemsilt University, Tissemsilt, Algeria*

[4]*Department of Physics, School of Chemical Engineering and Physical Sciences, Lovely Professional University, Phagwara, Punjab, India*

ABSTRACT

Heusler compounds are an exceptional family of intermetallic materials with compositions of 1:1:1 (commonly referred to as half-Heusler) or 2:1:1 with more than 1500 members. They are still a topic of ongoing research today; more than a century after Fritz Heusler first discovered them. The prediction of topological insulators is the most recent example. Novel features and possible application fields are continuously being discovered. Heusler alloys beginning with Cu_2MnAl have caught researchers' attention ever since their discovery in 1903. They are intermetallic compounds, often made up of some IIIeV group nonmagnetic element and a 3d transition metal. Using electron spins as information carriers, spintronics is one of the most promising next-generation information technologies. It has the ability to speed up data processing and have

Contemporary Advancements in Materials Technology: Novel Solutions to Complex Problems.
Shrikaant Kulkarni, Vipul Srivastava, & P. William (Eds.)

high circuit integration densities and low energy requirements. Spin creation and injection, long-distance spin transport, spin orientation manipulation, and spin orientation detection are only a few of the difficulties that spintronics must overcome. The basic electrical and magnetic properties of full-Heusler alloys such as X_2MnGe (X = Sc, Fe, Ni), Ti_2YAl (Y = Co, Fe, Mn), Co_2CrZ (Z = Al, Ga, In), and Co_2MnZ (Z = Al, Ge, Si, Ga) are studied in this chapter.

6.1 INTRODUCTION

The discovery that an alloy with the chemical formula Cu_2MnAl acts like a ferromagnetic, despite the fact that none of its constituent elements is magnetic on their own, was made by Fritz Heusler in 1903, and this discovery marked the beginning of one of the most fascinating material classes.[1] The numerous intriguing physical properties, including different spin behavior known as half-metallicity, temperature-dependent size, and shape effect popularly known as shape memory effect, etc., Heusler and half-Heusler alloys have recently been the subject of theoretical and experimental investigation. These alloys show promise for use in upcoming technological applications.

The half-metallic nature of the Heusler compounds' spin-split band structure is the source of this curiosity. Half-metallic ferromagnets (HMF) are materials that are a metal for spin-up and a semiconductor (or insulator) for spin-down electrons. Heusler compounds were previously thought of as prospective candidates to exhibit this behavior in 1983, when theoretical calculations suggested that the half-Heusler complex NiMnSb had energy gap.[2]

Heusler compounds actually cover lots of ternary and quaternary compounds that are currently known. They frequently adopt different magnetic ordering and most of them form cubic structures. Due to the discovery of half-metallicity, Heusler compounds and alloys have received a lot of interest in the last twenty years. Heusler alloys have the general chemical formula X2YZ, where X and Y are transition metals (TM) and Z is either a nonmetallic metal or nonmetal element.

Heusler systems come in two main categories. The first is the full-Heusler alloy, which crystallizes in two different types of structures: the first is the Cu_2MnAl structure [Fm-3m], which has four sets of interpenetrating fcc planes A, B, C, and D. The X atoms are located in the A and C sublattices with the Wyckoff coordinates (0, 0, 0) and (0.5, 0.5, 0.5), respectively, while the Y and Z atoms are located in the B and D sublattices with the Wyckoff coordinates (0.25, 0.25, 0.25) and (0.75, 0.75, 0.75). The local environment of the X atoms at the two separate sublattices is identical, but it has been rotated by 90°. The second full Heusler compound has the formula Hg_2CuTi, in which A, B, C, and D take position differently as compared to Cu2MnAl structure as X: A (0, 0, 0) and B (0.25, 0.25, 0.25), Y: C (0.5, 0.5, 0.5), and Z: D (0.75, 0.75, 0.75). The X atoms are labeled as X(1) and X(2) for positions A and B in the Hg_2CuTi-type structure, respectively.

The half-Heusler alloy with composition XYZ, which crystallizes in the Clb structure, is the second type of Heusler. This structure can be created by omitting one of the two equivalent (A, C) sites from the full Heusler structure. A promising area of nanoscale electronics called spintronic employs the electrons' spin, rather than their electric charge, to encode and process data.[3] Due to their strong spin polarization, half-metallic materials with complete spin polarization at the Fermi level are very appealing for use in spintronic applications. Due to the potential device applications for these materials, including nonvolatile magnetic random access memory (MRAM) and magnetic sensors, scientists are very interested in them.

The materials that can provide a strongly spin-polarized current are a crucial component of spintronic technology. In this case, half-metallic ferromagnets[2] are primarily appealing because they exhibit both metallic and insulating behavior for one of their spin channels simultaneously. Since they have a fully spin-polarized electronic density of states at the Fermi level, they may theoretically provide a fully spin-polarized current.[2]

The first compounds to be created were Full-Heusler ones.[4,5] Kübler et al. examined the mechanisms stabilizing the ferro- or antiferromagnetism in these compounds in a ground-breaking article.[6] Ab-initio electronic structure calculations were used by two Japanese research teams to make the first predictions about the existence of half-metallicity in the case of full-Heusler compounds: Co_2MnZ compounds (Z= Si and Ge) were studied by Ishida and collaborators[7] and Fe_2MnZ compounds were studied by Fujii and other collaborators.[8]

6.2 STRUCTURAL, ELECTRONIC, AND MAGNETIC PROPERTIES OF SAME HEUSLER ALLOYS

6.2.1 *X_2MNGE (X = SC, FE, NI)*

6.2.1.1 *STRUCTURAL PROPERTIES*

The present study results in structural analysis of Full-Heusler X_2MnGe (X = Sc, Fe, Ni),[9] [L21 structure; Cu_2MnAl and Hg_2CuTi -types]. The illustration of the two structures Cu_2MnAl and Hg_2CuTi types is presented in Figure 6.1. As discussed in Section 1, the atomic positions in the considered structures remain the same.

In Cu_2MnAl-type structure (L21 structure is X-Y-X-Z)

X atoms (Sc, Fe and Ni): A and C sublattices (0, 0, 0) and (0.5, 0.5, 0.5), respectively,

Y atoms (Mn): B sublattice (0.25, 0.25, 0.25) and
Z atoms (Ge): D sublattice (0.75, 0.75, 0.75).

In Hg_2CuTi-type structure (L21 structure is X-X-Y-Z)[10]

X atoms (Sc, Fe and Ni): A and B sublattices at (0, 0, 0) and (0.25, 0.25, 0.25), respectively,

Y atoms (Mn): C sublattice at (0.5, 0.5, 0.5),
Z atoms (Ge): D sublattice at (0.75, 0.75, 0.75).

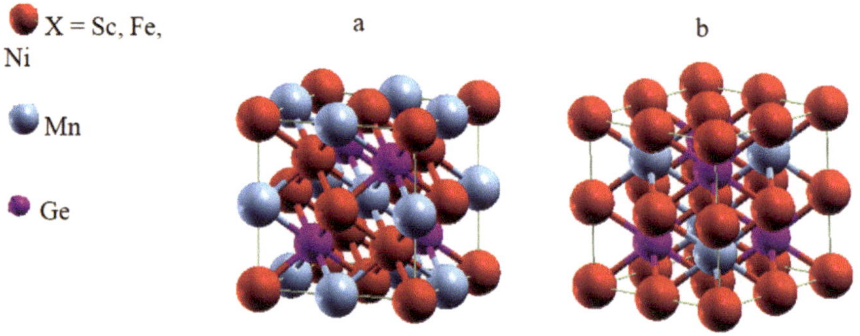

FIGURE 6.1 Structural representation of X_2MnGe (X=Sc, Fe, Ni) alloys in the two types of structures. (a) Hg_2CuTi. (b) Cu_2MnAl.

It has been determined that the quantity of 3d electrons on the X and Y atoms has a significant impact on where they want to be. The A and C sites are preferred by elements with more 3d electrons, while the B sites are more frequently occupied by atoms with less 3d electrons. In order to test this rule, optimizations were carried out in order to identify the equilibrium structural parameters and stable structure of X_2MnGe compounds. Figure 6.2 depicts that the Cu_2MnAl-type structure is the most stable and favorable structure (lowest energy in Cu_2MnAl-type).[9]

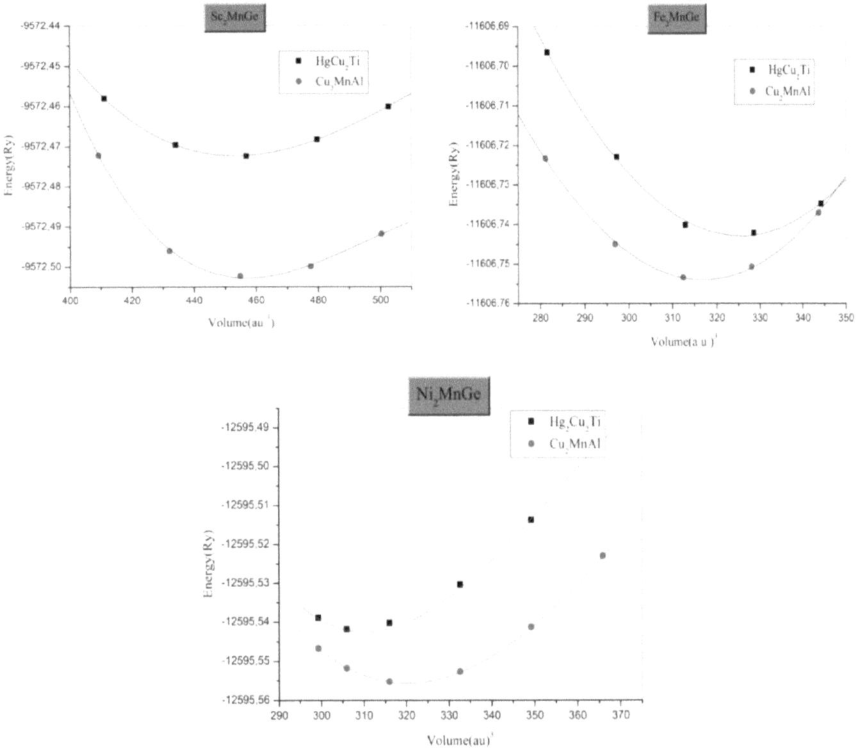

FIGURE 6.2 E-V optimization curve for X_2MnGe (X = Sc, Fe, and Ni).

6.2.1.2 *ELECTRONIC PROPERTIES*

Further, electronic structures of Sc_2MnGe, Fe_2MnGe, and Ni_2MnGe Heusler alloys have been calculated in two said structures. These Heusler alloys (HAs) exhibit ferromagnetic characteristics and both the spin states have been relooked to examine the material's behavior. We have therefore calculated electronic band structure (BS) and density of states (DOS) for different spin states. Here only DOS have been presented to visualize the electronic states. In Figure 6.3, we have presented total and partial DOS in Hg_2CuTi- and Cu_2MnAl-type structures for majority (spin up) and minority (spin down) states. For the very first HA, Sc_2MnGe, the inspection of Figure 6.3 depicts half-metallic nature in its Hg_2CuTi-type structure, while metallic nature in its Cu_2MnAl-type structure. These two different behaviors in two different structures are appealing, which can be understood by studying DOS diagrams. In its Hg_2CuTi-type structure for the majority (spin up) state, Ef is situated in the gap of valence band (VB) and conduction band (CB) indicating semiconducting nature, while for the minority (spin down) state, Ef crossed the electronic states indicating metallic nature. The overall Si_2MnGe HA shows HM behavior.[9] However, in its Cu_2MnAl-type structure for the majority (spin up) and minority (spin down) states, Ef crossed the electronic states indicating metallic nature. The overall Si_2MnGe HA shows metallic behavior.

For second HA, Fe_2MnGe have the Cu_2MnAl-type L21 structure. However, DOSs have been plotted for both the considered structures. In its Hg_2CuTi-type structure for the majority (spin up) and minority (spin down) states, Ef crossed the electronic states indicating metallic nature. The overall Fe_2MnGe HA shows metallic behavior. But in its Cu_2MnAl-type structure it shows HM behavior. The third HA, Ni_2MnGe, in this way exhibits overall metallic behavior, which is evident from the DOS diagram (Figure 6.3, right most diagram).

Three categories of the origin of the band gap in the crystals are recognized as covalent band gap (Eg), d-d band gap, and charge transfer band gap.[11,12] The first type of Eg can be seen in C1b structured HHAs, d-d band gap is seen in L21 structured FHAs, and is accountable for HM. Further, Galanakis et al.[13] have described how the d orbitals of Fe (high valance) covalently mixed with Cr (low valence) to create the band gap in Fe_2CrSi.

In the instance of the X_2YZ inverse Heusler compounds, such as Sc-based Heusler compounds, Skaftouros et al.[14] have provided compelling arguments for potential mixing between the d-bands of the TMs. They claim that the X(1)-Y atoms' mutual symmetry leads their d-bands to mix, resulting in 5 bonding d-states (2 eg and 3 t2g) and 5 nonbonding d-states (2 eu and 3 tu). The five X(1)-Y bonding d-states then mix with the d-bands of X(2) atom to produce new bonding and antibonding states.

6.2.1.3 *MAGNETIC PROPERTIES*

Slater and Pauling showed in two original articles that the addition of a V-electron in a binary magnetic alloy occupies spin-down states and lessens the total spin magnetic moment (MM) by 1 Bohr Magneton.[15,16] Interestingly, HMHAs exhibit a similar behavior as well. It has been demonstrated that the total spin magnetic in the unit cell, Mt, scales as a function of the

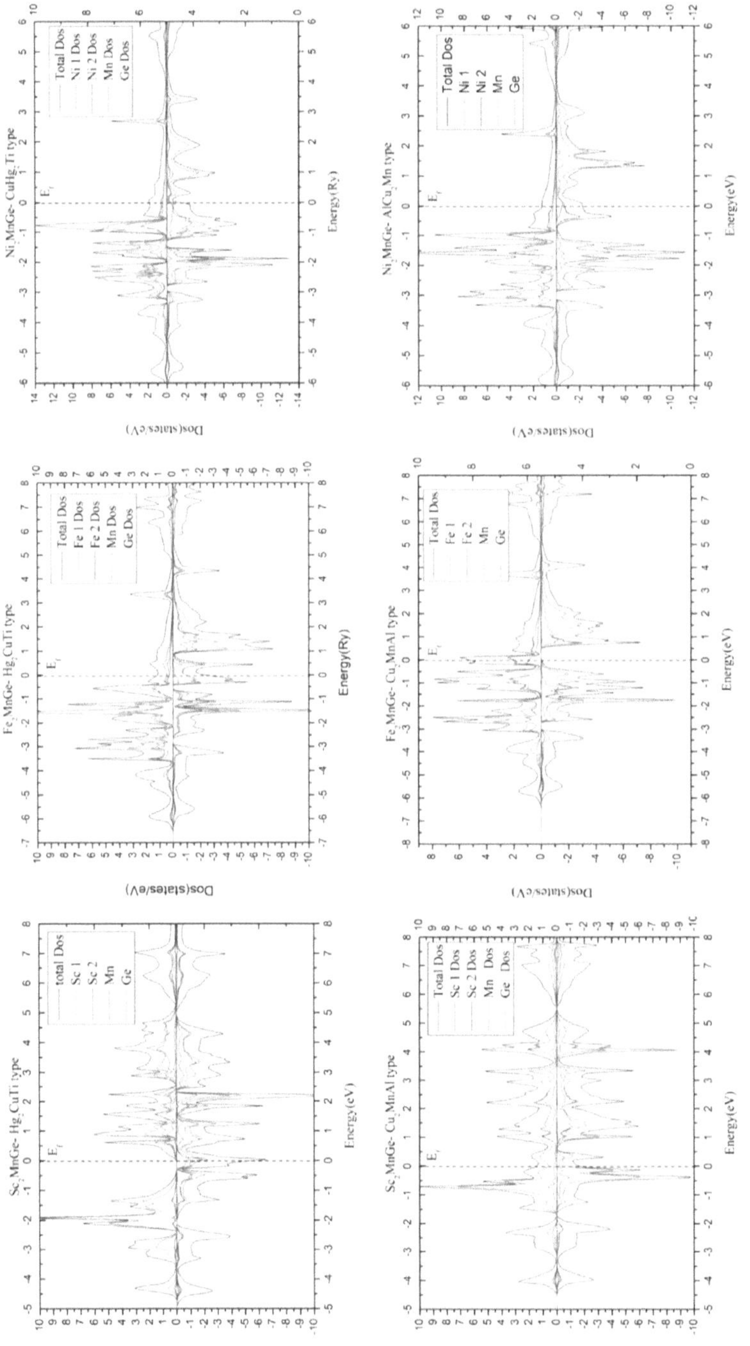

FIGURE 6.3 DOS of X_2MnGe (X= Sc, Fe, and Ni) in both the Cu_2MnAl and Hg_2CuTi structures.

total number of valence electrons, Zt, following the relation Mt = 18 – Zt,[17] whereas this relation changes for the L21 FH compound to Mt = Zt – 24.[18]

By directly connecting the electronic properties and the magnetic properties, the Slater–Pauling (SP) principles offer an efficient tool for the learning of HMF compounds. In HMF, one spin channel behaves as metal and the other behaves as semiconducting. The conduction electrons are therefore totally spin polarized at the Fermi level. Because of the materials' HM, the spin magnetization is always a complete number of MM. The Slater–Pauling rule is remarkably well followed by the Heusler complex Sc_2MnGe, which has 17 valence electrons and a total magnetic moment of 1 μ_B.[9] The same outcome was discovered.[19]

The estimated M^{Tot} for X_2MnGe compounds in both the structures is displayed in Table 6.2. The Heusler complex Sc_2MnGe, which has 17 valence electrons and an M^{Tot} of 1 μ_B, remarkably follows the Slater–Pauling rule. Referring to the paper,[29] the authors have discovered the same result concurred. The Heusler compound Fe_2MnGe has an M^{Tot} of 3μ_B and 27 valence electrons, according to Slater–Pauling. The Ni_2MnGe compound does not have a half-metal nature, as shown by the absence of an integer value of M^{Tot} for the compound in either of the two structures (Hg_2CuTi or Cu_2MnAl-type structure).

TABLE 6.1 The Magnetic Moments (in the unit of Bohr Magneton) for X_2MnGe (X=Sc, Fe, Ni).

Compound	Structure	M^{X1}	M^{X2}	M^{Mn}	M^{Ge}	$M^{interstitial}$	M^{Tot}
Sc_2MnGe	Hg_2CuTi	–0.58587	–0.61871	2.80063	–0.03087	–0.56523	0.99995
	Cu_2MnAl	–0.04738	–0.04738	2.60	–0.02201	0.04249	2.30
Fe_2MnGe	Hg_2CuTi	1.83255	2.49791	1.9614	–0.0596	–0.1031	6.129
	Cu_2MnAl	0.03305	0.03795	2.88854	0.00113	0.04249	3.00315
Ni_2MnGe	Hg_2CuTi	–0.04852	0.12763	2.99504	–0.06352	0.00094	3.11156
	Cu_2MnAl	0.20495	0.18721	3.54106	–0.02984	0.05035	3.95372

6.2.2 *TI_2YAL (Y = CO, FE, MN)*

6.2.2.1 *STRUCTURAL PROPERTIES*

The stoichiometry of the Heusler compounds is X_2YZ, where X and Y are TM and Z is the main group element. There are two different kinds of full-Heusler compounds that form cubic crystals: Cu_2MnAl and Hg_2CuTi.

In Cu_2MnAl-type structure

X atoms (Ti): A and C sublattices (0, 0, 0) and (0.5, 0.5, 0.5), respectively,
Y atoms (Co, Fe, Mn): B sublattice (0.25, 0.25, 0.25) and
Z atoms (Al): D sublattice (0.75, 0.75, 0.75).

In Hg_2CuTi-type structure

X atoms (Ti): A and B sublattices at (0, 0, 0) and (0.25, 0.25, 0.25), respectively,
Y atoms (Co, Fe, Mn): C sublattice at (0.5, 0.5, 0.5),
Z atoms (Al): D sublattice at (0.75, 0.75, 0.75).

The full-Heusler Ti_2YAl (Y=Co, Fe, Mn) with Cu_2MnAl and Hg_2CuTi structures is shown in Figure 6.4.

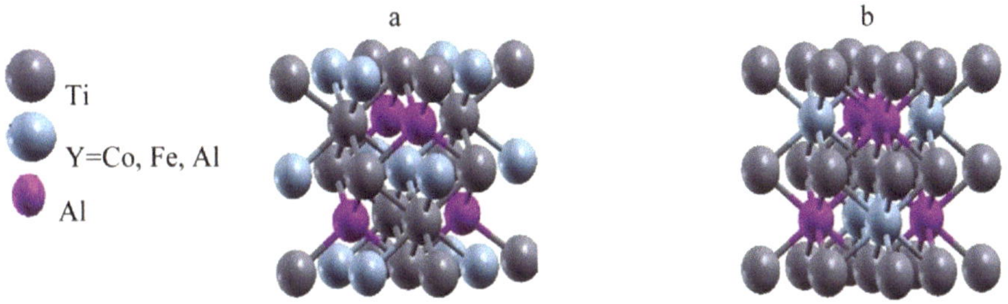

FIGURE 6.4 Structural representation of Ti_2YAl (Y=Co, Fe, Al) alloys in the two types of structures. (a) Hg_2CuTi, (b) Cu_2MnAl.

We are aware of no experimental lattice constant value that can be compared to ours. To ascertain the ground state properties, the estimated total energies vs volume are fitted to the empirical Murnaghan's equation of state.[20] The optimized lattice constants for the Ti_2YAl (Y = Co, Fe, Mn) alloys are listed in Table 6.2 and are in good agreement with the others' theoretical calculations as 6.14 Å,[22] 6.00 Å,[23] and 6.24 Å.[24]

6.2.2.2 ELECTRONIC PROPERTIES

The DOS for Ti_2YAl (Y=Co, Fe, Al) inverse HAs are shown in Figure 6.5. Ti_2CoAl, Ti_2FeAl, and Ti_2MnAl all have majority-spin bands that are metallic in terms of their electronic characteristics, while the minority-spin band shows energy gap near the Fermi level.[21] At the Fermi level, the semiconducting nature in the minority-spin band consequences in hundred percent spin polarization.[21] The HM property of the Ti_2CoAl and Ti_2FeAl alloys is confirmed by the presence of the majority-spin states at the Fermi level and a large band gap in the minority-spin state. In HM materials, the minority-spin band gap is a crucial component (Figure 6.5), and the explanation for the HM gap is conferred underneath. Three factors are typically responsible for the HM band gaps:[21, 25]

- Covalent band gap that can be found in half-Heusler structures with C1b,
- d-orbital : d-orbital band gap in FH alloys with a Cu_2MnAl structure,
- Charge transfer band gap,[25, 26] in double perovskites.[25]

TABLE 6.2 The Calculated Initial Properties for Ti_2ZAl (Z=Co, Fe, Mn).

Compound	Lattice parameter a (Å)	Bulk modulus, B (GPa)	Pressure derivative B'
Ti_2CoAl	6.08 [21]	144.6461	4.7252
Ti_2FeAl	6.07 [21]	134.1383	6.0915
Ti_2MnAl	6.16 [21]	102.4472	5.3872

Further, Galanakis et al.[27] have described how the d orbitals of Fe (high valance) covalently mixed with Cr (low valence) to create the band gap in Fe_2CrSi.

In Figure 6.5, we have presented total and partial DOS in Hg_2CuTi-type structure for the majority (spin up) and minority (spin down) states. For the very first HA, Ti_2CoAl, the inspection of Figure 6.5 depicts HM nature in its Hg_2CuTi-type structure. In its Hg_2CuTi-type structure for the majority (spin up) state, Ef crossed the electronic states indicating metallic nature, while for the minority (spin down) state Ef is situated in the gap of valence band (VB) and conduction band (CB) indicating semiconducting nature. The overall Ti_2CoAl HA shows HM behavior. In the same way, Ti_2FeAl also exhibits HM behavior. On the other hand, in its Hg_2CuTi-type structure for the majority (spin up) state, Ef is situated in the gap of valence band (VB) and conduction band (CB) indicating semiconducting nature, while for the minority (spin down) state Ef crossed the electronic states indicating metallic nature. The overall Ti_2FeAl HA shows HM behavior.

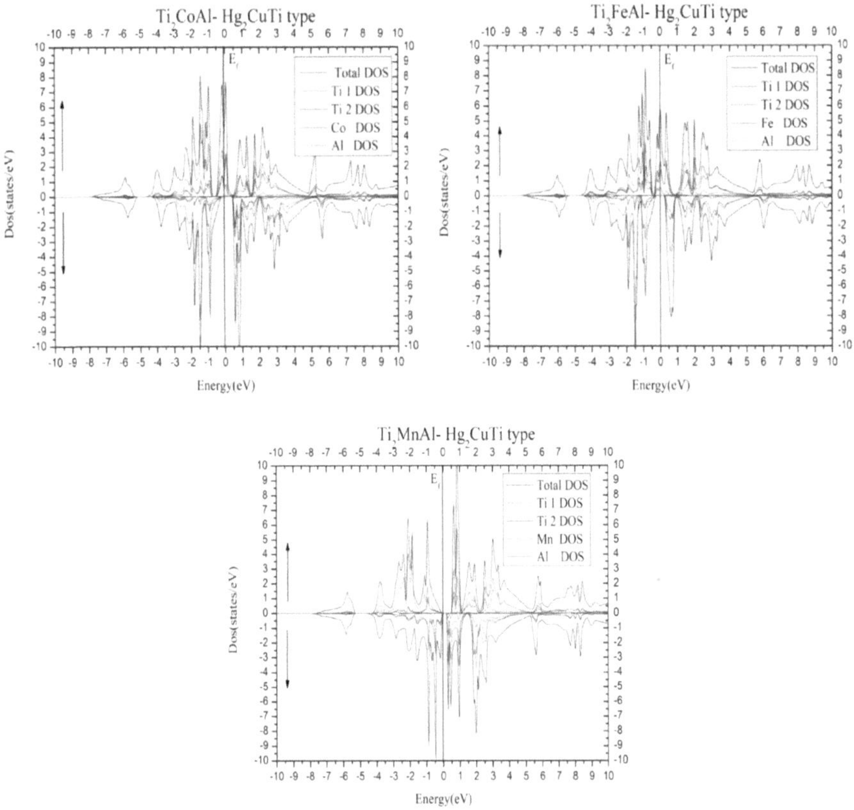

FIGURE 6.5 DOS of Ti_2YAl (Y=Co, Fe, Mn) in Hg_2CuTi-type structures.

6.2.2.3 MAGNETIC PROPERTIES

Numerous Ti_2-based FH alloys with the Hg_2CuTi-type structure can also be classified as half-metallic materials, according to the current study, the spin magnetic moment for each unit cell of the Ti_2CoAl alloy is 20 − 18 = 2 μ_B (Zt = 20), while the atomic MMs calculated and their values are presented in Table 6.3.[21] The Ti_2MnAl alloy's Mtot is an integer of 0 B[21] and is determined by the Slater–Pauling formula Mtot = Zt – 18 and the obtained values are depicted in Table 6.3.[21] The total magnetic moment of Ti_2FeAl is 1 μ_B,[21] while the obtained values are depicted in Table 6.3. Because the alloy possesses an integral total magnetic moment, Mtot, which complies with the Slater–Pauling criterion, it is evident that it has half-metallicity.

The two Ti atoms in a Ti_2MnAl alloy with a Hg_2CuTi-type structure have various local magnetic moments because of their various atomic surroundings. For instance, the Ti1 atom is closest to 4- Ti_2 atoms, 4- Al atoms, and 6- Mn atoms, while the Ti_2 atom is closest to four Ti1 atoms, 4- Mn atoms, and 6- Al atoms.

TABLE 6.3 The Magnetic Moments (in the unit of Bohr Magneton) Ti_2YAl (Y=Co, Fe, Al).

Compound	M^{Ti1}	M^{Ti2}	M^{X}	M^{Al}	$M^{interstitial}$	M^{Tot}
Ti_2CoAl	1.14807	0.65137	–0.29248	–0.00109	0.49423	2.00009
Ti_2FeAl	0.86238	0.56666	–0.83923	0.00531	0.40475	0.99987
Ti_2MnAl	–1.08091	–0.97271	2.60640	–0.01920	–0.53371	–0.00013

6.2.3 CO_2CRZ (Z=AL, GA, IN)

6.2.3.1 STRUCTURAL PROPERTIES

In this section, the considerable structures are presented in Figure 6.6 and are Cu_2MnAl and Hg_2CuTi types. The total energy calculations have been performed for Co_2CrZ (Z = Al, Ga and In) compounds in the considered structures and presented in Figure 6.7 using Birch–Murnaghan's equation of state.[20]

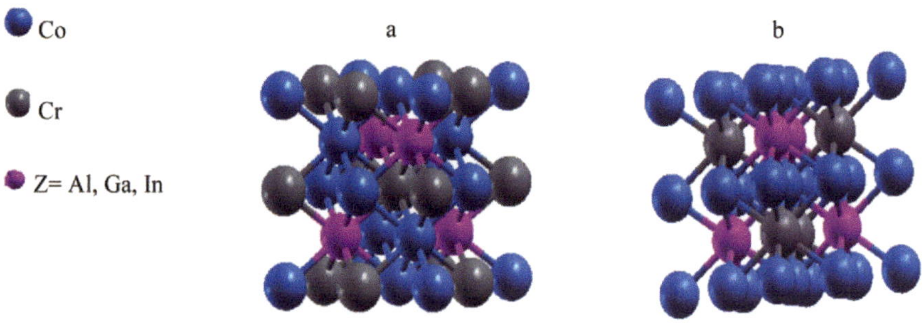

FIGURE 6.6 Structural representation of Co_2CrZ (Z = Al, Ga, In) alloys in the two types of structures. a) Hg_2CuTi, b) Cu_2MnAl.

The atoms of FH Co_2CrZ (Z = Al, Ga and In) occupy positions as shown in Figure 6.6. Further, Figure 6.7 depicts that the Cu_2MnAl-type structure is the most stable and favorable structure (lowest energy in Cu_2MnAl-type), which is consistent with nuclear charge consideration.[28]

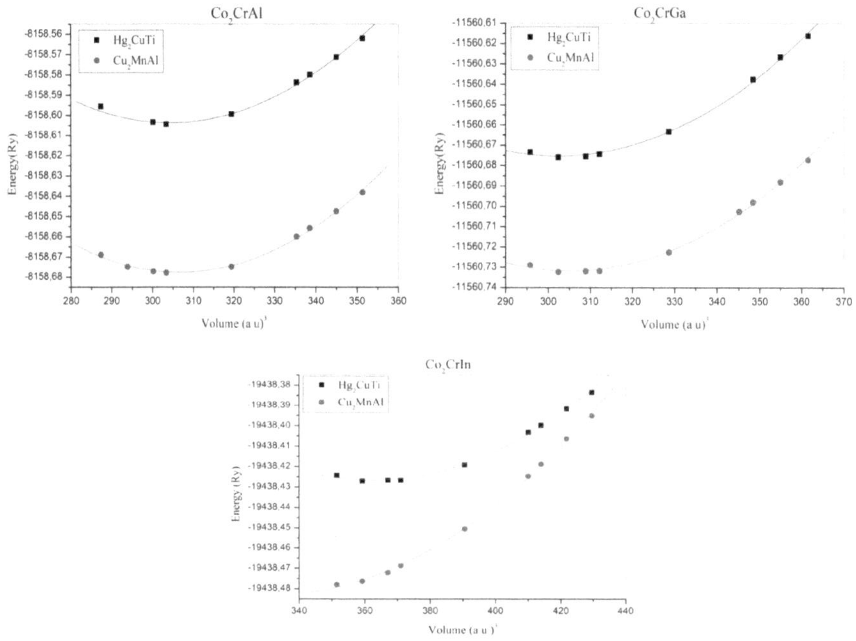

FIGURE 6.7 E-V optimization curve for Co_2CrZ (Z = Al, Ga, and In).

The calculated ground-state parameters of Co2CrZ (Z = Al, Ga, and In) materials in the two considered structures are presented in Table 6.4. These ground-state properties are listed in Table 6.4 at zero pressure.

TABLE 6.4 The Calculated Initial Properties for Co_2CrZ (Z=Al, Ga, In).

Compound	Structure	Lattice parameter, a (Å)	Bulk modulus, B (GPa)	Pressure derivative B'
Co_2CrAl	Hg_2CuTi	5.64	249.5985	7.5721
	Cu_2MnAl	5.66	223.7557	4.1664
Co_2CrGa	Hg_2CuTi	5.652	251.7494	6.4475
	Cu_2MnAl	5.665	239.7246	5.7381
Co_2CrIn	Hg_2CuTi	5.99	162.0538	6.0612
	Cu_2MnAl	5.91	234.7519	8.0378

6.2.3.2 ELECTRONIC PROPERTIES

In case of Co_2CrAl with Cu_2MnAl structure, the BS and DOS indicate that majority-spin (spin-up) states are available at Ef. The existence of these states at Ef shows metallic characteristic by those majority-spin (spin-up) states. On the other hand, the majority-spin (spin-up) states indicate semiconducting nature.[28] Overall Co_2CrAl with Cu_2MnAl structure exhibits HM nature. The similar results have been observed by Seema and Kumar.[29]

In Co_2CrGa with Cu_2MnAl structure, minority (spin down) state, Ef is situated in the gap of valence band (VB) and conduction band (CB) indicating semiconducting nature, while for the majority (spin up) state, Ef crossed the electronic states indicating metallic nature. The overall Co_2CrGa HA shows HM behavior. For Co_2CrIn both the spin channels indicate Ef crossed bands and depict metallic nature.[28]

When determining the electronic structure of the alloy for half-metallic materials, the minority-spin states are crucial. There is an indirect band gap in the Co_2CrAl alloy. This finding is consistent with earlier findings.[29] All the alloys of Hg_2CuTi exhibit a complete lack of half-metallic characteristics.

6.2.3.3 MAGNETIC PROPERTIES

The primary contributor to the total magnetic moment of Co_2CrAl, which has a total magnetic moment of 3 Bohr Magneton, is the chromium. Cobalt, aluminum, gallium, and indium atoms have modest magnetic moments, respectively, 0.83, 1.50, and 0.05 Bohr Magneton.[28] Our calculations yielded total magnetic moment values that are very near (particularly to integer values) and well in line with the SP rule.[28]

Remarkably, the behavior of half-metallic Heusler alloys is identical. In the case of NiMnSb, it was shown the total spin MM in the unit cell, Mt, given by Mt = Zt – 18.[1] Nevertheless, for the L21 FH, the link changes to Mt = Zt – 24.[30] The entire number of fully populated minority states, which equals 12 states, is divided into the 1 – s, 3 – p, and 8 – d states,[14] yielding the number "24" in the computation above.

These Slater–Pauling (SP) principles provide a potent tool for the study of half-metallic materials by establishing a direct connection between the magnetic properties and the electronic properties because (i) valence electron counting may be used to anticipate the magnetic belongings of an HM, and (ii) magnetic capacities can be used to confirm a compound's HM status.[31]

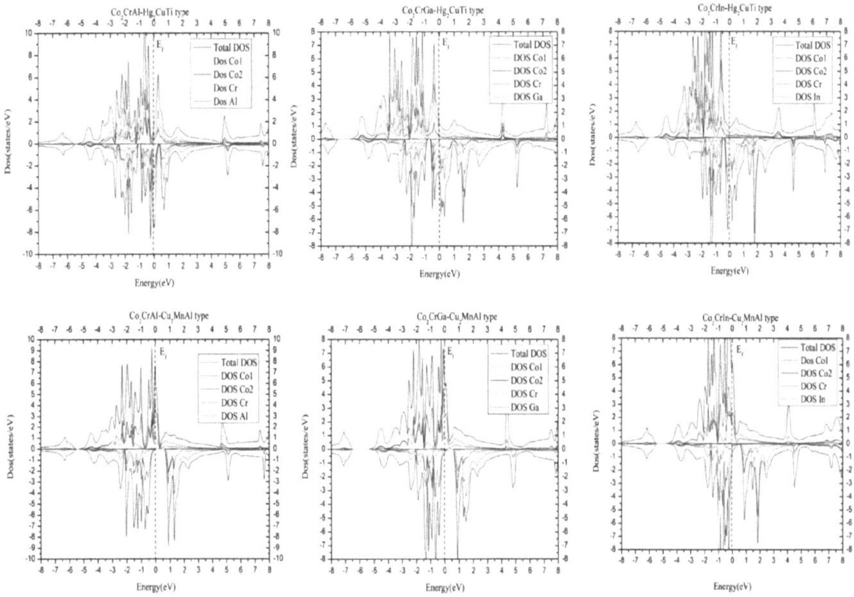

FIGURE 6.8 DOS of Co_2CrZ (Z = Al, Ga, In) in both Cu_2MnAl and Hg_2CuTi structures.

TABLE 6.5 The Magnetic Moments (in the unit of Bohr Magneton) for Co_2CrZ Compounds.

Compound	Structure	M^{Co1}	M^{Co2}	M^{Cr}	M^{X}	$M^{interstitial}$	M^{Tot}
Co_2CrAl	Hg_2CuTi	0.07751	0.47912	0.61065	0.00015	0.01469	1.18213
	Cu_2MnAl	0.83211	0.83118	1.50409	–0.05684	–0.11067	2.99986
Co_2CrGa	Hg_2CuTi	1.26088	1.72342	2.44757	–0.01045	0.03777	5.45920
	Cu_2MnAl	0.74534	0.74274	1.67262	–0.04598	–0.07343	3.04128
Co_2CrIn	Hg_2CuTi	1.41389	1.81031	3.00378	–0.02390	0.02620	6.23027
	Cu_2MnAl	0.45075	0.45282	2.40689	–0.04218	–0.03901	3.22926

6.2.4 CO_2MNZ (Z = AL, GE, SI, GA)

6.2.4.1 STRUCTURAL PROPERTIES

Co_2MnZ (Z = Al, Ge, Si, Ga) compounds crystallize in the L21 structure. In a similar way within GGA potential, the E-V diagrams are obtained by using Birch–Murnaghan's equation of state.[20]

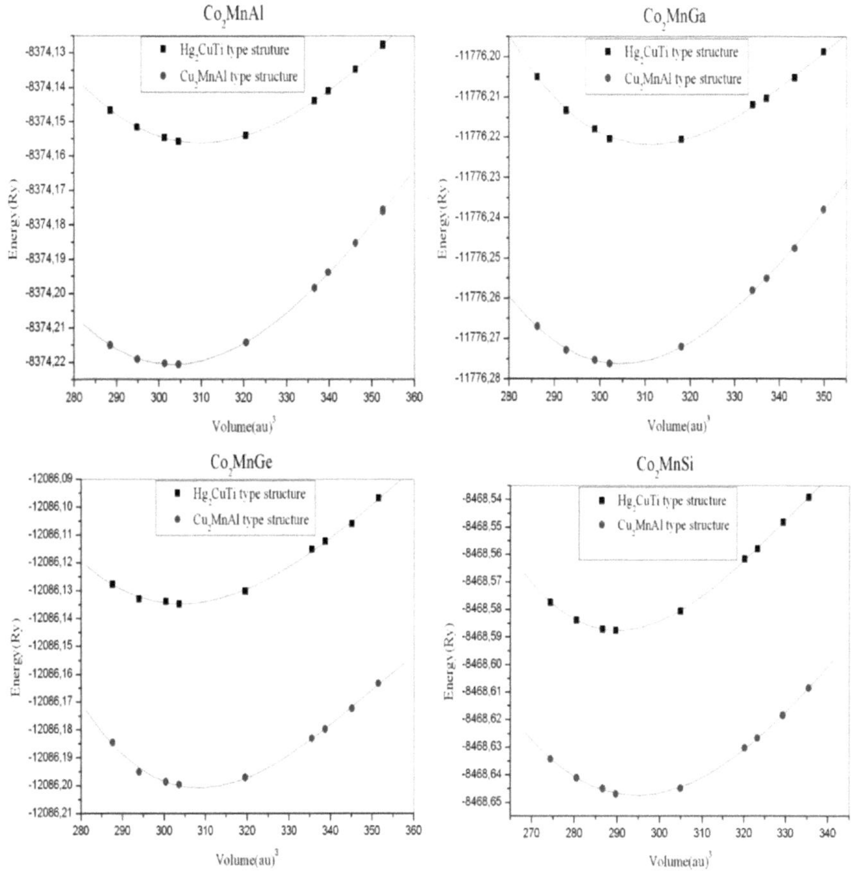

FIGURE 6.9 E–V optimization curve for Co_2MnZ (Z= Al, Ge, Si, Ga).

The connecting total energies with unit cell volumes are plotted in Figure 6.9 for the two structures. As one can understand, the Cu_2MnAl-type structure (SG_216) is favorable for these compounds, which is consistent with the nuclear charge consideration.[31] However, due to the highest energy[32] in this manuscript, antiferromagnetic states are not presented.

6.2.4.2 ELECTRONIC PROPERTIES

De Groot et al.[2] initially proposed the idea of a half-metal based on the estimates of the E–K diagram of the NiMnSb half-Heusler combination. De Groot et al.[2] initially proposed the idea of a half-metal based on the estimation of the E–K diagram of the NiMnSb half-Heusler compound. In such type of material an asymmetric spin-polarized E-K diagram is obtained in which one of the spin channels shows semiconducting performance while the other spin channel shows metallic performance. Consequently, the electrons of the CB have a full SP. Therefore, SP of the electrons, P at Fermi level can be defined by considering the following equation:

$$P = \frac{N^{\uparrow}(E_{\mathrm{F}}) - N^{\downarrow}(E_{\mathrm{F}})}{N^{\uparrow}(E_{\mathrm{F}}) + N^{\downarrow}(E_{\mathrm{F}})}$$

where $N^{\uparrow}(E_F)$ and $N^{\downarrow}(E_F)$ are the spin-dependent densities of states at E_F for the majority and minority states, respectively. Even below the magnetic transition temperature, the P=0 for paramagnetic (PM) or antiferromagnetic (AFM) materials. In ferromagnetic or ferrimagnetic materials, however, it has a finite value below the Curie temperature. The electrons at E_F are fully spin polarized (P = 100%) when either $N^{\uparrow}(E_F)$ or $N^{\downarrow}(E_F)$ equals zero.

TABLE 6.6 The Calculated Initial Properties for Co_2MnZ (Z= Al, Ge, Si, Ga).

Compound	Structure	Lattice parameter, a (Å)	Bulk modulus, B (GPa)	Pressure derivative B'
Co_2MnAl	Hg_2CuTi	5.69	175.7994	4.4381
	Cu_2MnAl	5.65	216.3541	4.8106
Co_2MnGe	Hg_2CuTi	5.65	211.5903	6.9013
	Cu_2MnAl	5.68	273.4231	10.1597
Co_2MnSi	Hg_2CuTi	5.56	319.6697	10.0093
	Cu_2MnAl	5.59	251.0557	4.2108
Co_2MnGa	Hg_2CuTi	5.69	193.4175	8.8298
	Cu_2MnAl	5.65	220.0236	5.6484

The DOS of these compounds for the spin-up (majority spin) and spin-down (minority spin) states in both the Cu_2MnAl and Hg_2CuTi-type structures were estimated in order to examine the electronic structure of Co_2MnZ (Z= Al, Ge, Si, Ga) compounds.

We discovered that the energy gap for Co_2MnAl in a Cu_2MnAl-type structure is found in the minority spin channel (spin down), whereas Fermi level (EF) cuts across bands in the majority spin channel, which is the situation in the spin up (↑).

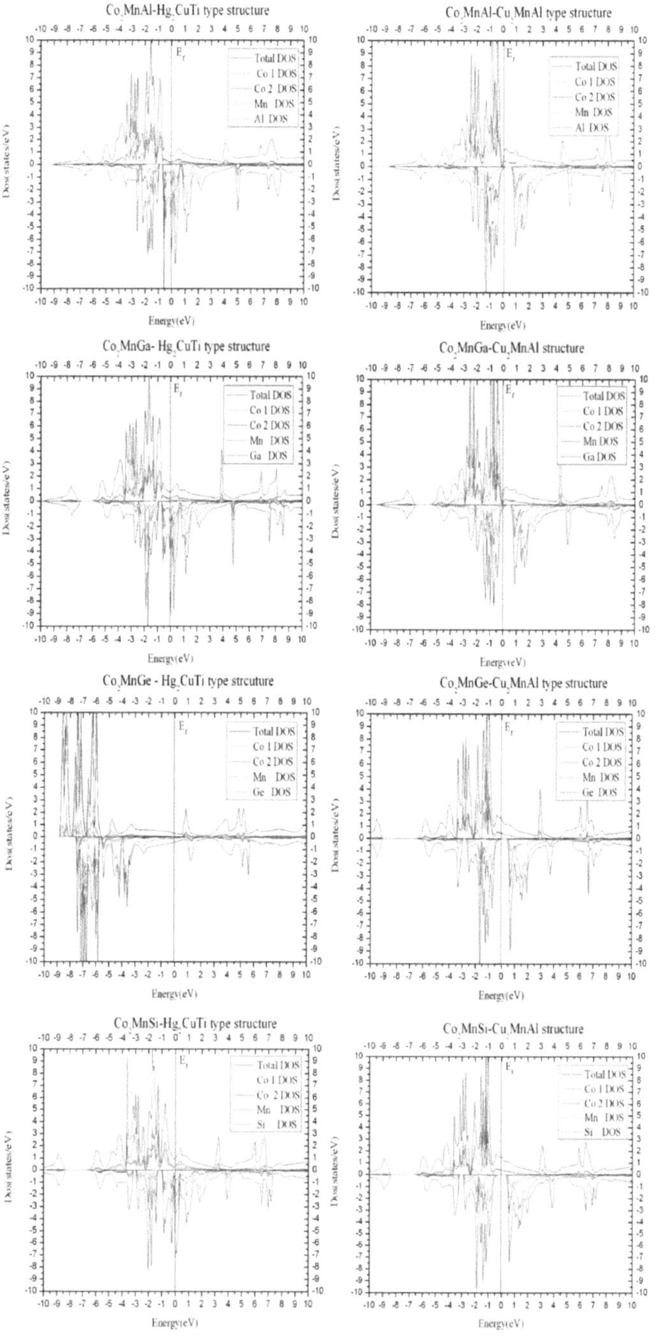

FIGURE 6.10 The calculated total and partial DOS for the Co_2MnZ (Z= Al, Ge, Si, Ga) in both the Cu_2MnAl and Hg_2CuTi-type structures.

Thus, (↑) have metallic properties, whereas the (↓) exhibit semiconducting properties because of the energy gap at the EF in the minority-spin band. We can observe that in the event of (↑), Co_2MnAl behaves like a metal with DOS concentrated at EF, whereas in the case of (↓), an energy gap appears around EF, in the range of 0.30 eV.

We can observe from Figure 6.10 an Eg in the minority spin band for the Co_2MnGe and Co_2MnSi in Cu_2MnAl type structure compounds, and the Fermi level situated in the gap, making these compounds excellent half-metals. We can see the absence of HM behavior for all compounds Co_2MnZ (Z=Al, Ge, Si, and Ga) in a Hg_2CuTi-type structure.

6.2.4.3 MAGNETIC PROPERTIES

For the Heusler compounds Co_2MnZ, we concised the computed magnetic moments per atom in Table 6.2. The results reveal that the Mn ion is the source of the total MM, whereas Al, Ge, Si, or Ga has a very low value. The total spin moment in half-metallic materials must be an integer.

TABLE 6.7 The Magnetic Moments (in the unit of Bohr Magneton) for Co_2MnZ (Z = Al, Ge, Si, Ga).

Compound	Structure	M^{Co1}	M^{Co2}	M^{Mn}	M^{Z}	$M^{interstitial}$	M^{tot}
Co_2MnAl	Hg_2CuTi	1.03688	1.77458	2.67893	–0.04607	–0.11911	5.32521
	Cu_2MnAl	0.69323	0.69306	2.83333	–0.06018	–0.13966	4.01978
Co_2MnGe	Hg_2CuTi	0.76176	1.72482	2.48638	–0.03785	–0.06666	4.86846
	Cu_2MnAl	0.99256	0.97246	3.11337	–0.03302	–0.04668	4.99868
Co_2MnSi	Hg_2CuTi	0.80338	1.72257	2.02711	–0.04467	–0.08770	4.42069
	Cu_2MnAl	1.03000	1.02986	3.03353	–0.04657	–0.04711	4.99971
Co_2MnGa	Hg_2CuTi	0.99605	1.79917	2.72743	–0.04329	–0.09627	5.38309
	Cu_2MnAl	0.64483	0.65355	2.90643	–0.04685	–0.09590	4.06207

The MM of the Co_2MnZ materials follow the formula Mt=Zt – 24, the generalized Slater–Pauling rule.

6.3 CONCLUSION

We have introduced the electronic structure and magnetic properties of Heusler alloys in this chapter. These alloys are intriguing mixtures of metallic ferromagnets and semiconductors. Understanding the gap in the minority band is crucial because it causes half-metallicity, which gives rise to many peculiar traits.

The electronic structure of Heusler alloys X_2MnGe (X = Sc, Fe, Ni) has been determined using first-principles FP-LAPW approach based DFT by considering Hg_2CuTi-type and Cu_2MnAl-type structures. The Sc_2MnGe FH combination is HM at the Hg_2CuTi-type structure, according to spin-polarize calculations. Fe_2MnGe compound has a metallic character in both Hg_2CuTi and Cu_2MnAl for the entire Heusler alloys, and Ni_2MnGe has a metallic character in Cu_2MnAl. The findings demonstrated that the Cu_2MnAl-type structure is more favourable based on the total energy for all compounds X_2MnGe (X = Sc, Fe, Ni). In a similar way Co_2MnZ (Z= Al, Ge, Si, and Ga) Heusler alloys have been investigated. The computed initial properties of the materials are promising with the related theoretical findings. Co_2MnZ (Z= Al, Ge, Si, or Ga) in a Cu_2MnAl-type structure displayed a half-metallic ferromagnetic activity, according to the electronic structure. In the Cu_2MnAl-type structure, the total magnetic moments are 4 μ_B, 5 μ_B, 5 μ_B, and 4 μ_B, respectively. These values are in accordance with the Slater–Pauling rule (Mt=Zt - 24). Co_2MnZ (Z=Al, Ge, Si, and Ga) appears to be a promising material for prospective spintronics applications.

KEYWORDS

- **Heusler alloys**
- **DFT theory**
- **half-metallic**
- **review**
- **ferromagnetic materials**
- **electronic structure**
- **ground state properties**

REFERENCES

1. Heusler, F.; Starck, W.; Haupt, E. *Verh DPG* **1903,** *5,* 220–223.

2. de Groot, R. A.; M¨uller, F. M.; Engen, P. G.; Buschow, K. H. J.:*Phys. Rev. Lett.* **1983,** *50,* 2024.
3. Shaughnessy, M.; Fong, C. Y.; Snow, R.; Yang, L. H.; Chen, X. S.; Jiang, Z. M. *Phys. Rev. B* **2010,** *82,* 035202.
4. Webster, P. J. *J. Phys. Chem. Solids* **1971,** *32,* 1221.
5. Ziebeck, K. R. A.; Webster, P. J. *J. Phys. Chem. Solids* **1974,** *35,* 1.
6. Kübler, J.; Williams, A. R.; Sommers, C. B. *Phys. Rev. B* **1983,** *28,* 1745.
7. Ishida, S.; Akazawa, S.; Kubo, Y.; Ishida, J. *J. Phys. F Met. Phys.* **1982,** *12,* 1111.
8. Fujii, S.; Asano, S.; Ishida, S. *J. Phys. Soc. Jpn.* **1995,** *64,* 185.
9. Dahmane, F.; Mesri, D.; Doumi, B.; Tadjer, A.; Abbar, B.; Yakoubi, A.; Boutaleb, M.; Aourag, H. *J. Supercond. Nov. Magn.* DOI 10.1007/s10948-015-2994-9.
10. Ahmadian, F. *J. Supercond. Nov. Magn.* **2013,** *26,* 1737–1743.
11. Fang, Q.-L.; Zhao, X.-M.; Zhang, J.-M.; Xu, K.-W. *Thin Solid Films* **2014,** *558,* 241–246.
12. Fang, C. M.; de Wijs, G. A.; de Groot, R. A. *J. Appl. Phys.* **2002,** *91,* 8340.
13. Galanakis, I.; Mavropoulos, P.; Dederichs, P. H. *J. Phys. D: Appl. Phys.* **2006,** *39,* 765.
14. Skaftouros, S.; ¨Ozdogan, K.; Galanakis, I. *Phys. Rev. B* **2013,** *87,* 024420.
15. Slater, J. C. *Phys. Rev.* **1936,** *49,* 931.
16. Pauling, L. *Phs. Rev.* **1938,** *54,* 899.
17. Galanakis, I.; Dederichs, P. H.; Papanikolaou, N. *Phys Rev. B* **2002,** *66,* 134428.
18. Galanakis, I.; Dederichs, P. H.; Papanikolaou, N. *Phys. Rev. B* **2002,** *66,* 174429.
19. Ahmadian, F.; Salary, A. *Intermetallics* **2014,** *46,* 243–249.
20. Murnaghan, F. D. *Proc. Natl. Acad. Sci. USA* **1944,** *30,* 244.
21. Dahmane, F.; Benalia, S.; Djoudi, L.; Tadjer, A.; Khenata, R.; Doumi, B.; Aourag, H. *J. Supercond. Nov. Magn.* DOI 10.1007/s10948-015-3109-3.
22. Bayar, E.; Kervan, N. Selc-uk Kervan and 2945–2948. *J. Magn. Magn. Mater.* **2011,** 323.
23. Liping, M.; Yongfan, S.; Yu, H. *J. Magn. Magn. Mater.* **2014,** *369,* 205–210.
24. Nan, Z.; Yingjiu, J. *J. Magn. Magn. Mater.* **2012,** *324,* 3099–3104.
25. Ahmadian, F. *J. Alloys. Compnd.* **2013,** *576,* 279–284.
26. Fang, C. M.; De Wijs, G. A.; De Groot, R. A. *J. Appl. Phys.* **2002,** *91,* 8340.
27. Galanakis, I.; Mavropoulos, P.; Dederichs, P. H. *J. Phys. D: Appl. Phys.* **2006,** *39,* 765.
28. Dahmane, F.; Mesri, D.; Tadjer, A.; Khenata, R.; Benalia, S.; Djoudi, L.; Doumi, B.,Boumia, L.; Aourag, H. *Modern Phys. Lett. B*, DOI: 10.1142/S0217984915502656.
29. Seema, K.; Kumar, R. *Appl. Phys. A* **2014,** *116,* 1199–1209.
30. Galanakis, I.; Dederichs, P. H.; Papanikolaou, N. *Phys. Rev. B* **2002,** *66,* 174429.
31. Xing, N.; Gong, Y.; Zhang, W.; Dong, J.; Li, H. *Comput. Mater. Sci.* **2009,** 45:489.
32. Dahmane, F.; Doumi, B.; Mogulkoc, Y.; Tadjer, A.; Prakash, D.; Khenata, R. *J. Supercond. Nov. Magn.* DOI 10.1007/s10948-015-3357-2.
33. Dahmane, F.; Mogulkoc, Y.; Doum, B.; Tadjer, A.; Khenata, R.; Bin Omran, S.; Rai, D. P. *J. Magn. Magn. Mater.* **2016,** *407,* 167–174.

CHAPTER 7

Development of Hall Effects: Derivation and Their Application in Materials Science

DIPAK BHATTARAI, DEERGH BAHADUR SHAHI, and MADHAV PRASAD GHIMIRE

Central Department of Physics, Tribhuvan University, Kirtipur, Kathmandu, Nepal

ABSTRACT

The Hall effect is a fundamental concept in materials science which is important for understanding the electrical and magnetic properties of materials. Edwin Hall in 1879 discovered the classical Hall effect which is widely used in material science to determine the charge carrier density and their type (electrons or holes). This information is crucial for the development of new materials for electronics, spintronics, and other related applications. It can also be used to measure the quality of materials and their uniformity, making it an important tool for material characterization and optimization. It was later observed that, in certain two-dimensional electron systems and strong magnetic fields, the Hall resistance became quantized, giving rise to the quantum Hall effect (QHE). Discovery of QHE in the 2D electronic system was the step to give topology a major role in condensed matter physics. The cause of the plateau in the Hall resistivity/conductance is due to the lack of scattering. At the plateau, the Landau levels are filled so there are no available states for particles to scatter into. They can only skip the orbit at the edges, showing no resistance and hence longitudinal resistivity becomes zero. In this situation, disorder is not protecting the plateau and magnetic field has broken the time reversal symmetry, so there must be another preserving factor which is the topological characteristic of the sample. To

Contemporary Advancements in Materials Technology: Novel Solutions to Complex Problems.
Shrikaant Kulkarni, Vipul Srivastava, & P. William (Eds.)

provide extremely accurate measurements of the Planck's constant and fine structure constant, quantization of resistance is used. Classically, the linear dependency of transverse resistivity with magnetic field is used to make accurate and sensitive magnetic field sensors due to a wide range of linearity from zero to a large value of magnetic field. The longitudinal resistivity is independent of the field and describes the transport characteristics of the sample along the direction of the current. This chapter contains the stepwise development of Hall effect, i.e., from classical Hall effect to quantum anomalous Hall effect (QAHE) with clear distinction between classical and quantum Hall effects. The lucid derivation of quantum Hall effect and stepwise explanation of the terms involved in it is the major content of this chapter. The conditions required for the observation of various Hall effects are explained properly. A comparison of different QHE device shows more stable and reproducible quantized resistance with a fundamental value of $\frac{h}{e^2}$ = 25813.807 Ω, also called the Klitzing constant. This value is found to be independent of microscopic details of the materials, dimension of the device, disorder and localized states of the Landau levels because the parameters involved in the expression of transverse resistivity are universal constants. QAHE are noted especially in magnetic topological insulators which can be observed in the absence of external magnetic field. In the QAH system, the electrons are confined to a two-dimensional plane and the magnetic field is used to generate a topological phase that has a quantized Hall conductance even in the absence of an external magnetic field. In QAHE, the edge states remain independent of the external bias or magnetic field. This development helps to develop dissipationless devices, spintronic devices, and many more. New hopes on novel quantum phenomenon such as magnetoelectric effect, majorana states are under extensive investigation after the realization of QAHE. Some experimental progress on QAHE and other classes of materials for it with the proper application in related area are covered in this chapter. The cause of proposing QAH state in materials such as graphene, transition metal oxides, topological insulators such as $MnBi_2Te_4$, Cr-doped $(Bi, Sb)_2Te_3$ are also covered in this chapter.

7.1 HALL EFFECT

The production of potential difference across the current (I) carrying conductor in a perpendicular magnetic field (B) because of the deflection of electron by Lorentz force is called Hall effect (by Edwin Hall in 1879) which works well for a small magnetic field where Hall resistance is linear with it according to the relation: $R_H = \frac{B}{net}$, where "n" is the electron density of a sample and "t" is the thickness.[1] The transverse voltage or Hall voltage V_H or V_{xy} is developed across the edges of the sample perpendicular to both current and the applied magnetic field as shown in Figure 7.1.

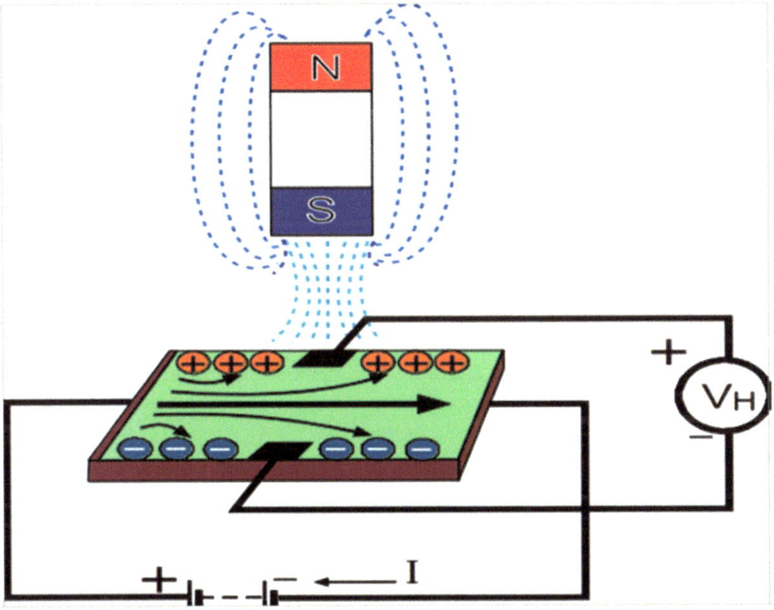

FIGURE 7.1 Schematic diagram of Hall effect.

From the classical Drude model, the transverse and longitudinal resistivities are given by

$$\rho_{xy} = \frac{B}{net} \tag{7.1}$$

and

$$\rho_{xx} = \frac{m}{ne^2\tau} = \frac{1}{ne\mu} = constant \tag{7.2}$$

where "τ" is relaxation time, "μ" is mobility, and "m" is carrier effective mass.[2]

Equation 7.1 can be applicable to calculate the charge density "n" and carrier type such as electron or hole. The linear dependency of resistivity ρ_{xy} with "B" is being used to make accurate and sensitive magnetic field sensors due to a wide range of linearity from zero to very large Tesla (T). These relations show that classically, transverse resistivity is linearly dependent on magnetic field, whereas ρ_{xx} is independent of the field and describes the transport characteristics of the sample along the direction of applied current as shown in Figure 7.2(a).

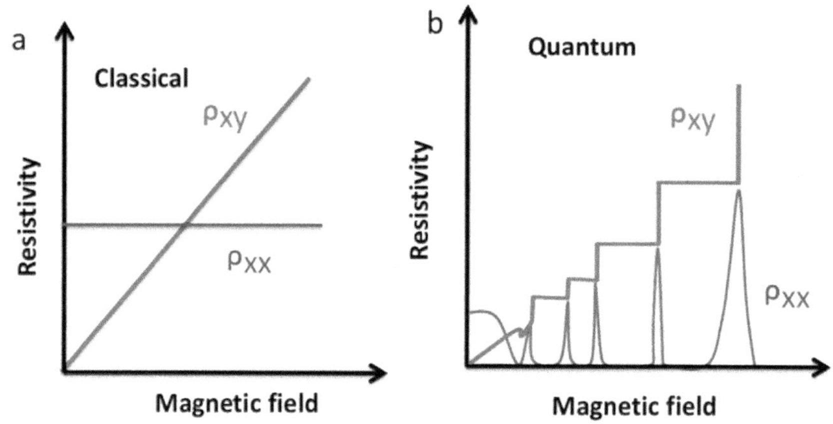

FIGURE 7.2 Resistivity versus magnetic field: (a) classical Hall effect and (b) quantum Hall effect.

7.2 QUANTUM HALL EFFECT

A century after the finding of the classical Hall effect, at extremely low temperatures ($T < 4$ K) and strong magnetic fields (1–10 T), Klaus von Klitzing and colleagues discovered in 1980 that the Hall resistance (R_H) is stepped-dependent on magnetic field strength in 2D electron systems. The simultaneous drop of the magnetoresistance ρ_{xx} to zero at sufficiently low temperature results in the plateaus.

To achieve this phenomenon, we need 2D electron gas (2DEG) system in which magnetic field is applied perpendicularly to the plane of the system at very low temperature environment.

At "B" = 0 T, electrons move according to the applied electric field but at "B" ≠ 0 T, electrons make circular path within the sample and localized in cyclotron orbit but at each edge of the sample, they make skipping orbits showing 1D channel generating chiral edge state as shown in Figure 7.3. This semi-classical explanation is further turned into a quantum mechanical way with the help of Landau's level.

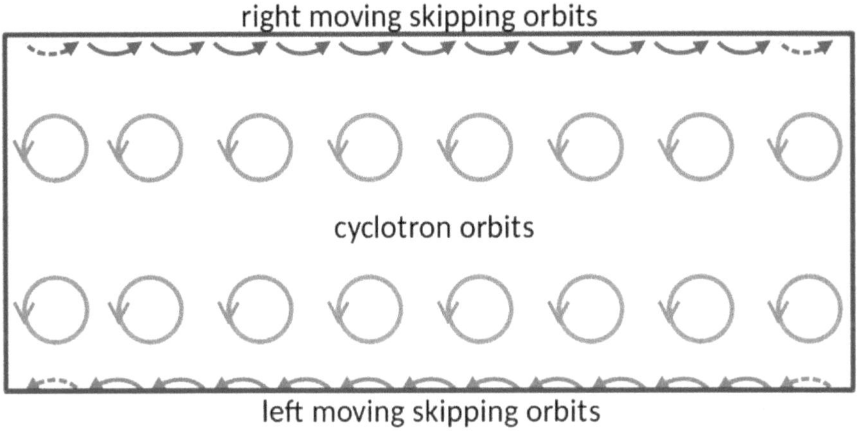

FIGURE 7.3 Chiral edge modes in 2D sample.

Klitzing et al. (1980) measured the Hall voltage of a 2DEG and found that Hall resistance is constant which depends on the universal constants such as speed of light and fine structure constant rather than the geometry of the device.[3] It is the quantization of Hall resistance precisely in integer multiples of $\frac{h}{e^2}$ as given in eq. 7.3.

$$\rho_{xy} = \left(\frac{h}{e^2}\right)\frac{1}{\gamma} \tag{7.3}$$

where γ is an integer up to an accuracy of 10^{-9}. The presence of this quantization is the striking feature of quantum Hall effect (QHE). Comparisons between different QHE devices revealed about the more stable and more reproducible quantized resistance having a fundamental value of $\frac{h}{e^2}$ = 25813.807 Ω, also called the Klitzing constant or quantum of resistivity, and is found to be independent of microscopic details of the materials, dimension of the device, disorder and localized states of the Landau levels because the parameters involved in eq. 7.3 are all universal constants due to which it is widely used in the field of metrology. In this situation, disorder is not protecting the plateau and magnetic field already has broken the time reversal symmetry. Thus, there must be another preserving factor which could be the topological characteristic of the sample.

Figure 7.2(b) shows the distinctive hallmarks of integer QHE, where Hall resistivity ρ_{xy} makes plateaus with different values of "*B*" and ρ_{xx} becomes zero at the plateau region showing peaks at the time of transition from one plateau to another. This is similar to the transition from metal to insulator, and vice versa. The cause of the plateau in the Hall resistivity/conductance is due to the lack of scattering of the carriers. At the plateau, the Landau levels are filled so there are no available states for particles to scatter into. They can only do the skipping orbit at the edges, which is hence dissipationless showing no resistance and hence $\rho_{xx} = 0$. It is the quantum mechanical phenomenon in which Landau levels (discrete energy levels) are generated at the density states of 2DEG.

One of the earliest determinations of magnetoresistance under QHE condition on a GaAs/AlGaAs sample was studied by Von Klitzing (2005) where at $B = 14\ T$, fundamental quantized resistance $\frac{h}{e^2}$ was observed and other plateaus were noticed at lower magnetic field for filling factors 2, 4, and 6. The plateaus for odd filling factors were not seen distinctly because their energy gaps arise from comparatively minor Zeeman splitting at each Landau level.[4] QHE is caused due to the quantized Landau levels, filling factor (i.e., no. of electrons over ground-state degeneracy) dependence on magnetic field along with the disorder effect. The Fermi level remains in between the extended states when QHE occurs. The gap for QHE is created by the localized states which can be measured through the temperature dependence of longitudinal resistance.

Discovery of QHE in the 2D electronic system[3,5] was the step to give topology a major role in condensed matter physics. Its experimental realization in the 3D electronic system is done by Tang et al. (2019) in bulk $ZrTe_5$ crystals by performing electric transport measurement at low temperature under the magnetic field[6] although its possibility was proposed decades ago.[7,8] They also noticed metal–insulator transition by further increasing the magnetic field with the observation of much increment of both longitudinal and Hall resistivity. Anomalous Hall effect was noticed with a larger value of Hall resistance in ferromagnetic substance exposed to a perpendicular magnetic field obtained from the combined result of spin–orbit coupling and magnetic polarization.

Integer QHE was discovered under strong magnetic field on a Si/SiO_2 field-effect transistor by von Klitzing. For the same, he received a Nobel prize in Physics in 1985. Another Nobel prize was awarded for the discovery of fractional QHE by Tsui, Stormer, and Laughlin in 1998, which stablished QHE in condensed matter physics. Magnetically induced Landau levels contribute the unique dissipationless chiral edge states responsible for QHE, but well-defined Landau levels require high magnetic field on high mobility sample. To overcome the difficulty of high magnetic field, first QHE model without Landau levels was proposed by Haldane on a graphene lattice having Dirac-cone-shaped gapeless band structure. The gape was opened at the Dirac point by breaking time reversal symmetry (TRS) with unit chern number.[9]

To give accurate measurements of the Planck constant and fine structure constant, quantization of resistance in QHE was used. The popularity of QHE is based on the importance of properties of charged particles in magnetic field which is being connected in different fields of materials science, high energy physics, astrophysics, and so on.

7.2.1 INTEGER QUANTUM HALL EFFECT

When a vertical magnetic field is acted in a slab carrying current, a Hall voltage appears perpendicular to both. The motion of an electron in the slab may be described in a single relaxation time approximation by

$$m\frac{d\vec{v}}{dt}+\frac{m\vec{v}}{\tau}=-e\left[\vec{E}+\frac{\vec{v}\times\vec{H}}{c}\right] \tag{7.4}$$

At the steady state, $\frac{d\vec{v}}{dt}=0$, so the above equation becomes

$$\frac{m\vec{v}}{\tau}=-e\left[\vec{E}+\frac{\vec{v}\times\vec{H}}{c}\right]$$

$$\text{or, } \vec{v}=\left(-\frac{e\tau}{m}\right)\left[\vec{E}+\frac{\vec{v}\times\vec{H}}{c}\right]$$

Here, we consider 2D example; thus

$$v_x=\mu\left[E_x+\frac{v_y\times H}{c}\right] \text{ and}$$

$$v_y=\mu\left[E_y-\frac{v_x\times H}{c}\right] \text{where, } \mu=\left(\frac{e\tau}{m}\right)=\text{mobility}$$

Using the value of v_y in v_x, we get

$$\text{or, } v_x=\mu\left[E_x+\frac{H}{c}\mu\left(E_y-\frac{v_xH}{c}\right)\right]$$

$$\text{or, } v_x=\mu\left[E_x+\frac{\mu H}{c}E_y-\frac{\mu H^2}{c^2}v_x\right]$$

$$\text{or, } v_x\left(1+\frac{\mu^2H^2}{c^2}\right)=\mu\left[E_x+\frac{\mu H}{c}E_y\right]$$

$$\therefore v_x=\frac{\mu\left(E_x+\frac{\mu H}{c}E_y\right)}{\left[1+\left(\frac{\mu H}{c}\right)^2\right]}$$

Here, $\frac{\mu H}{c}=\left(\frac{-e\tau}{m}\right)\frac{H}{c}=\left(\frac{-eH}{mc}\right)\tau=-\omega_c\tau$, where ω_c is the cyclotron frequency.

Therefore, the above equation becomes

$$v_x=\frac{\mu(E_x-\omega_c\tau E_y)}{1+\omega_c^2\tau^2}\ \&\ v_y=\frac{\mu(E_y+\omega_c\tau E_x)}{1+\omega_c^2\tau^2}$$

$$\therefore\begin{bmatrix}v_x\\v_y\end{bmatrix}=\frac{\mu}{1+\omega_c^2\tau^2}\begin{bmatrix}1&-\omega_c\tau\\\omega_c\tau&1\end{bmatrix}\begin{bmatrix}E_x\\E_y\end{bmatrix}$$

The current vector is given by $\vec{j}=-ne\vec{v}$

$$\therefore\begin{bmatrix}j_x\\j_y\end{bmatrix}=-\frac{ne\mu}{1+\omega_c^2\tau^2}\begin{bmatrix}1&-\omega_c\tau\\\omega_c\tau&1\end{bmatrix}\begin{bmatrix}E_x\\E_y\end{bmatrix}$$

$$\text{or, }\begin{bmatrix}j_x\\j_y\end{bmatrix}=\frac{\sigma_0}{1+\omega_c^2\tau^2}\begin{bmatrix}1&-\omega_c\tau\\\omega_c\tau&1\end{bmatrix}\begin{bmatrix}E_x\\E_y\end{bmatrix}$$

where $\sigma_0=-ne\mu=\frac{ne^2\tau}{m}$ = conductivity

since $J=\sigma E$

$$\therefore\sigma=\frac{\sigma_0}{1+\omega_c^2\tau^2}\begin{bmatrix}1&-\omega_c\tau\\\omega_c\tau&1\end{bmatrix}$$

where $$\sigma_{xx}=\sigma_{yy}=\frac{\sigma_0}{1+\omega_c^2\tau^2}\ \&\ \sigma_{xy}=-\sigma_{yx}=-\frac{\sigma_0\omega_c\tau}{1+\omega_c^2\tau^2}\qquad(7.5)$$

After finding the inverse of σ, we get

$$\rho=\frac{1}{\sigma_0}\begin{bmatrix}1&\omega_c\tau\\-\omega_c\tau&1\end{bmatrix}$$

$$\Rightarrow\rho_{xx}=\rho_{yy}=\frac{1}{\sigma_0}\ \&\ \rho_{xy}=\frac{\omega_c\tau}{\sigma_0}=-\rho_{yx}$$

From eq. 7.5,

$$\sigma_{xy} = -\frac{\sigma_0 \omega_c \tau}{1+\omega_c^2 \tau^2} = -\frac{\sigma_0 \omega_c \tau / \sigma_0^2}{\frac{1}{\sigma_0^2} + \left(\frac{\omega_c \tau}{\sigma_0}\right)^2}$$

$$\therefore \sigma_{xy} = -\frac{\rho_{xy}}{\rho^2_{xx} + \rho^2_{xy}} \qquad (7.6)$$

Similarly, $\sigma_{xx} = \frac{\rho_{xx}}{\rho^2_{xx} + \rho^2_{xy}} \left(\because \rho_{xy} = \frac{\omega_c \tau}{\sigma_0} \text{ and } \rho_{xx} = \frac{1}{\sigma_0} \right)$ (7.7)

It is observed that ρ_{xx} vanishes when ρ_{xy} remains constant at a plateau. In general, it is considered that a system with $\rho_{xx} = 0$ is a perfect conductor, i.e., there is no dissipation of energy. If $\rho_{xy} = 0$, from eq. 7.7, we obtain the conventional relation between conductivity and resistivity, i.e., $\rho_{xx} = \frac{1}{\sigma_{xx}}$. However, if $\rho_{xy} \neq 0$, both σ_{xx} and σ_{xy} exist and an interesting result is noticed.

From eq. 7.7, when $\rho_{xx} = 0 \Rightarrow \sigma_{xx} = 0$. It shows that the system is a perfect conductor and a perfect insulator at the same time which seems as controversy. The fact that $\rho_{xx} = 0$ and $\sigma_{xx} = 0$ at the same time simply shows that the current flowing direction is normal to the electric field. $\sigma_{xx} = 0$ implies that the relaxation time τ will be infinite from eq. 7.5, showing no scattering and hence current is flowing perpendicular to the field.

The electric current may be allowed to flow only in the x-direction. Under this condition $j_y = 0$ then

$$E_x = -\frac{E_y}{\omega_c \tau}$$

$$\& \ j_x = \frac{\sigma_0}{1+\omega_c^2 \tau^2} \left(-\frac{E_y}{\omega_c \tau} - \omega_c \tau E_y \right)$$

$$= \frac{-\sigma_0 E_y}{1+\omega_c^2 \tau^2} \left(\frac{1+\omega_c^2 \tau^2}{\omega_c \tau} \right)$$

$$= \frac{-\sigma_0 E_y}{\omega_c \tau}$$

So, Hall conductivity is given by

$$\sigma_H = \frac{j_x}{E_y} = \frac{-\sigma_0 E_y}{\omega_c \tau} \frac{1}{E_y} = \frac{-\sigma_0}{\omega_c \tau} = \frac{1}{\rho_{yx}}$$

$$= -\frac{ne^2\tau / m}{\left(\frac{eH}{mc}\right)\tau}$$

$$= -\frac{nec}{H}$$

$$= \frac{1}{\rho_{xy}}$$

It is noted that in contrast to σ_0, this conductivity is dependent on the sign of the charge. This helps to identify the types of charge carriers (i.e., either electron or hole).

Instead of the condition $j_y = 0$, if the condition $E_y = 0$ is imposed, the Hall conductivity can be written as

$$\sigma_{yx} = \frac{j_y}{E_x} = -\sigma_H$$

According to eq. 7.5, σ_{xx} does not vanish for a finite density. However, if $\sigma_{xx} = 0$ then σ_{xy} is given by

$$\sigma_{xy} = \sigma_H = \frac{j_x}{E_y} = -\frac{nec}{H}$$

The last equality is based on the classical equation of motion. It shows that the absolute magnitude of σ_{xy} is proportional to electron density "*n*."

The Hall coefficients of bulk metals, especially alkali metals, agree usually with the theoretical expectation. Measurements of the magnetoconductivity of surface electron started around in 1966 and those of the Hall conductivity around 1975. These measurements supplied important information about the dimensionality and Coulomb effects. While interest in 2D electron systems was rapidly growing, a very unusual phenomenon of the electrons in silicon MOSFETs was invented by Klitzing et al. (1980), during the course of studying electron localization, the Hall resistance took on plateau values for finite intervals of gate voltage where the magnetoresistance dropped to zero. Note that electron density is changed by gate voltage in silicon MOSFETs.

The magnetoconductivity σ_{xx} in the plateau region stays nearly zero; it can be 10^{-7} times smaller than its maximum value. The phenomenon is now called the Integer quantum Hall effect (IQHE) which is illustrated schematically as shown in Figure 7.4. The upper graph illustrates the Hall conductivity (σ_{xy}) versus electron density "*n*." The lower curve illustrates the magnetoconductivity. This graph may be considered to represent the density of states plotted against the Fermi energy where the dashed line represents about the classical expectation.

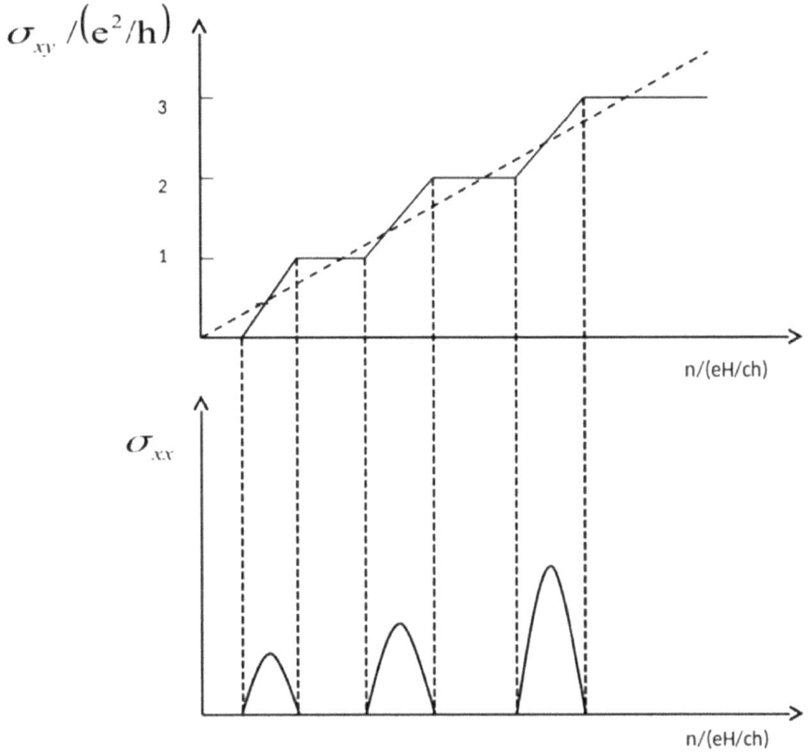

FIGURE 7.4 Schematic representation of IQHE.

7.2.2 *FRACTIONAL QUANTUM HALL EFFECT*

The fractional quantum Hall effect (FQHE) is a phenomenon that occurs in 2D systems, such as thin films or semiconductor devices when a high magnetic field is applied on it. It is characterized by the quantized value of the Hall resistance into some fraction of the quantum of resistance, $\frac{h}{e^2}$.

It is a result of the electronic interaction with the magnetic field, which leads to the formation of quantized energy levels known as Landau levels (LL). When the Fermi level remains inside the two LL, the system exhibits a quantized Hall resistance which is a fraction of $\frac{h}{e^2}$. Tsui, Stormer, and Laughlin got the 1998 Nobel Prize in Physics for the discovery of the fractional Hall effect. They discovered that the plateaus observed at some fractional filling factor values also by replicating the Klitzing's previous experiment under large magnetic fields. The fractional Hall effect has since been studied extensively for the discovery of new states of matter like FQH state, which is a type of topological phase of matter that has potential applications in quantum computing and other areas of technology.

In the FQH effect, the electrons in the 2DEG form a collective behavior, known as a "quantum fluid," in which the electrons can only exist at certain discrete energy levels called the Landau levels. The magnetic field normal to the plane quantizes the in-plane motion to Landau levels having energies $E_i = (i + \frac{1}{2})\ \hbar\omega_c$, where ω_c refers to the cyclotron frequency. The number of available states (d) in each LL is linearly related to the magnetic field (B) according to the relation $d = \frac{2eB}{h}$. The spin of electron further splits the LL into two having $\frac{eB}{h}$ states per unit area.[10] The electron population in the 2D system at low temperature and in higher magnetic field is obtained from the LL filling factor $\nu = \frac{n}{d} = \frac{n}{eB/h}$. The characteristic resistances of a 2D electron system versus magnetic field at 85 mK temperature were observed by Eisenstein and Stormer (1990).[11]

The filling factor ν is shown as numbers to show the sequence of filling LL with electrons. Hall resistance R_{xy} exhibits plateaus quantized to $\frac{h}{\nu e^2}$ at $\nu = i =$ integer as the hallmarks of IQHE and at ν = some fractional number showing the footprint of FQHE. The IQHE is because of the quantization conditions for noninteracting 2D electrons whereas FQHE can be explained by many body problems. It is due to the highly correlated motion of large no. of electrons in 2D under magnetic field. Some of the fractions observed are $\frac{1}{3}, \frac{1}{5}, \frac{1}{7}, \frac{2}{3}, \frac{2}{5}, \frac{2}{7}, \frac{2}{9}, \frac{3}{5}, \frac{3}{7}$ etc. It is noticed that the purer the sample, the greater is the no. of observed fractions.[12] Under certain conditions, when the electron density is low and the magnetic field is sufficient then electrons can form FQH states, in which the electrons behave as if they have a fraction of an electron charge. These states are characterized by quantized Hall conductance values that are fractions of the conductance quantum $\frac{e^2}{h}$.

7.3 SPIN HALL EFFECT

The spin Hall effect (SHE) refers to the phenomenon in which an electric current flowing through a material generates a spin polarization in the transverse direction, perpendicular to the current flowing direction as shown in Figure 7.5. Specifically, when a current flows through a material having strong SOC, such as a heavy metal or a semiconductor, the spin angular momentum of the moving electrons can become separated from their charge and collection of spins on opposite sides of the sample takes place. In 1971, its first prediction was done by Dyakonov and Perel[13] and recaptured in 1999 by Hirsch.[14] The first SHE signal was detected experimentally in nonmagnetic material GaAs which was reported in 2004.[15]

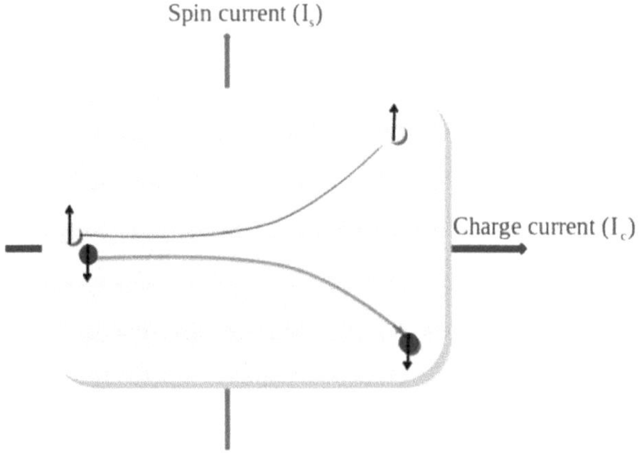

FIGURE 7.5 Schematic diagram of SHE.

This effect is found to have important applications in spintronics where the spin of electrons rather than their charge is utilized for information processing, storage, quantum computing, and energy harvesting. The generation and manipulation of spin currents can be used to control the magnetization of a magnetic device which could lead to more efficient and faster spintronic devices.[16]

Conditions to obtain the SHE

- The material must have a strong SOC to lead to the accumulation of spin angular momentum.

- An electric current must be applied to the material. This can be achieved by applying a voltage across the sample, which will cause electrons to flow through the material.
- The sample must be designed in such a way that the accumulation of spin currents can be detected. This can be done using a variety of techniques, such as measuring the voltage across the sample or using a magnetic sensor to detect changes in the magnetization of the material.
- The material must have large spin Hall angle that describes the ability of conversion of a charge current to spin current.
- The sample geometry must be suitable for observing the spin Hall effect. Typically, a thin film or a narrow wire is used to observe the effect, as this allows for the collection of spin currents at the sample's edges.
- The experimental setup must be able to measure the spin accumulation. This can be done using techniques like spin Hall magnetoresistance.
- Finally, the sample must be carefully prepared and characterized to ensure that any observed effects are not due to other factors, such as impurities or defects in the material.

7.4 QUANTUM SPIN HALL EFFECT

The quantum spin Hall effect (QSHE) is a phenomenon that occurs in some special two-dimensional (2D) materials with strong SOC strength, in which an electric current flows without losing energy due to the topological protection of edge states with opposite spin directions. It is characterized by the presence of two dissipationless edge channels that run along the edges of the material, one for each spin orientation. These edge channels can transport spin-polarized electrons without scattering, even in the presence of defects or impurities in the material.[17] It is a totally quantum phenomenon that results from the material's electronic band structure's topology. Originally, graphene was expected to show QSHE but due to weak SOC in carbon, it could not be observed.[18]

Figure 7.6 shows the schematic picture of insulators with different surface or edge states. For the case of QSHE, though a bulk band gap is present (see Figure 7.6 bottom), the spin-polarized edge states remain and carry the current of up-spin and down-spin electrons. QSHE arises when two edge states with opposite spins hybridize to form a single, spin-polarized state that is robust against back scattering. This means that the state cannot be scattered back into the bulk of the material by impurities or defects, leading to a highly conductive state resistant to dissipation.

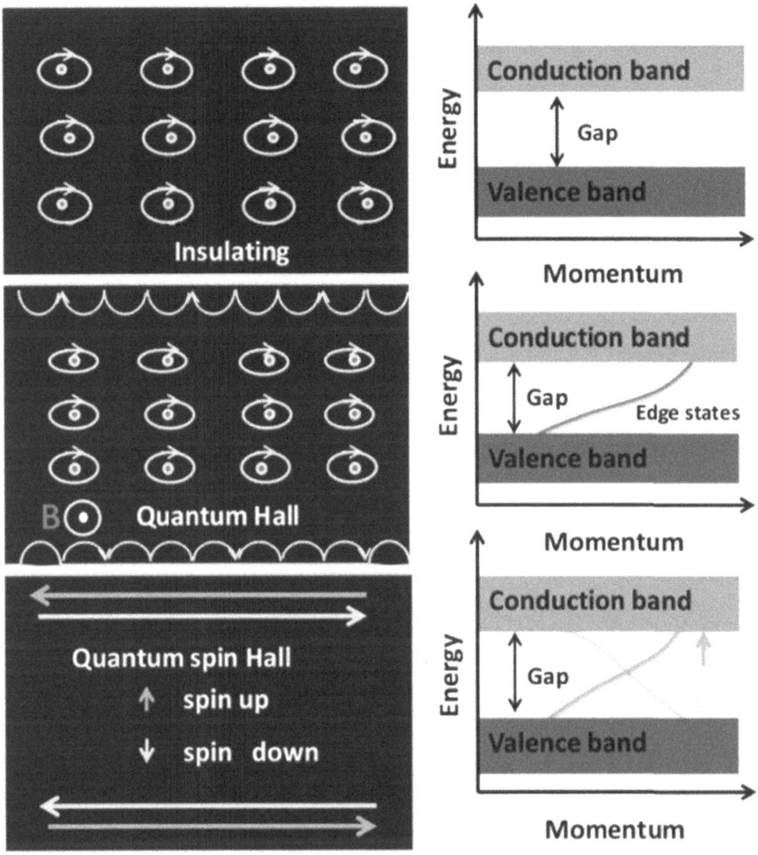

FIGURE 7.6 Schematic diagram showing three topologically distinct states.

QSHE are specifically noticed in three-dimensional topological insulators (TI) that host 2D surface state. A TI behaves as an insulator in its bulk but conducts electricity on its surfaces. For the first time, Bernevig et al. (2006) predicted QSHE in thin layer of semiconducting material HgTe/CdTe 2D-quantum wells[19] which was later confirmed by Konig et al. (2007) from experiment by observing the quantized conductance of $\frac{2e^2}{h}$ at 30 mK.[20]

Conditions to obtain the QSHE

- The material with nontrivial band topology is most suitable, such as TI which gives rise to gapless surface/edge modes.

- The material with strong SOC is preferable for the creation of spin-polarized edge modes and to split the bands for QSHE.
- The Fermi energy of the material should lie within the bulk bandgap of the electronic band structure to ensure that the edge modes are well-defined and do not overlap with the bulk electronic states.
- The sample should be designed in such a way that the edge modes can be detected. This can be done using a variety of techniques, such as measuring the conductance of the edge modes or using a magnetic sensor to detect the spin polarization at the edge modes.

This phenomenon has important implications for future electronics, spintronics, quantum computing as it provides a pathway for the development of low-power, high-speed electronic devices that are robust against disorder and other types of perturbations.

7.5 QUANTUM ANOMALOUS HALL EFFECT

Quantum anomalous Hall effect (QAHE) is defined as the quantized version of Hall effect identified in materials without any external magnetic field.[21] It is found to originate from the topological properties that occur in magnetic topological insulators (TI), in which a spin-polarized current flows without any charge current, leading to the quantization of Hall conductance. In magnetic TI, the presence of an intrinsic magnetic field from the doped atoms, the up- and down-spin bands in the electronic band structure can shift relative to each other, leading to a nonzero Chern number and the emergence of quantized conductance in Hall measurement. The Chern number, which is a topological invariant describes the relationship between the electronic band structure's topology and the Hall conductance quantization. This topological invariant is related to the number of times the electronic band structure wraps around the Brillouin zone in momentum space. In the case of the QAHE, the Chern number is quantized to an integer value leading to the Hall conductance quantization. F.D.M. Haldane first suggested the Haldane model, which realized quantized Hall conductance without magnetic field, in 1988.[9] It is the theoretical possibility to realize a QAH state. Following this, a novel QAH state was proposed by Rui et al. (2010) in Bi_2Se_3 by doping 3d magnetic atoms.[22] To realize QAHE, we need chern insulator (2D ferromagnetic insulator with nonzero chern number). When a TI thin film's time-reversal symmetry is broken by ferromagnetism, the QAHE is observed.[22,23] A 2D TI is expected to show QSHE in which each edge will only have one topologically protected spin channel so the helical edge states that cause QSHE will transform into chiral edge states that cause QAHE.[24] A bulk insulator, the 3D TI has 2D surface states on all of its surfaces.[25] In 3D TI, the gap of Dirac surface band is opened by ferromagnetism and makes it as a QH system. Interestingly, a ferromagnetic TI was found which is fully independent of the applied external magnetic field. Following this, after three years of theoretical prediction, the first experimental observation of QAHE was done by Chang et al.[26] wherein chromium is doped to the (Bi, Sb)$_2$Te$_3$ thin films to make magnetic TI. Further verification was done by doping 3d elements in (Bi, Sb)$_2$Te$_3$.[27–29]

In a magnetic TI, the intrinsic magnetization from the magnetic atom breaks the time-reversal symmetry which leads to the surface band gap opening in TI. When the magnetization from the doped atom is finite that it opens the surface band gap within the Fermi level, the electronic states in the material become topologically nontrivial, with a nonzero Chern number. This causes gapless edge states carrying a quantized spin Hall conductance (which is independent of the shape and size of the sample) at the Fermi level.

Conditions to obtain the QAHE

- The material should host the emergence of gapless, chiral edge states that are protected against backscattering, i.e., magnetic TI.
- The material should have a large magnetic exchange field induced either by doping the material with transition metal impurities or by breaking TRS with the use of magnetic field.
- Finite strength of SOC is essential to link the spin and electronic momentum for the formation of the topologically protected edge states.
- The magnetic ordering in the material should be uniform. Disorder or defects disrupt the magnetic ordering which can lead to the destruction of the QAHE.

Overall, QAHE is a fascinating phenomenon that arises due to the interplay between the magnetic exchange field, the SOC, and the topology of the electronic band structure in few materials. In QAHE, the edge states remain independent of the external bias or magnetic field. New hope on novel quantum phenomenon like magnetoelectric effect, majorana states etc. are arisen and extensively being studied after the realization of QAHE. At present, this topic is gaining interest in the condensed matter physics and materials science community due to possible application as dissipationless devices, spintronic devices, quantum computing, energy harvesting, etc.

Few Experimental Observations of QAHE

To achieve QAHE at higher temperature, various practices have been done after its first experimental observation on a thin sheet of magnetic TI made by doping chromium on (Bi, Sb)$_2$Te$_3$, at very low temperature <30 mK and large magnetic field.[26] Kou et al. (2014) investigated the QAHE from the thin sheet of $(Cr_{0.12}Bi_{0.26}Sb_{0.62})_2Te_3$ at low temperature experimentally.[28] Bestwick et al. (2015) reported an almost ideal QAHE in thin film of 3D TI after doping with ferromagnetic atoms.[30] Mogi et al. (2015) noticed the significant increase of QAHE temperature with magnetic doping in (Bi,Sb)$_2$Te$_3$ films.[31] Alloying is a successful strategy for enhancing the magnetic properties of magnetic metals, so this technique is applied to promote ferromagnetic coupling by codoping various magnetic impurities on diluted magnetic semiconductors to obtain QAHE at higher temperature due to improved ferromagnetic order and electronic structure.[32] Combination of magnetic doping and modulation doping methods can raise the QAH temperature.

7.6 SUMMARY

In this chapter, we give an overview of the development of several versions of Hall effects starting from 1879 until 2013, derive and give their important applications in materials science.

To briefly summarize, Hall effect is a phenomenon that describes the generation of an electric field perpendicular to both electric current and a magnetic field applied to a conducting material. There are several types of Hall effects, including the classical Hall effect, anomalous Hall effect, spin Hall effect, quantum Hall effect, quantum spin Hall effect, and quantum anomalous Hall effect.

The classical Hall effect was discovered by American physicist Edwin Hall in 1879. It was later observed by Klaus von Klitzing and colleagues in 1980 in the 2D electron gas system. Under the application of strong magnetic field the Hall resistance is found to be quantized, giving rise to quantum Hall effect. The spin Hall effect refers to the phenomenon in which an electric current flowing through a material generates a spin polarization in the transverse direction, perpendicular to the flow of current. The first spin Hall effect signal was detected experimentally in 2004 in material GaAs. Based on the proposal of Bernevig et al. (2006), quantum spin Hall effect was realized experimentally by Konig et al. (2007) in HgTe/CdTe (2D TI) quantum wells. This became a milestone to understand the topological features of matter with the invention of topological superconductors and insulators. Furthermore, following the prediction of quantum anomalous Hall effect by Yu et al. (2010) in magnetic doped topological insulators Bi_2Te_3, Bi_2Se_3, and Sb_2Te_3, the first experimental confirmation was possible by doping elements like Cr or Fe in Bi_2Te_3 samples by Chang et al. (2013). This novel discovery gave the paving path for dissipationless devices fabrication without external magnetic field.

Figure 7.7 depicts the schematic picture of several Hall effects. One can notice that an external magnetic field is required for classical Hall and quantum Hall effect, whereas intrinsic magnetization is necessary for anomalous Hall effect and quantum anomalous Hall effect. In a similar way, spin–orbit coupling is an essential parameter to tune spin Hall effect and quantum spin Hall effect.

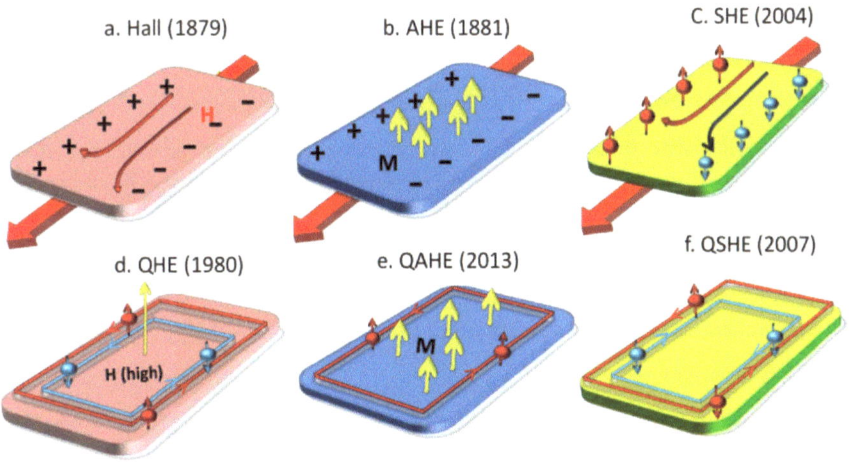

FIGURE 7.7 Schematic diagram representing various Hall effects with the discovery date. Here, H and M denote the magnetic field (external) and intrinsic magnetization respectively.

Use and applications: The classical Hall effect is commonly used to measure magnetic field strength and to determine the type of charge carrier in a material. Anomalous Hall effect is observed in materials with magnetic impurities, and is used to study the magnetic properties of these materials. The spin Hall effect is used in spintronics to generate and detect spin currents, while the quantum Hall effect is used to study the behavior of electrons in two-dimensional materials. Applications of Hall effects include current sensors, magnetic field sensors, and magnetic memory devices. With the ongoing development in Hall effects, it provides a pathway for low-power, high-speed electronic devices.

KEYWORDS

- **quantum Hall effect**
- **topological materials**
- **Landau levels**
- **spintronics**
- **dissipationless devices**

REFERENCES

1. Hall, E. H. On a New Action of the Magnet on Electric Currents. *Am. J. Math.* **1879,** *2* (3), 287–292.
2. Ashcroft, N. W.; Mermin, N. D. *Solid State Physics*; Cengage Learning, **2022**.
3. Klitzing, K. V.; Dorda, G.; Pepper, M. New Method for High-Accuracy Determination of the Fine-Structure Constant Based on Quantized Hall Resistance. *Phys. Rev. Lett.* **1980,** *45* (6), 494.
4. Von Klitzing, K. Developments in the Quantum Hall Effect. *Philos. Trans. R. Soc. A: Math., Phys. Eng. Sci.* **2005,** *363* (1834), 2203–2219.
5. Tsui, D. C.; Stormer, H. L.; Gossard, A. C. Two-Dimensional Magnetotransport in the Extreme Quantum Limit. *Phys. Rev. Lett.* **1982,** *48* (22), 1559.
6. Tang, F.; Ren, Y.; Wang, P.; Zhong, R.; Schneeloch, J.; Yang, S. A.; Zhang, L. Three-Dimensional Quantum Hall Effect and Metal–Insulator Transition in $ZrTe_5$. *Nature* **2019,** *569* (7757), 537–541.
7. Halperin, B. I. Possible States for a Three-Dimensional Electron Gas in a Strong Magnetic Field. *Jpn. J. Appl. Phys.* **1987,** *26* (S3–3), 1913.
8. Kohmoto, M.; Halperin, B. I.; Wu, Y. S. Diophantine Equation for the Three Dimensional Quantum Hall Effect. *Phys. Rev. B* **1992,** *45* (23), 13488.
9. Haldane, F. D. M. Model for a Quantum Hall Effect Without Landau Levels: Condensed-Matter Realization of the "Parity Anomaly". *Phys. Rev. Lett.* **1988,** *61* (18), 2015.
10. Stormer, H. L.; Tsui, D. C.; Gossard, A. C. The Fractional Quantum Hall Effect. *Rev. Mod. Phys.* **1999,** *71* (2), S298.
11. Eisenstein, J. P.; Stormer, H. L. The Fractional Quantum Hall Effect. *Science* **1990,** *248* (4962), 1510–1516.

12. Jain, J. K. Theory of the Fractional Quantum Hall Effect. *Phys. Rev. B* **1990,** *41* (11), 7653.
13. Dyakonov, M. I.; Perel, V. I. Current-Induced Spin Orientation of Electrons in Semi-Conductors. *Phys. Lett. A* **1971,** *35* (6), 459–460.
14. Hirsch, J. E. Spin Hall Effect. *Phys. Rev. Lett.* **1999,** *83* (9), 1834.
15. Kato, Y. K.; Myers, R. C.; Gossard, A. C.; Awschalom, D. D. Observation of the Spin Hall Effect in Semiconductors. *Science* **2004,** *306* (5703), 1910–1913.
16. Hoffmann, A. Spin Hall Effects in Metals. *IEEE Trans. Magnet.* **2013,** *49* (10), 5172–5193.
17. Maciejko, J.; Hughes, T. L.; Zhang, S. C. The Quantum Spin Hall Effect. *Annu. Rev. Condens. Matter Phys.*, **2011,** *2* (1), 31–53.
18. Yao, Y.; Ye, F.; Qi, X. L.; Zhang, S. C.; Fang, Z. Spin-Orbit Gap of Graphene: First-Principles Calculations. *Phys. Rev. B* **2007,** *75* (4), 041401.
19. Bernevig, B. A.; Hughes, T. L.; Zhang, S. C. Quantum Spin Hall Effect and Topological Phase Transition in HgTe Quantum Wells. *Science* **2006,** *314* (5806), 1757–1761.
20. König, M.; Wiedmann, S.; Brüne, C.; Roth, A.; Buhmann, H.; Molenkamp, L. W.; Zhang, S. C. Quantum Spin Hall Insulator State in HgTe Quantum Wells. *Science* **2007,** *318* (5851), 766–770.
21. Liu, C. X.; Zhang, S. C.; Qi, X. L. The Quantum Anomalous Hall Effect: Theory and Experiment. *Annu. Rev. Condens. Matter Phys.* **2016,** *7,* 301–321.
22. Yu, R.; Zhang, W.; Zhang, H. J.; Zhang, S. C.; Dai, X.; Fang, Z. Quantized Anomalous Hall Effect in Magnetic Topological Insulators. *Science* **2010,** *329* (5987), 61–64.
23. Nomura, K.; Nagaosa, N. Surface-Quantized Anomalous Hall Current and the Magnetoelectric Effect in Magnetically Disordered Topological Insulators. *Phys. Rev. Lett.* **2011,** *106* (16), 166802.
24. Liu, C. X.; Qi, X. L.; Dai, X.; Fang, Z.; Zhang, S. C. Quantum Anomalous Hall Effect in $Hg_{1-y}Mn_yTe$ Quantum Wells. *Phys. Rev. Lett.* **2008,** *101* (14), 146802.
25. Fu, L.; Kane, C. L.; Mele, E. J. Topological Insulators in Three Dimensions. *Phys. Rev. Lett.* **2007,** *98* (10), 106803.
26. Chang, C. Z.; Zhang, J.; Feng, X.; Shen, J.; Zhang, Z.; Guo, M.; Xue, Q. K. Experimental Observation of the Quantum Anomalous Hall Effect in a Magnetic Topological Insulator. *Science* **2013,** *340* (6129), 167–170.
27. Checkelsky, J. G.; Yoshimi, R.; Tsukazaki, A.; Takahashi, K. S.; Kozuka, Y.; Falson, J.; Tokura, Y. Trajectory of the Anomalous Hall Effect Towards the Quantized State in a Ferromagnetic Topological Insulator. *Nat. Phys.* **2014,** *10* (10), 731–736.
28. Kou, X.; Guo, S. T.; Fan, Y.; Pan, L.; Lang, M.; Jiang, Y.; Wang, K. L. Scale Invariant Quantum Anomalous Hall Effect in Magnetic Topological Insulators Beyond the Two-Dimensional Limit. *Phys. Rev. Lett.* **2014,** *113* (13), 137201.
29. Kandala, A.; Richardella, A.; Kempinger, S.; Liu, C. X.; Samarth, N. Giant Anisotropic Magnetoresistance in a Quantum Anomalous Hall Insulator. *Nat. Commun.* **2015,** *6* (1), 7434.
30. Bestwick, A. J.; Fox, E. J.; Kou, X.; Pan, L.; Wang, K. L.; Goldhaber-Gordon, D. Precise Quantization of the Anomalous Hall Effect Near Zero Magnetic Field. *Phys. Rev. Lett.* **2015,** *114* (18), 187201.
31. Mogi, M.; Yoshimi, R.; Tsukazaki, A.; Yasuda, K.; Kozuka, Y.; Takahashi, K. S.; Tokura, Y. Magnetic Modulation Doping in Topological Insulators Toward Higher-Temperature Quantum Anomalous Hall Effect. *Appl. Phys. Lett.* **2015,** *107* (18), 182401.
32. Andriotis, A. N.; Menon, M. Defect-Induced Magnetism: Codoping and a Prescription for Enhanced Magnetism. *Phys. Rev. B* **2013,** *87* (15), 155309.

PART III
Novel Materials for their Performances

CHAPTER 8

Transparent Conducting Oxides Thin Film

SHRUTI BAKSHI and SUMAN RANI

Department of Physics, School of Chemical Engineering and Physical Sciences, Lovely Professional University, Punjab, India

ABSTRACT

Transparent conducting oxides (TCO's) have less absorption of light in the range of visible spectrum and high electrical properties and transmission. In 1907, the first transparent conducting oxide, i.e., cadmium oxide was explained by Badekar. Transparent conducting oxides (TCO's) are a semiconductor with a broadband gap and a high concentration of free electrons in the conduction band. Their energy value is higher than that of visible light. Display devices are transparent and conducting material. For this study increased electron carrier concentration produces electromagnetic radiation absorption in both the visible and infrared sections of the spectrum with the former being more essential. Transparent conducting oxides have vast applications such as solar cells, low e windows, liquid crystals, touch screens, automobile windows, defogging, dye-sensitive solar cells, organic-related solar cells, photovoltaic cells, carbon nanotubes, random metallic networks, perovskites, and LED. Generally, it is present in n-type and enhances the ability of the material. Mostly utilized TCOs are ITO, zinc oxide, aluminum zinc oxide, and P-type films. In this chapter, we focus on the transparent conducting oxide thin film. They are mostly fabricated with thin-film technologies and are also utilized in the areas of optoelectronic devices such as display devices, solar cells.

Contemporary Advancements in Materials Technology: Novel Solutions to Complex Problems.
Shrikaant Kulkarni, Vipul Srivastava, & P. William (Eds.)

8.1 INTRODUCTION

Transparent conducting films (TCF) have been the focus of extensive research in recent years due to their widespread use in various optoelectronic applications, including touchscreens, solar cells, and displays. The latest research in TCF is focused on developing new materials and deposition techniques that can overcome the limitations of traditional TCF materials, such as indium tin oxide (ITO).

Graphene, a two-dimensional compound consisting of carbon atoms packed in a hexagonal lattice, is one of the excellent material for TCF applications. Graphene has excellent conductivity and transmission electrical conductivity, high optical transmittance, and mechanical flexibility, making it a potential alternative to ITO. Researchers are exploring various deposition techniques, including chemical vapor deposition and inkjet printing, to deposit graphene-based TCFs with high conductivity and transparency.

Another emerging TCF material is Metal Mesh, which consists of a pattern of thin metallic wires deposited onto a transparent substrate. Metal Mesh TCFs offer excellent conductivity and flexibility and can be deposited using low-cost techniques such as roll-to-roll printing. However, metal mesh TCFs have the disadvantage of reduced optical transmittance due to the presence of the metallic wires.

Researchers are also exploring hybrid TCF materials that combine the benefits of different TCF materials. For example, hybrid TCFs made up of ITO and silver nanowires (AgNWs) have been shown high conductivity and transmission while maintaining mechanical flexibility. Additionally, researchers are exploring the use of conductive polymers, for example, Poly(3,4-ethylenedioxythiophene) (PEDOT), as a TCF material due to their high conductivity and transparency.

In addition to developing new TCF materials, researchers are also investigating new deposition techniques, such as Aerosol Jet Printing and Electrohydrodynamic Printing, to deposit TCFs with high precision and uniformity. These techniques offer the potential to produce TCFs with high conductivity and transparency while reducing material waste and production costs.

The latest research in TCF is focused on developing new materials and deposition techniques that can overcome the limitations of traditional TCF materials, such as ITO. Graphene, Metal Mesh, and hybrid TCF materials are emerging as potential alternatives to ITO, while new deposition techniques offer the potential to produce TCFs with high precision and uniformity. These advances in TCF research will help to drive the achievements of new optoelectronic devices and technologies in the future.

Metal oxides [MO] having high level of transmission and electronic conductivity are known as transparent conducting oxides.[1] They are caused by either material defects or external dopants whose impurities exist near conduction band. A transparent conducting oxide TCO explained a bond between conduction of the material and transmission and also explains equilibrium to their properties.[2] They show wavelength lies in between near infrared and infrared. TCFs having band gap lie in between ultraviolet and they rarely absorb visible light and seem transparent to naked eye. In these materials having high transmission and electrical conductivity equal metals.[3] In 1907, the first transparent conducting oxide, i.e. cadmium oxide was explained by "Badekar."[4] In terms of practical development, very less amount of work has been done due to their low kind of transparency. After that some of the materials have been prepared such as oxides of zinc (ZnO), aluminum-doped zinc oxide (AZO), SnO_2. Out of this kind of material, the best film has been formed by using indium tin oxide (ITO).[5] But ITO is limited in range and quite costly. So it is necessary to find any another material for optoelectronic devices such as ZnO, ZnS, ZnSe, ZnTe SnO_2 TiO_2 CeO_2, and Ga_2O_3.

In a semiconductor material, the electrical conduction process takes place when electrons go from the valence band to the conduction band. Photons (higher in their energy level) required broad band gap. Hence, widening of band gap, i.e. (Eg > 3.0 Ev) can enhance the difference between the bands (such as valence band and conduction band) and decreases the excitation of electrons in conduction band.[6] In transparent conducting oxides, optical band gap of the material explains the transparency. For getting n-type conductivity, electrons are emitted from valence band to near conduction band.

Transparent conducting thin films (TCTFs) are thin layers of material that are capable of conducting electricity while remaining optically transparent. There are several materials that can be used to make TCTFs, for example, zinc oxide (ZO), fluorine-doped tin oxide (FTO), and indium tin oxide (ITO). ITO is currently the most widely utilized material because of its transmission and conductivity, but its supply is limited and it can be expensive.

The process of making TCTFs involves depositing a thin layer on the material having transmission and conductivity onto a substrate, for example, glass substrate or silicon substrate and quartz, using different method or technique such as sol gel, spray pyrolysis, sputtering, and chemical vapor deposition. The thickness of the film lies in between nanometers (nm) to micrometers (um), and its transmission and electrical conductivity can be tuned by focussing on the deposition conditions.

The main challenge in the development of TCTFs is finding a balance between high conductivity and high transparency. When the thickness of the film is increased to improve conductivity, it can become less transparent due to increased light scattering. Conversely, when the thickness is decreased to improve transparency, the conductivity may suffer.

Despite these challenges, TCTFs have found widespread use in modern electronics and are expected to play a key role in the development of emerging technologies, for example, flexible displays and wearable electronics. Researchers are actively exploring new materials and deposition techniques to improve the performance and cost-effectiveness of TCTFs, making them an exciting area of research in materials science and engineering. They have a variety of applications, including touch screens, solar cells, and flexible electronics. Below is a brief review of the literature related to TCFs.

Materials for TCFs: There are several materials that can be used to make TCFs, including carbon nanotubes, indium tin oxide (ITO), graphene, and nanowires of metals. Many studies have focused on comparing the properties of these materials, such as their conductivity, transparency, and stability.

Fabrication methods: There are also several methods that can be used to fabricate TCFs, including sputtering, chemical vapor deposition, and solution-based techniques. Studies have examined the properties of TCFs made using different fabrication methods, as well as the scalability and cost-effectiveness of these methods.

Performance optimization: Several studies have focused on optimizing the performance of TCFs for specific applications. For example, researchers have developed TCFs with improved stretchability for flexible electronics, and TCFs with improved light trapping for solar cells.

Stability and durability: The stability and durability of TCFs are important factors to consider, as they can affect the performance and lifespan of devices that use them. Many studies have examined the stability and durability of TCFs under different conditions, such as exposure to moisture, heat, and mechanical stress.

Transparent conductive films: There are several materials that can be used to make TCFs. Some of the most common materials for TCFs include:

Indium tin oxide: ITO is the widely utilized material for TCFs because of its excellent combination of optical transparency and electrical conductivity.

Zinc oxide: ZnO is a promising alternative to ITO because of its high conductivity and lower cost.

Aluminum-doped zinc oxide (AZO): AZO is a type of ZnO that has been doped with aluminum to increase its electrical conductivity.

Tin-doped indium oxide: This is a type of ITO that has been doped with tin to reduce the use of indium and lower the cost.

Graphene: It is a two-dimensional material with excellent conductivity and high optical transparency.

Carbon nanotubes (CNTs): CNTs are cylindrical structures made of carbon atoms and have high electrical conductivity and optical transparency.

Metal nanowires: Metal nanowires, such as silver and gold, can be used to make TCFs with high conductivity and transparency.

Conducting polymers: Conducting polymers, for example polyaniline and PEDOT:PSS, can be used to make TCFs with high flexibility and mechanical stability.

Overall, the choice of material for TCFs depends on the specific application requirements, such as transparency, conductivity, cost, and flexibility. Researchers are continuously exploring new materials and developing innovative techniques to improve the performance and lower the cost of TCFs.

8.1.1 INDIUM TIN OXIDE

ITO is an attractive material because it has a high transmission and conductivity lies in the visible spectrum. This means that it can conduct electricity while remaining optically transparent, which is essential for many electronic applications. In addition, it is also relatively stable, chemically inert, and can be easily deposited onto a substrate.

The most common method for depositing ITO is by sputtering, which involves bombarding a target made of the ITO material with high-energy ions. This causes the target to release ITO atoms, which then deposit onto the substrate. Other methods such as spray pyrolysis and chemical vapor deposition can also be used to deposit ITO.

ITO is an indium tin oxide well known TCO. It is basically a ternary composition of indium, tin, and varying compositions of oxygen. In ITOs, it is 74% weightage of indium, 18% weightage of Sn, and 8% weightage of oxygen. It is highly utilized transparent conducting oxide. When the thickness and concentration of charge carriers of ITOs are enhanced, then the conductivity of the film increases but transparency starts decreasing.

One of the major drawbacks of ITO is its limited availability and high cost due to the scarcity of indium, which is a rare earth element. In addition, ITO is brittle and can crack under mechanical stress, which limits its use in flexible electronic devices.

To address these issues, researchers are exploring alternative TCO materials such as aluminum-doped zinc oxide, zinc oxide, and fluorine-doped tin oxide, which offer comparable or even superior electrical and optical properties to ITO, while also being more cost-effective and flexible.

In conclusion, indium tin oxide is a well-established TCO material that has played a critical role in the achievements of modern electronic devices. However, due to its limited availability, high cost, and brittleness, researchers are actively exploring alternative TCO materials that can offer comparable or superior performance at a lower cost.

It is highly utilized in optical coating, organic LEDs, touch panels, solar cells, as an IR reflector for low-e-windows. It works as an associated contact layer in films of solar cells and heterojunction with intrinsic thin layer.

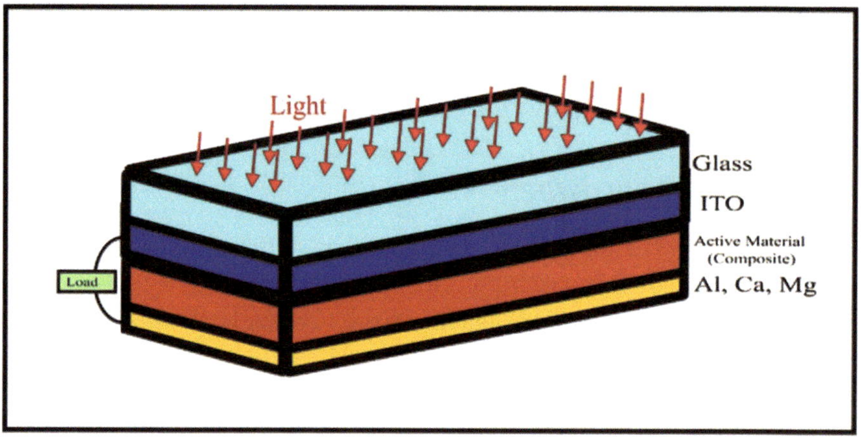

FIGURE 8.1 Use of ITO in solar cell.

8.1.2 ALUMINUM ZINC OXIDE

AZO is an aluminum-doped zinc oxide also well known as TCO material. Most preferable material after ITO is (AZO). For the formation of transparent conductive electrode for super substrate aluminum-doped zinc oxide, thin-film solar cells have been utilized. The layer of AZO is conductive and transparent to incoming light and allows the transportation of electrons. It requires a rough surface, which enhances current and efficiency of solar cells. The morphological properties of aluminum-doped zinc oxide depend upon its initial thickness. To get effective surface texturing properties in thin-film Si solar cell applications, the AZO layer must be thick enough (700 nm).[8]

Aluminum-doped zinc oxide is a transparent conducting oxide material that is a promising alternative to ITO. AZO is a ternary compound made up of zinc oxide (ZnO) and aluminum oxide (Al_2O_3), with a typical composition of 1–2% aluminum doping.

AZO is attractive as a TCO material because it has many advantages over ITO. Firstly, it is more cost-effective and abundant, as zinc and aluminum are readily available and inexpensive compared to indium. Secondly, AZO is more mechanically robust and flexible, making it suitable for use in flexible electronic devices. Thirdly, AZO can be deposited using less costly techniques such as spray pyrolysis, sol-gel, making it a more scalable and accessible technology.

AZO has a high electronic conductivity because of the presence of free electrons, which are created by doping with aluminum. The high optical transmittance of AZO is due to its wide bandgap, which allows light to pass through the material. The combination of these properties makes AZO an excellent material for use in optoelectronic devices, including solar cells and displays.

One of the challenges of using AZO as a TCO material is controlling its doping level, which can affect its electrical and optical properties. Additionally, AZO has a tendency to form a nonconducting Al_2O_3 layer on its surface, which can limit its conductivity. Researchers are actively exploring ways to overcome these challenges by optimizing the deposition process and developing new doping methods.

In conclusion, aluminium-doped zinc oxide (AZO) is a promising TCO material that offers several advantages over traditional TCO materials for example indium tin oxide (ITO). Its high transmission, conductivity, cost-effectiveness, and flexibility make it an attractive candidate for the use of optoelectronic applications.

8.2 ELECTRICAL AND OPTICAL PROPERTIES OF TRANSPARENT CONDUCTING MATERIALS

Transparent conducting oxides (TCOs) have been recognized as superior materials for the optoelectronic industry because of their tremendous combination of optical transparency and electrical conductivity. This synergistic balance of properties makes TCOs ideal for applications such as solar cells, touchscreens, and light-emitting diodes (LEDs). Indium tin oxide (ITO) is currently a leading TCO due to its remarkable transparency, exceeding 85%, and its high conductivity, ranging from ~10^3 to 10^4 $(\Omega{\cdot}cm)^{-1}$. But ITO is limited in range and quite expensive. So it is necessary to find another material for optoelectronic devices such as ZnO, ZnS, ZnSe, ZnTe, SnO_2, TiO_2, CeO_2, Ga_2O_3 etc. Table 8.1 shows the list of materials with optical and electrical properties of TCO materials.

TABLE 8.1 Electrical and Optical Properties of Transparent Conducting Materials.

S. no	Transparent conducting material	Substrate	Sintering temp. (oC)	Resistivity (Ω cm)	Transmission (%)	Optical Band gap (eV)	References
1.	ITO	Glass	300°C	4×10^{-4} to 2×10^{-4}	85%	3.5–4.3	9
2.	Ga_2O_3	Corning glass-squared substrate	300°, 350°C, 400°C	6×10^{-6}	100%	~4.9	10
3.	TiO_2	Glass	(400°C,450°C, 500°C, 550°C)	0.0802	70%	Indirect 3.43–3.04	11
4.	ZnO	Glass, Silicon	500°C	$\sim 10^{6}$	89.7%	3.11–3.18	12
5.	CeO_2	Glass	(300–500°C)	$3.44–6.06 \times 10^{5}$	70%	3.28–3.52	13
6.	SnO_2	Glass	400°C	2.5×10^{-3}	Greater than 80%	4.03	14
7.	ZnS	Glass	450°C	1.5×10^{5} $–20.84 \times 10^{5}$	35–75%	3.48–3.92	15
8.	AZO	Soda lime glass	650	$(2–4) \times 10^{-4}$	85 %	3.276 to 3.299	16
9.	Cu-Zn-S	Soda lime	370–470°C	$18.25–392.92 \times 10^{-2}$		–	17
10.	ZnO	Si and Glass	200–500°C	$10^{-2} – 10^{-5}$	70–85%	3.3	18
11.	ZnS	Glass	500°C	–	64 %	3.9	19
12.	Ag doped Cu-Zn-S	Glass, n, p type of silicon	280°C	0.66	66%	2.29–2.09	20
13.	ZnO	Glass	350°C	$1.92 \times 10^{+2}$	80%	2.71 eV to 3.04	21
14.	ZnO :Al	Glass	550°C	3.3×10^{-3}	80 to 90%	3.21 to 3.273	22
15.	ZnO	Quartz, Silicon	100–800°C	55	80 to 90%	3.5	23
16.	Gd :ZnO	Glass	100–400°C	$1.7–1.8 \times 10^{-3}$	82.8%	3.18	22
17.	Gd doped and ZnO :Al	Glass	–	8.4×10^{-3} $–10.6 \times 10^{-3}$	90%	3.38–3.58	24
18.	Indium doped ZnO	Glass	300–450°C	$(8–9) \times 10^{-4}$	85%	3.3	25

8.3 EXPERIMENTAL METHODS AND MATERIALS

The fabrication of transparent conducting films (TCFs) involves the creation of thin films that are both conductive and transparent. The fabrication process typically involves depositing a thin layer of conductive material onto a glass or different substrate using one of several deposition techniques. The choice of deposition technique and the type of material used can impact the performance of the resulting TCF.

Deposition techniques: There are many methods of deposition that can be utilized to form TCFs, including sputtering, chemical vapor deposition (CVD), solution-based method, for example, inkjet printing and spin-coating method. Sputtering is a popular technique for fabricating TCFs because it can produce films with high conductivity and good transparency. CVD is another technique that is commonly used, particularly for the deposition of graphene and other two-dimensional materials. Solution-based techniques are often used for fabricating flexible TCFs, which require a more flexible substrate.

Conductive materials: The choice of conductive material used in the fabrication of TCFs can impact the performance of the resulting film. ITO has traditionally been the famous conductive material for TCFs because of its transmission and conduction. The scarcity of indium and its high cost have led to the development of alternative materials, including graphene, silver nanowires, carbon nanotubes, and conductive polymers.

Substrates: The substrate used in the fabrication of TCFs can also impact the performance of the resulting film. Glass is a common substrate due to its high transparency, but some of the flexible substrates such as polyethylene naphthalate (PEN) and polyethylene terephthalate (PET) are becoming increasingly popular for flexible TCFs.

Postprocessing: After deposition, TCFs may undergo postprocessing steps to improve their performance. For example, annealing can be used to enhance the electrical conductivity of the film by decreasing defects in the material. Surface treatments such as plasma treatment can also be used to enhance the adhesion of the TCF to the substrate.

Overall, the fabrication of TCFs involves a complex process that requires careful selection of deposition technique, conductive material, substrate, and postprocessing steps. As new materials and techniques are developed, the fabrication process for TCFs is likely to continue to evolve and improve, leading to better performing and more versatile TCFs.

For the preparation and synthesis of nanoparticles, a distinct method is used such as solution method, plasma, electrochemical, spray pyrolysis, dip coating, vapor phase compression method, and mechanical alloying method. All the methods can produce large amount of nanomaterial, but sol gel shows more industrial properties as compared to other methods.

8.3.1 SOL-GEL METHOD

It is like a wet chemical method for producing distinct nanomaterials such as [MO] nanoparticles. In this process, chemical precursor undergoes reaction to form a sol-like structure which combines together to form a gel-like structure. This method is used for making gels, ceramics, and thin films. When it is present in their drying state, gel broke down into small granules. Material that we mostly used in this process are of inorganic configuration.

Various steps are followed in sol-gel method such as hydrolysis, condensation, aging, drying, and calcination. In this method, a silica solution is first prepared and then a gel formation process occurs. Due to the dispersion colloidal particles, a sol is formed. After that we obtained a dry gel which is made up of 3D silica network. In the process of drying remained solvent is removed. At the end of the process sintering, densification, growth mechanism is performed to increase its mechanical properties and stability of the structure.

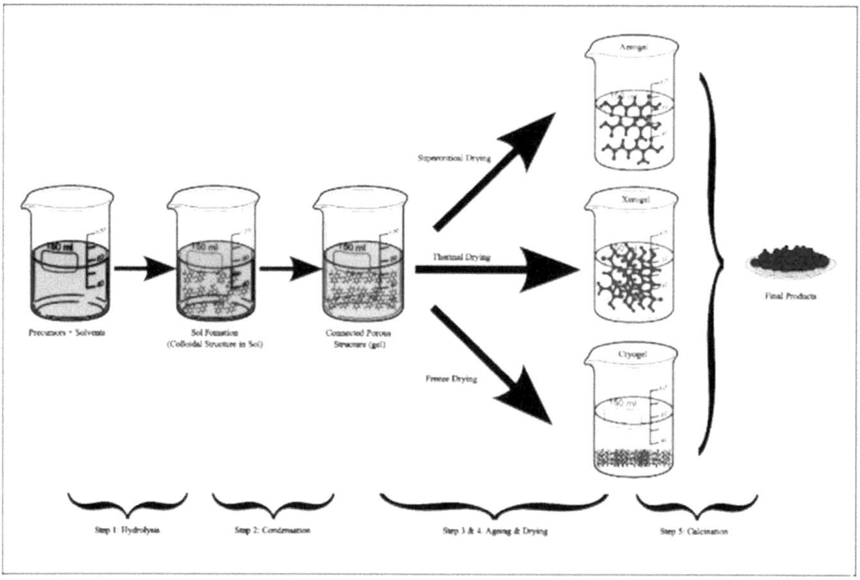

FIGURE 8.2 Sol-gel method.

8.3.2 SPIN COATING

It is a method that is utilized to make a uniformly distributed thin film on a glass substrate or silica substrate. In this procedure, a limited amount of prepared solution in a dropper is put into the center position of a glass substrate. The glass substrate rotates for a particular period of time such as 2000 rpm and acceleration 10 m/s^2 and the prepared solution uniformly spreads on a glass substrate with the help of a centrifugal force. The tool that is mostly used for spin coating is known as spin coater or normally called spinner. The rotation process continued until the prepared solution spread all over the glass substrate.

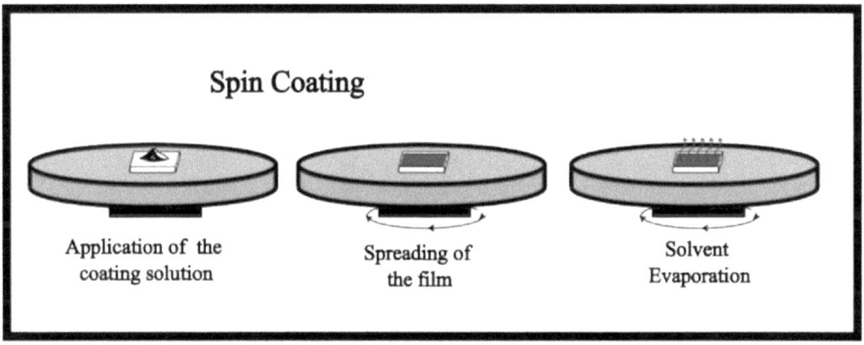

FIGURE 8.3 Spin coating process.

8.3.2.1 STEPS INVOLVED IN THE SPIN COATING PROCESS

1. The viscosity of the material must be balanced up to an optimum range so that the material on the glass substrate is seldom cohesive.
2. The drop of a material or solution should be poured perfectly at the center of the glass.
3. To avoid sprinkling of the solution, the rate of rotation should be increased.
4. The angular speed of a spin coater should be increased to get a satisfied thin film.
5. For getting a homogenous thin film, spinning speed and time must be high.
6. At a constant rate, the substrate should be spinned; otherwise, the solvent must be evaporated easily.

8.3.2.2 USES OF SPIN COATING

- In the industrial sector especially semiconductor industries, spin coating is highly utilized for the formation of thin films.
- It produces a very high level of thickness such as 10 nanometers.
- In this instrument steps are very easy.
- It is very reliable for the thin film.
- It is used in the process of photolithography.

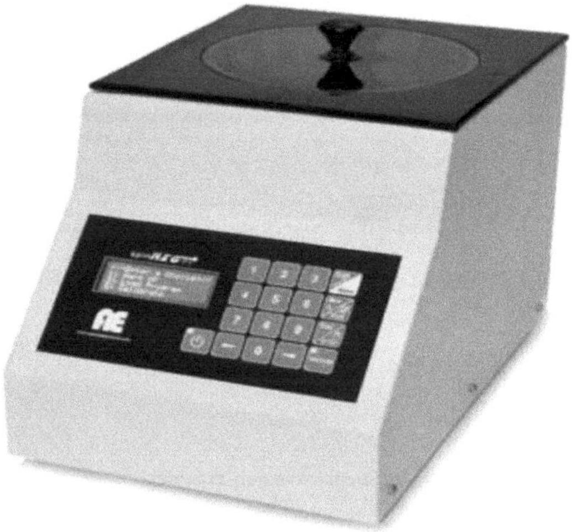

FIGURE 8.4 Experimental setup: diagram of spin coater.

8.3.3 DIP COATING

Dip coating is one of the prominent procedures for the creation of thin films. It is an oldest technique to prepare thin films. The basic flow is constant, thickness governed by the interaction between viscous, capillary (surface tension), and gravity forces. The thicker the layer deposited, the faster the substrate is extracted. The technology behind this process is based on slowly dipping a substrate into a solution. When the liquid part of the solution is dried, a glass substrate is left with a solid thin coating. In this method of preparing thin film, adhesion and cohesion forces in between the substrate play an important role.

In this method, a clean substrate (such as glass, silica, and quartz) is dipped into a prepared solution and then extracted over a continuous period of time at a constant speed under favorable conditions such as temperature and atmospheric conditions. Hence, the thin film is formed by this method which is amorphous in nature and used an annealing step to achieve crystallization.

8.3.4 SPRAY PYROLYSIS

In the spray pyrolysis method, precursor solution is very important. The atomic cloud aerosol produces large droplets because of ultrasonic spray method that analyzes small droplets. It affects their surface morphology.

Spray pyrolysis is extremely efficient, cost-effective, and easy-to-use. The thin films formed have a significant surface area of substrate covering and also have uniformity. Due to low quality value of thin film, spray pyrolysis is not that much useful. The vapors are produced because of temperature difference that prevents the source from binding to the substrate.

8.3.5 CHEMICAL VAPOR DEPOSITION (CVD)

Chemical vapor deposition is a process utilized in the formation of high-quality coatings and thin films. It is a technique in which a thin film is deposited on a surface of substrate introducing a reactive gas mixture into a reaction chamber at elevated temperatures and pressures.

During this process, the substrate is heated at a high temperature, typically in the range of 600 to 1200° Celsius, which causes the gas mixture to decompose and react on the surface of the substrate.

It can be deposited to a wide range of materials such as ceramics, metals, polymers, and semiconductors. The process is utilized in the semiconductor industry for the production of coatings and thin film utilized in microelectronics, optoelectronics, and photovoltaics.

One of the advantages of CVD is that it allows for precise control over the thickness and composition of the deposited film. The process can be tailored to produce films with specific properties, for example, high purity level, less defect density, and high uniformity.

There are several different types of CVD processes, including atmospheric-pressure, plasma-enhanced and low-pressure CVD (LPCVD). Each of these processes has its own advantages and disadvantages and is used for different applications.

Despite its many benefits, CVD has some limitations. The process can be expensive and requires specialized equipment, and the high temperatures used can limit the types of substrates that can be used. Additionally, the process can be time-consuming, and the deposition rate is generally slower than other techniques of deposition, such as electroplating or sputtering.

In conclusion, it is a powerful method for the deposition of high-quality thin films and coatings. It is widely used in the semiconductor industry and has many potential applications in other fields. With continued research and development, CVD is likely to play an increasingly important role in the production of advanced materials and devices in the future.

8.4 APPLICATIONS

Transparent conducting oxides have vast applications in terms of energy efficiency such as solar cells, liquid crystals, touch screens, automobile window deicing, defogging, dye-sensitive solar cells, low e windows, organic-related solar cells, photovoltaic cells, carbon nanotubes, random metallic networks, perovskites, and LEDs. Generally, it is present in n-type and enhances the ability of the material. Mostly utilized TCOs are ITO, ZnO, aluminum zinc oxide, and P-type films. They have diverse industrial applications. Some of the most common applications of TCO thin films include:

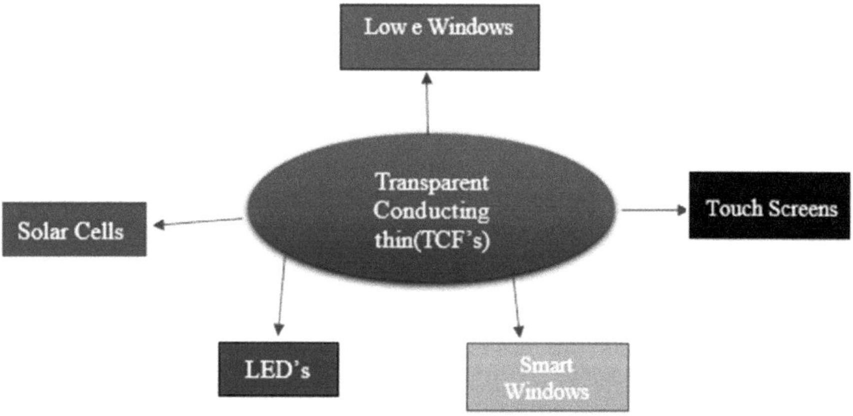

FIGURE 8.5 Application of TCFs.

- **Solar cells:** In terms of photovoltaic energy conversion, solar cell is a very essential device that produces electric energy from light energy. Thin films of TCO are utilized as front electrodes in solar cells. They allow light to pass through to the active layer while conducting the current generated by the photovoltaic effect. The most commonly used solar cell material is semiconductor as conversion process involves absorption of light which results in electron–hole pair and charge carrier separation. For the separation of charge carrier, p–n junction is widely used. The comprehensive knowledge of p–n junction gave you suggestion about how to improve solar cells in terms of their efficiency, production value etc.

- **LED:** In the light-emitting diode, when current passes through, it produces light. In the LED, when electrons recombine with electron–hole pair, it produces light in the form of photons. White light is produced by various semiconductor devices or layer of light-emitting phosphor on the semiconductor devices. It is used for TV back lighting, display devices, used in the automotives and dimming of lights. It utilized very less amount of energy. It required a low operational voltage. It does not require any warm-up time. TCO thin films are used as transparent electrodes in LED lighting. They improve the efficiency of light extraction by increasing the amount of light that can escape the LED chip.
- **Flat panel displays:** Thin films of TCOs are utilized as transparent electrodes in flat panel displays, such as LCD, OLED screens. They provide a uniform electric field and allow light to pass through without interfering with the display's color and contrast.
- **Electrochromic devices:** TCO thin films are used as the conducting layer in electrochromic devices. When a voltage is applied to these devices, they show the reversible change in the optical properties of a material such as the tinting of windows in response to sunlight.
- **Smart windows:** TCO thin films are used as the conductive layer in smart windows that can switch from transparent to opaque in response to an applied voltage. This technology is used in buildings to control the amount of sunlight and heat entering the building.
- **Touch screens:** In between 1965 and 1967, historians agree that the first touch screen was designed by E.A. In a 1968 publication, the inventor provided a detailed explanation of touch screen technology for air traffic management. It is basically a computer display screen that behaves as an input device. When a finger or Stylus touches a touch screen and it detects the events and sends it to a controller for processing, it shows images or words that the user can interact with by touching them. Registered touch screen depends upon three technologies such as capacitive, resistive, and surface acoustic wave.

TCO thin films are used as transparent electrodes in touchscreen displays of smartphones, tablets, and laptops. They allow for precise detection of touch and gestures while maintaining high optical transparency.

a. **Resistive:** It has a very thin metallic layer that is more conductive and resistive; due to this it causes a change in the electric current sent to the controller. It is much more affordable, not affected by dust and water.
b. **Surface acoustic waves (SAW):** In surface acoustic waves, ultrasonic waves pass through the screen. When touched the screen waves are absorbed; touch point is recorded and delivered to the controller.
c. **Capacitive:** An electrical charged material has been applied on this screen. When you touch the screen, the capacitance starts changing and location is to be identified and delivered to the controller.

- **Smart windows:** Most of the countries are trying to enhance their energy consumption. Few of them researchers focused on less power consumption. So the new advanced technologies are Smart Windows. In Smart Windows cutting-edge window technology is present which stops light from passing through them or easily pressing a knob. This kind of technology saves billion of dollar money on cooling and lightning. In the United States of America, researcher tells only about 2 percent of energy consumed on residential windows.
- **Low E windows:** Low E windows are widely spread in the market and it installed window treatment. It is used as a low transmission filter in which one or more than one semitransparent such as Ag or AI layer are placed in between transmissive dielectric layers such as TiO_2, Ta_2O_5, or transparent conducting oxide such as zinc oxide ZnO. It can be placed on flexible plastic and sandwiched between two glass pans. Transmissions such as visible and infrared depend upon the design of lower e windows and solar e coating. When the number of silver layers enhances reflection of heat increases and transmission decreases.

Overall, TCO thin films are essential in many applications that require both high electrical conductivity and optical transparency. Their versatility and unique properties make them valuable components in the development of next-generation electronic devices.

KEYWORDS

- **transmission**
- **conductivity**
- **transparent conductive oxides**
- **photovoltaic cells**
- **solar cells**
- **display devices**

REFERENCES

1. Granqvist, C. G.; Hultåker, A. Transparent and Conducting ITO Films: New Developments and Applications. *Thin Solid Films* **2002,** *411* (1), 1–5.
2. Coutts, T. J.; Young, D. L.; Li, X. Characterization of Transparent Conducting Oxides. *Mrs. Bull.* **2000,** *25* (8), 58–65.
3. Jayathilake, D.; Peiris, T. N. Overview on Transparent Conducting Oxides and State of the Art of Low-Cost Doped ZnO Systems. *SF J. Mater. Chem. Engl.* **2018,** (1), 1004.
4. Mattox, D. M.; Mattox, V. H.; Eds. *50 Years of Vacuum Coating Technology and the Growth of the Society of Vacuum Coaters*; Society of Vacuum Coaters, 2007.
5. Coutts, T. J.; Young, D. L.; Gessert, T. A. Modeling, Characterization, and Properties of Transparent Conducting Oxides. In *Handbook of Transparent Conductors*; Springer: Boston, MA. Coutts, 2011; pp 51–110.
6. Dixon, S. C.; Scanlon, D. O.; Carmalt, C. J.; Parkin, I. P. n-Type Doped Transparent Conducting Binary Oxides: An Overview. *J. Mater. Chem. C* **2016,** *4* (29), 6946–6961.
7. Horley, P. P.; Jiménez, L. L.; García, S. P.; Quintana, J. Á.; Vorobiev, Y. V.; Bon, R. R.; Hernández, J. G. Thin Film Solar Cells: Modeling, Obtaining and Applications. *Appl. Solar Energy* **2013,** *95*.
8. Yan, X.; Li, W.; Aberle, A. G.; Venkataraj, S. Textured AZO for Thin-Film Si Solar Cells: Towards Understanding the Effect of AZO Film Thickness on the Surface Texturing Properties. *Procedia Eng.* **2016,** *139*, 134–139.
9. Kim, H.; Gilmore, A. C.; Pique, A.; Horwitz, J. S.; Mattoussi, H.; Murata, H.; Chrisey, D. B. Electrical, Optical, and Structural Properties of Indium–Tin–Oxide Thin Films for Organic Light-Emitting Devices. *J. Appl. Phys.* **1999,** *86* (11), 6451–6461.
10. Juárez-Amador, L. I.; Galván-Arellano, M.; Andraca-Adame, J. A.; Romero-Paredes, G.; Kennedy-Magos, A.; Peña-Sierra, R. Electrical, Optical and Structural Characteristics of Gallium Oxide Thin Films Deposited by RF-Sputtering. *J. Mater. Sci.: Mater. Electron.* **2018,** *29* (18), 15726–15731.
11. Hanini, F.; Bouabellou, A.; Bouachiba, Y.; Kermiche, F.; Taabouche, A.; Hemissi, M.; Lakhdari, D. Structural, Optical and Electrical Properties of TiO2 Thin Films Synthesized by Sol–Gel Technique. *IOSR J. Eng. (IOSRJEN)* **2013,** *3* (11), 21–28.

12. Kennedy, J.; Murmu, P. P.; Leveneur, J.; Markwitz, A.; Futter, J. Controlling Preferred Orientation and Electrical Conductivity of Zinc Oxide Thin Films by Post Growth Annealing Treatment. *Appl. Surface Sci.* **2016,** *367*, 52–58.
13. Suresh, R.; Ponnuswamy, V.; Mariappan, R.; Kumar, N. S. Influence of Substrate Temperature on the Properties of CeO2 Thin Films by Simple Nebulizer Spray Pyrolysis Technique. *Ceram. Int.* **2014,** *40* (1), 437–445.
14. Abdelkrim, A.; Rahmane, S.; Abdelouahab, O.; Abdelmalek, N.; Brahim, G. Effect of Solution Concentration on the Structural, Optical and Electrical Properties of SnO2 Thin Films Prepared by Spray Pyrolysis. *Optik* **2016,** *127* (5), 2653–2658.
15. Derbali, A.; Attaf, A.; Saidi, H.; Benamra, H.; Nouadji, M.; Aida, M. S.; Ezzaouia, H. Investigation of Structural, Optical and Electrical Properties of ZnS Thin Films Prepared by Ultrasonic Spray Technique for Photovoltaic Applications. *Optik* **2018,** *154,* 286–293.
16. Jun, M. C.; Koh, J. H. Optical and Structural Properties of Al-Doped ZnO Thin Films by Sol Gel Process. *J. Nanosci. Nanotechnol.* **2013,** *13* (5), 3403–3407.
17. Emegha, J. O.; Olofinjana, B.; Ukhurebor, K. E.; Adegbite, J. T.; Eleruja, M. A. Electrical Properties of Semiconducting Copper Zinc Sulphide Thin Films. *Curr. Appl. Sci. Technol.* **2022,** *22,* 1–9.
18. Herissi, L.; Hadjeris, L.; Moualkia, H.; Abdelmalek, N.; Attaf, N.; Aida, M. S.; Bougdira, J. Realization and Study of ZnO Thin Films Intended for Optoelectronic Applications. *New Technol. Mater* **2011,** *1*, 39–43.
19. Balachander, M.; Saroja, M.; Venkatalachalam, M.; Kumar, V.; Shankar, S. Structural and Optical Properties of Zinc Sulfide Thin Film Prepared by Sol-Gel Spin Coating Method. *Int. J. Chem. Concepts* **2016,** *2* (2), 65–69.
20. Khalil, A. A. I.; Abd El-Gawad, A. S. H.; Gadallah, A. S. Impact of Silver Dopants on Structural, Morphological, Optical, and Electrical Properties of Copper-Zinc Sulfide Thin Films Prepared Via Sol-Gel Spin Coating Method. *Opt. Mater.* **2020,** *109*, 110250.
21. Amroun, M. N.; Salim, K.; Kacha, A. H. Molarities Effect on Structural Optical and Electrical Properties of Nanostructured Zinc Oxide Deposited by Spray Pyrolysis Technique. *Int. J. Thin. Fil. Sci. Tec.* **2021,** *10* (1), 67–73.
22. Ammaih, Y.; Lfakir, A.; Hartiti, B.; Ridah, A.; Thevenin, P.; Siadat, M. Structural, Optical and Electrical Properties of ZnO: Al Thin Films for Optoelectronic Applications. *Opt. Quantum Electron.* **2014,** *46*, 229–234.
23. Cho, S. Effect of Growth Temperature on Structural, Electrical, and Optical Properties of Gd-Doped Zinc Oxide Films. *Phys. Status Solidi (a)* **2014,** *211* (3), 709–713.
24. Lin, W.; Ma, R.; Shao, W.; Liu, B. Structural, Electrical and Optical Properties of Gd Doped and Undoped ZnO: Al (ZAO) Thin Films Prepared by RF Magnetron Sputtering. *Appl. Surface Sci.* **2007,** *253* (11), 5179–5183.
25. Major, S. A. K. L.; Banerjee, A.; Chopra, K. L. Highly Transparent and Conducting Indium-Doped Zinc Oxide Films by Spray Pyrolysis. *Thin Solid Films* **1983,** *108* (3), 333–340.

CHAPTER 9

Thermoelectric, Mechanical, and Structural Properties of Lead-Free Halide Double Perovskite $Cs_2AgFeCl_6$

Y. AL-DOURI[1,2,3], RIYADH A. AL-SAMARAI[4], and VIPUL SRIVASTAVA[5]

[1]*College of Health and Medical Techniques, Al-Bayan University, Baghdad, Iraq*

[2]*Department of Mechanical Engineering, Faculty of Engineering, Piri Reis University, EflatunSk, No: 8, 34940 Tuzla, Istanbul, Turkey*

[3]*Nanotechnology and Catalysis Research Centre, University of Malaya, 50603 Kuala Lumpur, Malaysia*

[4]*Electromechanical Engineering Department, Engineering College, University of Samarra, Samarra, Iraq*

[5]*School of Chemical Engineering & Physical Sciences, Lovely Professional University Phagwara, Punjab, India*

ABSTRACT

The studies of perovskite have received a lot of interest for multiapplication for advanced technologies. The thermoelectric, mechanical and structural properties of $Cs_2AgFeCl_6$ lead-free halide double perovskite have been investigated by employing Boltzmann transport scheme within spin-polarized full-potential linearized augmented plane wave (FP-LAPW) method. The potential of exchange-correlation utilizes within approximation of generalized gradient approximation (GGA) and modified Becke-Johnson (mBJ). The physical properties are elaborated. The ferromagnetic phase for cubic structure is recrystallized. The $Cs_2AgFeCl_6$ shows p-type ferromagnetic properties.

Contemporary Advancements in Materials Technology: Novel Solutions to Complex Problems.
Shrikaant Kulkarni, Vipul Srivastava, & P. William (Eds.)

The 5 μB has a contribution to Fe atom to have total magnetic moment. The $Cs_2AgFeCl_6$ has ductile and stable elastic properties. Also, the $Cs_2AgFeCl_6$ has thermoelectric properties at the range; 100–900 K using equations of Boltzmann transport. The $Cs_2AgFeCl_6$ is applicable for multi-field at low and high temperatures.

9.1 INTRODUCTION

$Cs_2AgFeCl_6$ is a type of lead-free halide double perovskite with potential applications in optoelectronics and solar cells. Understanding its thermoelectric, mechanical and structural properties is required for optimizing its performance and designing new devises. Structural properties refer to the atomic arrangement of crystal lattice of the material. $Cs_2AgFeCl_6$ has a cubic crystal structure with the space group Fm-3m, which means that its lattice has a face-centered cubic symmetry.[1,2] The unit cell of $Cs_2AgFeCl_6$ contains 32 atoms, including 2 cesium atoms, 2 silver atoms, 2 iron atoms, and 12 chlorine atoms. $Cs_2AgFeCl_6$ has a halide double perovskite crystal structure, which consists of alternating layers of corner-sharing octahedra of AgCl6 and FeCl6. The Cs+ ions are located between these layers and are coordinated by 8 Cl-ions. The crystal structure is highly symmetric, with a cubic symmetry, and the unit cell has a volume of approximately 546 $Å^3$. The crystal structure of $Cs_2AgFeCl_6$ is stable at room temperature and has been characterized using techniques such as X-ray diffraction, Raman spectroscopy, and transmission electron microscopy.[3–5] the materials response to external forces indicates for mechanical properties or stresses. $Cs_2AgFeCl_6$ has a high bulk modulus and shear modulus, which means that it is relatively stiff and resistant to deformation. Its Young's modulus, which describes its resistance to stretching or compression, is also relatively high. These mechanical properties make $Cs_2AgFeCl_6$ a promising material for use in mechanical applications such as sensors and actuators.[6,7] The mechanical properties of $Cs_2AgFeCl_6$ have been studied using computational techniques like density functional theory (DFT). These studies suggest that the material has a high bulk modulus (over 100 GPa) and a high shear modulus (over 50 GPa), which indicates that it is relatively stiff and resistant to deformation. Its Young's modulus has been estimated to be around 60 GPa, which is comparable to other perovskite materials. These mechanical properties make $Cs_2AgFeCl_6$ a potential candidate for use in mechanical applications that require high stiffness and resistance to deformation, such as sensors and actuators.[8–10]

Thermoelectric properties refer to materials ability for converting thermal to electrical energies and vice versa. $Cs_2AgFeCl_6$ has a high Seebeck coefficient, which is a measure of its ability to generate a voltage in response to a temperature difference. This is applicable for using in thermoelectric generators, which can convert waste heat into electricity. The $Cs_2AgFeCl_6$ has thermoelectric properties, which has been investigated experimentally using measurements of thermal conductivity, electrical conductivity and Seebeck coefficient[11] The material has a high Seebeck coefficient, which indicates that it can generate a large voltage in response to a temperature difference. The $Cs_2AgFeCl_6$ has relatively low electrical conductivity, which can help to minimize heat loss due to electrical resistance. The thermal conductivity of $Cs_2AgFeCl_6$ is also low, which can help to maintain a large temperature gradient across the material. These properties make $Cs_2AgFeCl_6$ a potential candidate for use in thermoelectric generators that can convert waste heat into electricity.[12] The thermal stability of $Cs_2AgFeCl_6$ is an important property for its potential use in high-temperature applications. The material has been found to be stable up to temperatures of around 250°C, after which it undergoes a phase transition to a different crystal structure. The thermal stability of $Cs_2AgFeCl_6$ has been characterized using thermogravimetric analysis (TGA) and differential scanning calorimetry (DSC). The environmental stability of $Cs_2AgFeCl_6$ is an important consideration for its use in practical applications. The material is relatively stable in air and water, but can be degraded by exposure to light. The degradation of $Cs_2AgFeCl_6$ has been studied using techniques such as X-ray diffraction and photoluminescence spectroscopy. The electronic properties of $Cs_2AgFeCl_6$ have been studied using a combination of experimental and theoretical methods. The material has been found to have a direct band gap of around 2.3 eV, which makes it a potential candidate for use in optoelectronic applications such as photovoltaics and light-emitting diodes. The $Cs_2AgFeCl_6$ has analyzed its electronic structure using density functional theory (DFT) calculations and X-ray photoelectron spectroscopy (XPS).[13–15]

In summary, $Cs_2AgFeCl_6$ is a lead-free halide double perovskite with a cubic crystal structure and promising mechanical and thermoelectric properties. Its unique combination of properties makes it a potential candidate for use in a range of applications, including optoelectronics, solar cells, sensors, and thermoelectric generators. Overall, $Cs_2AgFeCl_6$ is a promising material for a range of applications, including optoelectronics, solar cells, sensors, and thermoelectric generators. Further research is needed to fully understand and optimize its properties for these applications.

9.2 TECHNIQUES OF CALCULATIONS

9.2.1 STRUCTURAL PROPERTIES

The structural properties of $Cs_2AgFeCl_6$ can be calculated using techniques such as density functional theory (DFT) calculations, X-ray diffraction (XRD), Raman spectroscopy, and transmission electron microscopy (TEM). DFT calculations can provide information on the electronic structure, crystal structure, and lattice vibrations of the material. XRD has been utilized to analyze the structural feature and lattice parameters of the material. Raman spectroscopy can be used to study the lattice vibrations and symmetry of the material. TEM can be used to obtain high-resolution images of the crystal structure and defects in the material.[16]

9.2.2 MECHANICAL PROPERTIES

The mechanical properties of $Cs_2AgFeCl_6$ can be calculated using techniques such as DFT calculations and molecular dynamics (MD) simulations. DFT calculations can be used to calculate the elastic constants, bulk modulus, and shear modulus of the material. MD simulations can be used to study the mechanical behavior of the material at the atomic level, including the deformation and fracture of the material under stress.[17]

9.2.3 THERMOELECTRIC PROPERTIES

The thermoelectric properties of $Cs_2AgFeCl_6$ can be calculated using techniques such as DFT calculations and Boltzmann transport calculations. DFT calculations can be used to calculate the electronic structure and transport properties of the material, including the electronic thermal conductivity, electrical conductivity and Seebeck coefficient. Boltzmann transport calculations can be used to calculate the thermoelectric properties of (ZT) for the material, which is a measure of its thermoelectric efficiency. The thermal properties of $Cs_2AgFeCl_6$ can be studied using techniques like thermogravimetric analysis (TGA) and differential scanning calorimetry (DSC). DSC is to measure the specific heat and heat flow in terms of temperature. TGA is to measure the weight loss of the material as a function of temperature, which can provide information on the thermal stability and decomposition of the material.[18]

9.2.4 ELECTRONIC PROPERTIES

The electronic properties of $Cs_2AgFeCl_6$ can be studied using a combination of experimental and theoretical techniques. For example, X-ray photoelectron spectroscopy (XPS) can be used to study the chemical composition and electronic structure of the material. UV-Vis spectroscopy can be used to measure the optical absorption and photoluminescence properties of the material. DFT calculations can be used to calculate the electronic structure of the material, including the density of states, band structure, in addition to optical properties such as the absorption spectrum.[19]

9.2.5 THERMAL PROPERTIES

The thermal properties of $Cs_2AgFeCl_6$ can be studied using thermogravimetric analysis (TGA) and differential scanning calorimetry (DSC). DSC can be used to measure the heat flow and specific heat of the material as a function of temperature. TGA is to measure the material weight loss as a function of temperature, which can provide information on the thermal stability and decomposition of the material.[20]

Environmental stability

The environmental stability of $Cs_2AgFeCl_6$ can be studied using techniques such as Fourier transform infrared (FTIR) spectroscopy, X-ray diffraction (XRD) and photoluminescence spectroscopy. XRD can be used to monitor changes in the crystal structure of the material as a function of time and exposure to different environmental conditions. Photoluminescence spectroscopy and FTIR spectroscopy can be used to measure changes in the optical and vibrational properties of the material as a function of environmental conditions.[21]

Overall, a combination of theoretical calculations and experimental techniques can be used to study the properties of $Cs_2AgFeCl_6$ and optimize its properties for practical applications. In summary, a range of experimental and theoretical techniques can be used to study the properties of $Cs_2AgFeCl_6$, including electronic, optical, thermal, and environmental properties. These techniques can provide valuable information on the properties of the material and can be used to optimize its properties for a range of practical applications.[22]

9.3 STRUCTURAL STUDIES

The structural studies of lead-free halide double perovskite $Cs_2AgFeCl_6$ involve analyzing the crystal structure of the material using experimental techniques such as X-ray diffraction (XRD) and neutron diffraction, as well as theoretical calculations such as density functional theory (DFT). XRD is a powerful technique that is commonly used for determining the structural properties of materials. XRD measurements involve passing X-rays through a crystal and measuring the angles and intensities of the diffracted X-rays. The diffraction pattern provides information on the arrangement of atoms in the crystal lattice, including the lattice parameters, unit cell volume, and symmetry of the crystal. XRD measurements can be used to determine the crystal structure of $Cs_2AgFeCl_6$, which is a cubic double perovskite with a space group of Fm-3m.[23]

Neutron diffraction is another powerful technique that can be used to study the crystal structure of materials, particularly those that contain light elements such as hydrogen. Neutron diffraction involves passing a beam of neutrons through a crystal and measuring the positions and intensities of the scattered neutrons. The diffraction pattern provides information on the positions of atoms in the crystal lattice, including the positions of light elements such as hydrogen. Neutron diffraction measurements can be used to refine the crystal structure of $Cs_2AgFeCl_6$ and obtain more accurate atomic positions.[24]

DFT calculations are a theoretical technique that can be used to calculate the electronic structure and crystal structure of materials. DFT calculations involve solving the Schrödinger equation for the electrons in the crystal lattice, using an approximation for the exchange-correlation energy of the electrons. DFT is utilized for predicting the structure of crystal for $Cs_2AgFeCl_6$, as well as calculate the lattice parameters, bond lengths, and angles in the crystal lattice. DFT calculations can also provide insight into the electronic structure of the material, including the band structure, density of states, and electronic properties such as the dielectric constant and polarizability. Overall, the structural studies of lead-free halide double perovskite $Cs_2AgFeCl_6$ involve a combination of experimental and theoretical techniques to determine the crystal structure and analyze its properties. The information obtained can be utilized for optimizing the properties of the material for practical applications, such as in optoelectronics and photovoltaics.[25–31]

The $Cs_2AgFeCl_6$ has accurate geometry and ground-state properties of B and B sites at different sub-lattice positions are achieved by optimized crystal structure. Basing on convention of Hermann-Maguin, the double perovskite $Cs_2AgFeCl_6$ of lead-free has cubic structure as shown in Figure 9.1 for space group (Fm-3m) (225). The atomic positions for Cl are 24e (0.25, 0, 0), for Fe are 4b (0.5, 0.5, 0.5), for Ag are 4a (0, 0, 0) and for Cs are 8c (0.25, 0.25, 0.25).[32] The Birch-Murnaghan's equation of state[33] is employed to compute total energy/ unit cell at different volumes of optimized crystal's geometry and structure. For GGA, the minimum ground state energy is configured by unit cell structure into ferromagnetic (spin-polarized) and non-magnetic (no spin-polarization) configurations. The energy corresponds to ferromagnetic phase lower than for non-magnetic phase as illustrated in Figure 9.2.

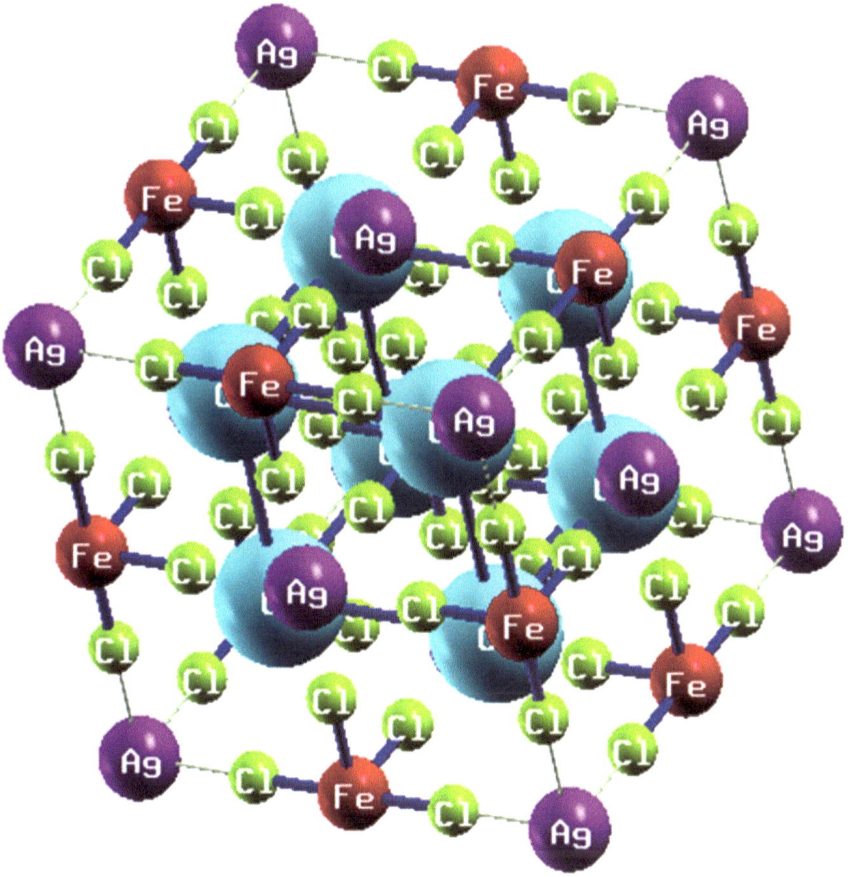

FIGURE 9.1 Crystal structure of double perovskite $Cs_2AgFeCl_6$.

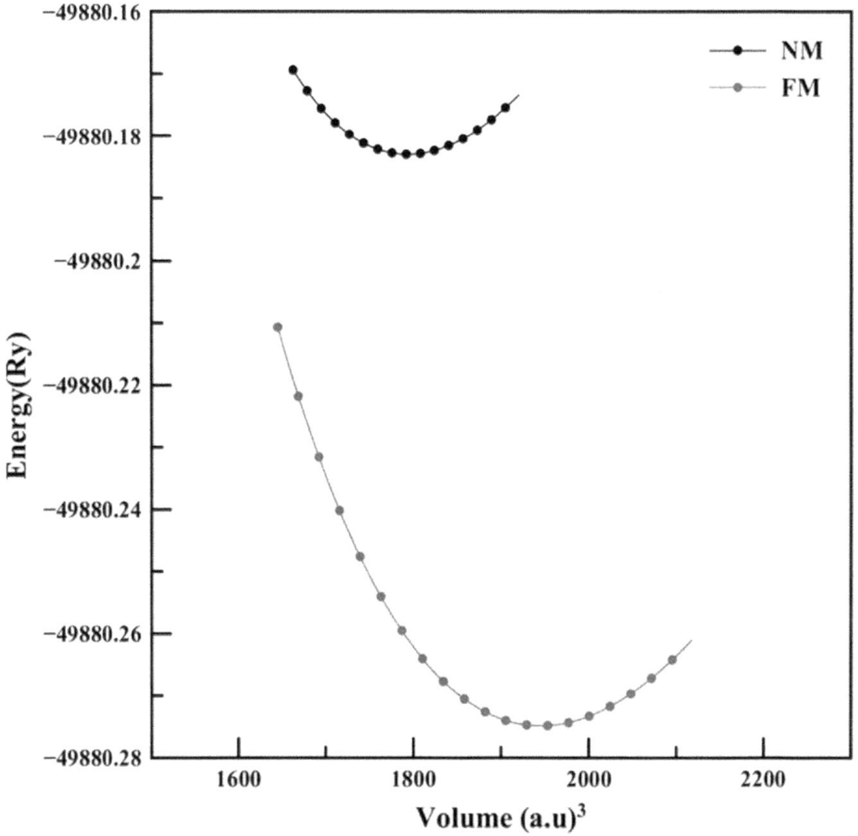

FIGURE 9.2 Total energy as a function of volume for $Cs_2AgFeCl_6$ double perovskite compound in ferromagnetic (FM) and non-magnetic (NM) phase.

Table 9.1 presents specific parameters of minimum energy (Ry), bulk modulus (B), pressure derivative of bulk modulus (B') and lattice constant (Å). It is seen that the calculated lattice constant of $Cs_2AgFeCl_6$ agrees with experimental value.[20] Anyway, our results forecast and support future works due to not available experimental or theoretical data in the literature for assessment. In addition, for perovskite, 0.25 is a deviated value of computed oxygen parameter x from an ideal value. The computed Goldschmidt tolerance factor () to determine the stability of cube is presented in Table 9.1.

$$\tau = 0.707 \frac{r_{Sr} + r_O}{\left(\frac{r_{Co} + r_{Ti}}{2}\right) + r_O} \tag{9.1}$$

where Caesium atom ionic radius refers to rCs, Chlorine atom ionic radius refers to rCl, silver atom ionic radius refers to rAg, and Iron atom ionic radius refers to rFe. Generally, for perovskites, the ranged value is 0.9–1.0, if $t < 0.9$ and/or $t > 1.0$ is observed for any cubic structure deviation.[34] The computed tolerance factor value of $Cs_2AgFeCl_6$ is 0.92, basing on effective ionic radii table of Shannon.[35]

TABLE 9.1 Structural Parameters, Tolerance Factor (τ), and Enthalpy of Formation (ΔH) of the $Cs_2AgFeCl_6$ Compound within Fm-3m Space Group Configurations.

	Present work	**Other**
Method	GGA	Exp
a(Å)	10.4848	10.2023 [20]
V $(a.u)^3$	1944.5413	1061.93 [20]
B (GPa)	30.5343	–
B'	4.1805	–
E_0: Minimum total energy per unit cell (Ry)	–49880.274869	–
Tolerance factor (τ)	0.92	–
ΔH (meV)	–1.12	–

To confirm the stability of thermodynamics, the formation enthalpy (*H*) is computed for double perovskite $Cs_2AgFeCl_6$ compound utilizing:

$$\Delta H = E_{tot}\left(Cs_2AgFeCl_6\right) - 2E_{Cs} - E_{Ag} - E_{Fe} - 6E_{Cl} \tag{9.2}$$

where the total energy refers to Etot $Cs_2AgFeCl_6$ and equivalent values of energy for the mentioned six atoms in their competent intrinsic lattices refer to ECs, EAg, EFe and ECl. Table 9.1 displays estimated formation energy to be 1.12 meV for illustrating an energetic stable.

9.4 ELASTIC AND MECHANICAL STABILITY

The mechanical and elastic stability of lead-free halide double perovskite $Cs_2AgFeCl_6$ can be analyzed using a range of experimental and theoretical techniques. The mechanical properties of materials refer to their resistance to deformation or failure under applied stresses or strains, while the elastic properties refer to the material's ability for returning after being deformed to its original shape.

Experimental techniques to measure mechanical properties including nanoindentation and mechanical testing, such as compression, tension, and bending tests. These tests involve applying a controlled stress or strain to the material and measuring its response, such as the deformation or failure mode. The elastic properties of materials can be determined using techniques such as ultrasonic measurements, Brillouin scattering, or resonant ultrasound spectroscopy.

Theoretical calculations can also be used to predict the mechanical and elastic properties of materials. Density functional theory (DFT) calculations can be used to calculate the elastic constants of the crystal lattice, which describe the relationship between the stress and strain of the material. These elastic constants can be used to calculate other mechanical properties of the material, such as its Poisson's ratio, shear modulus and bulk modulus. Molecular dynamics simulations can also be used to simulate the mechanical behavior of materials under applied stresses or strains.

Studies on the mechanical and elastic stability of $Cs_2AgFeCl_6$ have shown that the material exhibits good mechanical properties, with a high hardness and modulus of elasticity. The elastic constants of the crystal lattice have been calculated using DFT calculations, and the results indicate that $Cs_2AgFeCl_6$ has a high bulk modulus, which suggests good mechanical stability. These are suggested that $Cs_2AgFeCl_6$ is applicable to multi-field requiring good mechanical and elastic properties, such as in electronic devices and structural materials.

The elastic properties studies are crucial in understanding restoration capability of material's structure undergone to distortion pressure. They are close and matching to elastic properties, which have wide spectrum of engineering and industrial applications. The WEIN2K software package is utilized[36] for computing elastic properties Cij (C11, C12, C44) for investigating mechanical studies for double perovskite $Cs_2AgFeCl_6$.

$$C_{11} > 0;\ C_{11}\text{-}C_{12} > 0;\ C_{44} > 0;\ C_{11} + 2C_{12} > 0.$$ [37,38]

Born has referred to their importance of mechanical stability. Table 9.2 presents the elastic constants. Therefore, it can be said that the mentioned materials are stabled elastically at room temperature and pressure condition. Moreover, the mentioned elastic properties are related to the mechanical properties utilizing;[39,40]

$$E = \frac{9BG}{3B+G} \tag{9.3}$$

$$B = \frac{C_{11} + 2C_{12}}{3} \tag{9.4}$$

$$v = \frac{3B - 2G}{2(3B + G)} \tag{9.5}$$

$$A = \frac{2C_{44}}{C_{11} - C_{12}} \tag{9.6}$$

where E refers to Young's moduli and B refers to Bulk's moduli, refers to Poisson's ratio and A refers to Zener's Anisotropy factor as displayed in Table 9.2. The B value is weak, resulting in deformation of volumetry under external stress application. This deformation illustrates the compressibility of alloy.

TABLE 9.2 Elastic Constants C_{11}, C_{12}, C_{44} in (GPa), Bulk Modulus B (GPa), Shear Modulus G (GPa), Young's Modulus E (GPa), Poisson's Ratio v, Zener Anisotropy Factor A, Pugh's Ratio (B/G), Cauchy's Pressure (C_{12}-C_{44}) and Melting Temperature T_m (K) for $Cs_2AgFeCl_6$.

Alloy	C_{11}	C_{12}	C_{44}	B	G	E	v	B/G	C_{12}-C_{44}	A	T_m
$Cs_2AgFeCl_6$	34.8	16.86	11.71	22.84	10.53	27.37	0.3	2.17	5.15	1.3s	758.7067

The E represents a material stiffness, its value is higher than B to suggest a compressive force of resist uniaxial[41] There is importance for Poisson's ratio (υ). Regarding to Frantsevich's rule,[42] it has a nature of ductility with a value bigger than 0.26; if it is small, it will be a brittle. Table 9.2 presents υ equals 0.3 to its ductility. The Pugh's ratio is an additional criterion for determining its ductility or brittle character,[43] it equals 1.75. If it is bigger than 1.75, it has a ductility nature; if contrary, it has a brittle character. Table 9.2 displays B/G to be 2.17 referring to the mentioned material is ductile.

The (C_{12}–C_{44}) relates to Cauchy's pressure[44] for indicating to material's brittleness and ductility. If the value is negative, it is brittle, otherwise, it is ductile. Table 9.2 reports clearly the ductility nature. The Zener's anisotropy factor (A) shows the degree of solid's elastic anisotropy, which has 1 for isotropic character. While, the anisotropy has a divergence from unity. Table 9.2 indicates that the materials are extremely anisotropic. We have computed a melting temperature utilizing longitudinal elastic constant C_{11} in:[45]

$$T_m(K) = [553(K) + (5.911)C_{11}]\text{GPa} \pm 300\text{K} \tag{9.7}$$

The observed values are 758.7067 ± 300 K that is meaning it can be hard-done by different applications for a wide range of temperature. Therefore, it is not available yet any data of elastic properties in the literature. Now, our investigations of elastic and mechanical properties have actually different industrial and engineering applications.

9.5 MAGNETIC AND ELECTRONIC STUDIES

The electronic and magnetic properties of lead-free halide double perovskite $Cs_2AgFeCl_6$ can be analyzed using a range of experimental and theoretical techniques. These properties are important for understanding the material's behavior in electronic and magnetic devices and can be used to optimize the material's properties for specific applications. Experimental techniques for measuring the electronic properties of materials include photoemission spectroscopy, electrical conductivity measurements, and optical spectroscopy. These techniques can provide information on the material's band structure, density of states, and electronic properties such as the dielectric constant and polarizability. Magnetic properties of materials can be measured using techniques such as magnetometry, which measures the response of a material to an applied magnetic field.

Theoretical calculations can also be used to predict the electronic and magnetic properties of materials. Density functional theory (DFT) calculations can be used to calculate the band structure of the material, which describes the energy levels of the electrons in the crystal lattice. These calculations can also predict the magnetic properties of the material, such as its magnetic moment and magnetic ordering. Studies on the electronic and magnetic properties of $Cs_2AgFeCl_6$ have shown that the material has a narrow bandgap and to make a promising candidate, it is required for dielectric constant for using in photovoltaic devices. The magnetic properties of the material have also been studied, and it has been found to exhibit antiferromagnetic ordering at low temperatures. The magnetic moment of $Cs_2AgFeCl_6$ has been calculated using DFT calculations, and the results suggest that the material has a low magnetic moment, which indicates weak magnetic interactions between the Fe ions. Overall, the electronic and magnetic properties of $Cs_2AgFeCl_6$ can be analyzed using a combination of experimental and theoretical techniques, and the information obtained of the mentioned studies to be used to optimize the properties of the material for practical applications.

The electronic band structure is necessary for characterizing any material as insulator, semiconductor, metal, or semimetal. So, the spin-polarized calculation is used for electronic band structure of double perovskite $Cs_2AgFeCl_6$ of lead-free for realizing its potential application as illustrated in Figure 9.3. The GGA is ineffective due to d-state electrons existence. So, mBJ potential is included for establishing more accurate band structure. The Iron (Fe) atom is available in our double perovskite and its lattice structure. For the effect of Jahn-Teller (J-T), crystal field theory, Hund's rule and Pauli

Exclusion Principle, the orbital of Fe-d is divided into t_{2g} and eg orbitals, which could split into d_{xy}, d_{yz}, d_{zx} and $dx^2–y^2$, dz^2.[28] The t_{2g} states being are triply degenerate and eg states are doubly degenerate. The t_{2g} and eg states are partitioned by splitting the gap:

$$\Delta E_{\mathrm{CF}} = E_{\mathrm{eg}} - E_{\mathrm{t2g}} \tag{9.8}$$

The *d* orbital constitutes of electronic band profile of perovskite structure. For both GGA+mBJ and GGA, the band structure high symmetry for Brillouin zone (BZ) in spin-up and spin-down states is suggesting a behavior of semiconductor.

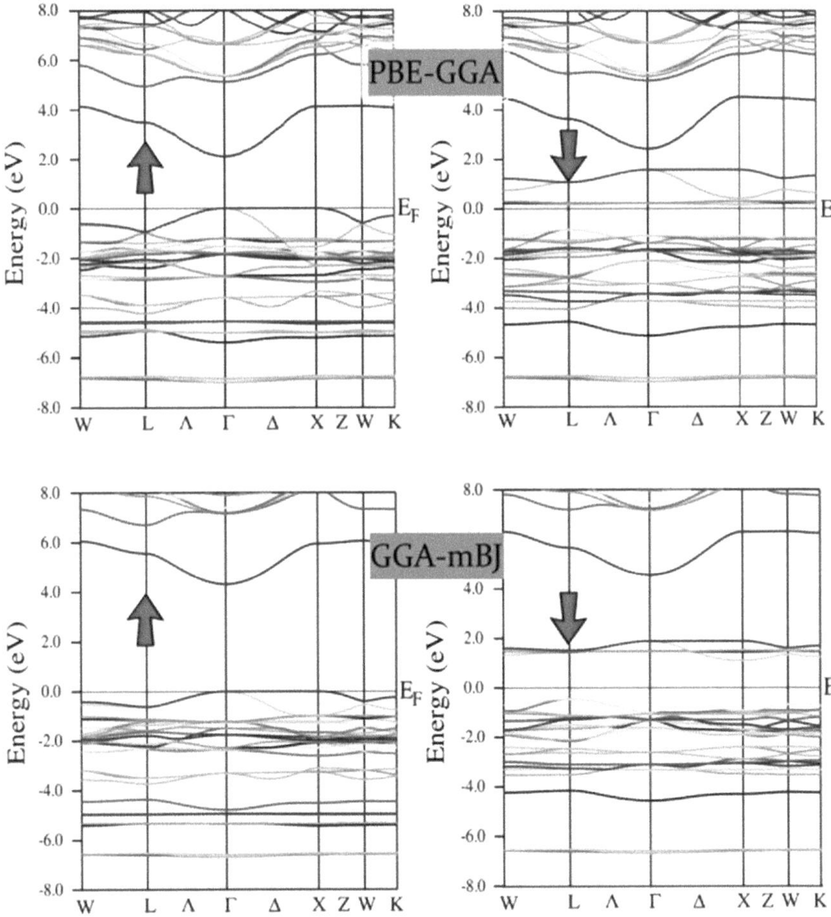

FIGURE 9.3 P-type semiconducting band profiles of $Cs_2AgFeCl_6$ in GGA and GGA-mBJ approaches respectively.

The energy bandgap computed via mBJ is different than utilizing GGA for spin-up, which equals 4.32 eV, 2.11 eV, respectively at point Γ. Anyway, for spin down, it can a direct bandgap equals 1.09 eV for GGA at point L and indirect bandgap equals 1.55 eV at points L → X. Moreover, GGA-mBJ includes Fe-partially *d* states and displaying an observed wide bandgap, which is attributing to energy state shifted of Fermi level (E_f). Therefore, for GGA-mBJ and GGA, the separation between conduction and valence band is resulted in semiconductor of p-type and spin-polarized effect. Figure 9.4 shows the total density of states that can be utilized for further investigation. In addition, GGA-mBJ is employed for calculating partial density of states (PDOS) of lead-free double perovskite $Cs_2AgFeCl_6$ for determining the occupation of elements to confirm the semiconductor nature. So, the investigations of band structures and density plots imply to maintain the nature of semiconductors.

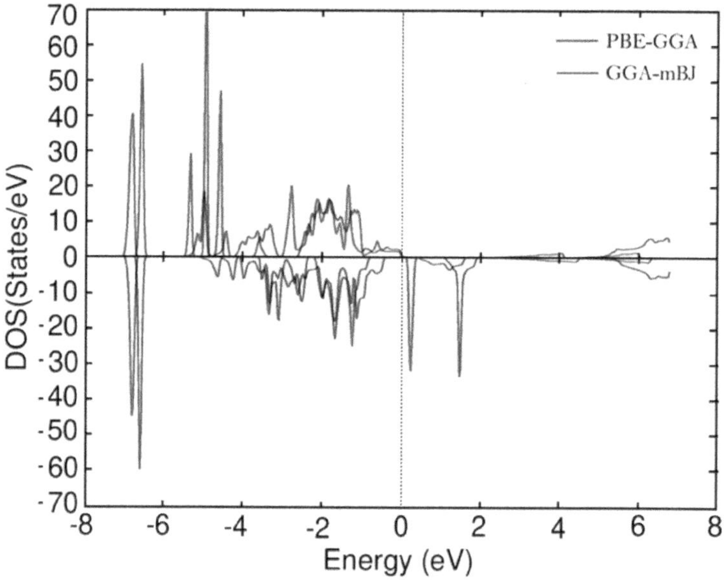

FIGURE 9.4 Total density of states for $Cs_2AgFeCl_6$ using GGA and GGA-mBJ approaches.

From spin up (Figure 9.5), it can be observed a band structure is extended by –7 eV, which attributes to 6p state of Cs atom. The energy band range, –5.26 to –3.1 eV is resulted of a mixture of Ag-4d, Fe-dt2g, Fe-deg and Cl-3p. the valence energy region at –2.6 to 0 eV is contributed to Ag-4d and Cl-3p states hybridization. While, for spin down, the Cl and Ag electrons are contributing to major peaks of VB, while Fe-dt2g state shows a major peak of conduction band. Additionally, we have aimed to find the magnetic properties that determine their abilities in spin-based technological domains, focusing on the most interesting applications such as spintronics, spin injectors, random memory shape devices and spin filters.

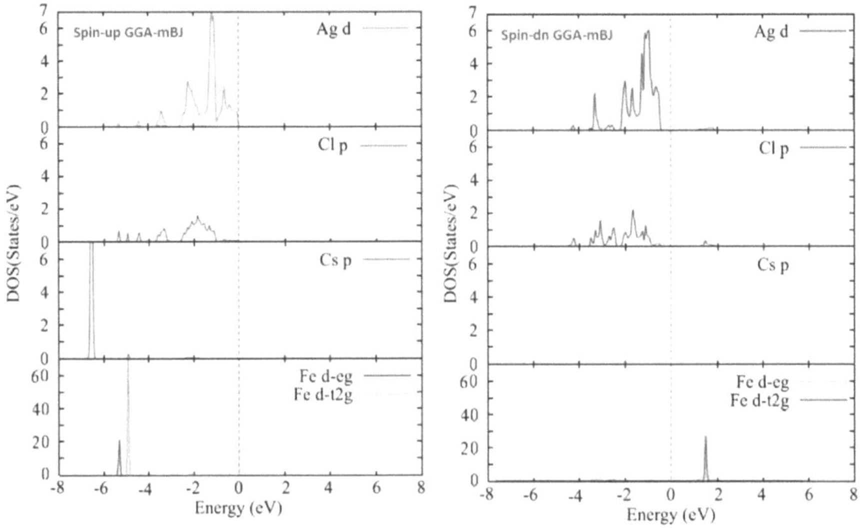

FIGURE 9.5 Representation of partial density of states of $Cs_2AgFeCl_6$ within GGA-mBJ approach.

The GGA and GGA-mBJ are employed to investigate magnetic properties. The calculated interstitial and total magnetic moments are displayed in Table 9.3. Our materials are ferromagnetic, with 5 μB magnetic property that is a fractional magnetic moment summation for different interstitial atoms and sites. The Fe atoms contribute to the total magnetic moment including a partial moment; 4.095 B (GGA) (mBJ) and 3.918 B (GGA). The nature of FM is predominately because of Iron atoms. The orbitals of unfilled Fe-3d are a magnetic principal source. It is suggested that the positive atoms have ferromagnetic character aligned with others.

9.6 THERMOELECTRIC STUDIES

The thermoelectric properties of lead-free halide double perovskite $Cs_2AgFeCl_6$ can be analyzed using a range of experimental and theoretical techniques. These properties are important for understanding the material's behavior in thermoelectric devices, which convert heat into electrical energy, or vice versa. Experimental techniques for measuring the thermoelectric properties of materials include the Seebeck coefficient, electrical conductivity, and thermal conductivity measurements. These measurements can be used to calculate the material's figure of merit, ZT, which describes the material's efficiency in converting heat into electrical energy.

TABLE 9.3 Calculated Partial, Interstitial and Total Magnetic Moment and Band Gap (eV) for $Cs_2AgFeCl_6$ within GGA and mBJ (in Bohr Magneton μ_B).

Compound	Method	Magnetic moment (μB)						Band gap (eV)	Band gap (eV)
		Interstial	Cs	Ag	Fe	Cl	Total		
$Cs_2AgFeCl_6$	GGA	0.2716	0.0025	0.0852	3.9184	0.1197	4.9991	2.11 (↑) 1.09 (↓)	1.18
	mBJ	0.1103	0.0007	0.0773	4.0953	0.1192	5.0000	4.32 (↑) 1.55 (↓)	2.14

Theoretical calculations can also be used to predict the thermoelectric properties of materials. The Boltzmann transport equation can be used to model the behavior of charge carriers in the material, and the resulting electrical and thermal conductivities can be used to calculate the Seebeck coefficient and figure of merit. DFT calculations can also be used to calculate the electronic structure of the material, which is important for understanding its thermoelectric behavior. Studies on the thermoelectric properties of $Cs_2AgFeCl_6$ have shown that the material has a high Seebeck coefficient, which indicates a high potential for thermoelectric energy conversion. However, the electrical conductivity of the material is relatively low, which limits its efficiency. The thermal conductivity of $Cs_2AgFeCl_6$ has also been tested, where it has suggested to have a low thermal conductivity, which is desirable for thermoelectric applications.

Theoretical calculations have been performed to predict the thermoelectric properties of $Cs_2AgFeCl_6$, and the results suggest that the material has a moderate figure of merit, which is comparable to other lead-free thermoelectric materials. These findings suggest that $Cs_2AgFeCl_6$ is a promising candidate for use in thermoelectric devices, and further optimization of its properties may lead to improved performance. The thermoelectric materials (TE) properties display a performance of intriguing to transform the wasted temperature. The thermoelectric efficiency is calculated by its intrinsic characteristics utilizing ZT = S2σT/κ. The power factor refers to S2σ, temperature refers to T and total thermal conductivity refers to κ including both electronic (e) and lattice (l) contributions, i.e., κ = κe + κl[46–50] Low thermal conductivity, High electrical conductivity and large Seebeck coefficient are required to materials having thermoelectric quality. The Perovskites have a wide interest of thermoelectric materials. The semiconductors have two spin channels. So, for investigating the transport properties of halide double perovskite $Cs_2AgFeCl_6$, it is recommended to use Boltzmann theory of semi-classical (CBT) into BoltzTraP code.[31] The thermoelectric parameters are reported at the range; 100–900 K.

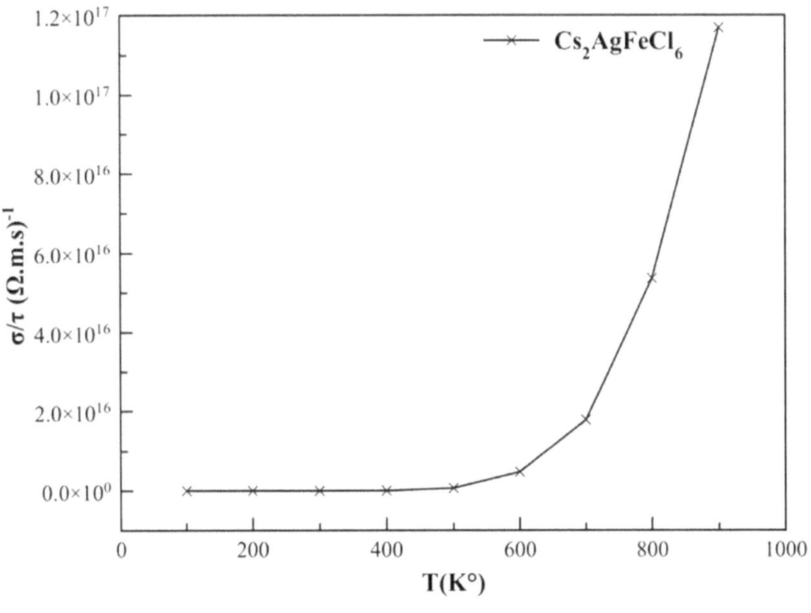

FIGURE 9.6 Variation of electrical conductivity (σ/τ) as a function of temperature for the $Cs_2AgFeCl_6$ compound.

The computed electrical conductivity graphical variation is σ/τ, where relaxation time refers to τ, (Fig. 9.6). It determines the holes/electrons number accessible to conduction band. The electrical conductivity value is 16.82×10^3 $(\Omega ms)^{-1}$, reaching to 1.17×10^{17} $(\Omega ms)^{-1}$ at at 100 K and 900 K, respectively. Moreover, when temperature increases, the electron receives enough thermal energies for moving forward to conduction band to form hole-electron pair escorting to increasing in electrical conductivity. For thermal conductivity ($\kappa e/\tau$), the electronic part only has been calculated due to classical theory-based BoltzTraP code limitation. Figure 9.7 illustrates the increasing from 136.79 W/mKs to 4.28×10^{13} W/mKs at at 100 K and 900 K, respectively. The variation of Seebeck coefficient (S) is shown in Figure 9.8 to show as T increases, S decreases to 575.88 μV/K at 900 K. the optimum value is 2848.21 μV/K at 100 K. The resulted values of S are positive referring to p-type of conduction level. Figure 9.9 illustrates the power factor (PF) to evaluate the thermoelectric efficiency of wide range of temperature. The correlation between PF and T is directly, which is increasing from 1.36×10^{14} ($W.m^{-1}.K^{-2}s$) to 3.49×10^{25} ($W.m^{-1}.K^{-2}s$) at 100 K and 900 K, respectively. Also, Figure 9.10 shows the merit factor (ZT) fluctuation with a wide range of temperature. The correlation between ZT and T is inversely. Our material has high ZT values (0.99, 0.92) at 100 K and 900 K, respectively due to its high S value and low thermal conductivity.[51–53]

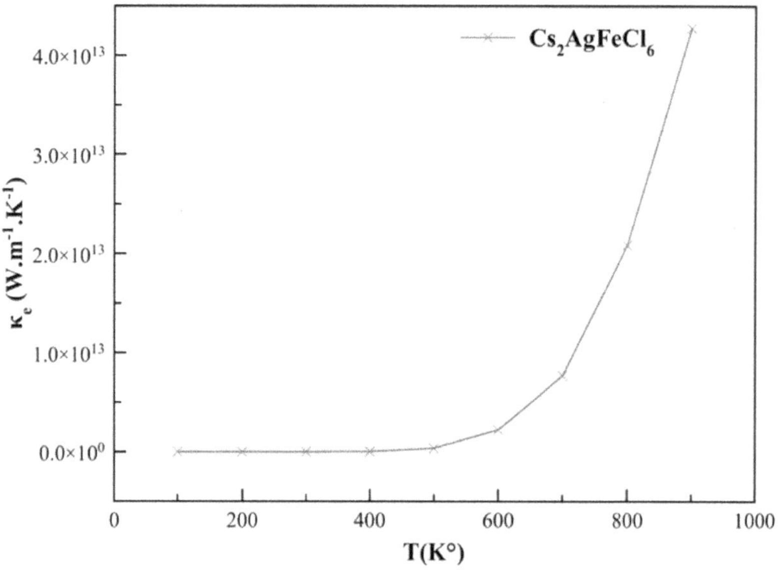

FIGURE 9.7 Thermal electronic conductivity variation (κe) as a function of temperature for the $Cs_2AgFeCl_6$ compound.

9.7 CONCLUSIONS

DFT has guided us to a success, the transport, magneto-electronic, elastic and structural properties for lead-free halide double perovskite

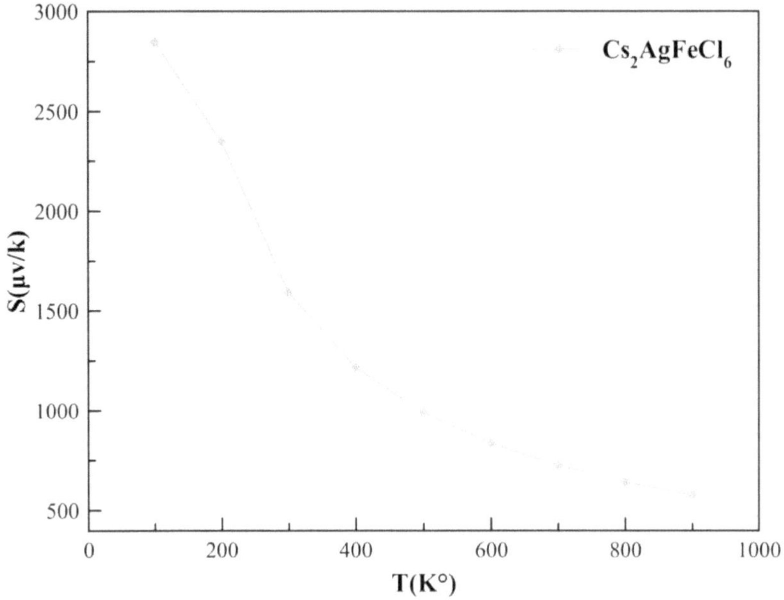

FIGURE 9.8 Variation of Seebeck coefficient (*S*) in terms of temperature for the $Cs_2AgFeCl_6$ compound.

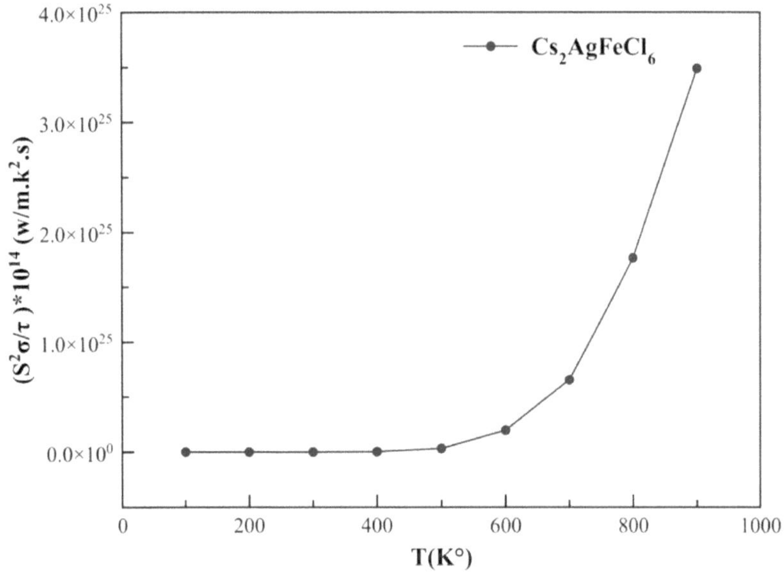

FIGURE 9.9 Variation of the power factor (PF) as a function of temperature for the $Cs_2AgFeCl_6$ compound.

$Cs_2AgFeCl_6$ were investigated. The GGA and TB-mBJ were employed for exchange correlation. The tolerance factor depiction and description of formation energy for structural stability of $Cs_2AgFeCl_6$ are stabled in cubic structure including lattice constant, 10.4848 Å. The lattice constant was agreed with other data. The researched elastic properties including mechanical stability criteria, Cauchy pressure, Pugh's ratio and Poisson's ratio referring to ductility character of $Cs_2AgFeCl_6$. Also, the electronic results present nature of semiconducting for spins. The magnetic properties were determined by their lattice parameters that reveal an important contribution of Iron (Fe) atom. Hence, the total magnetic moment was 5 μB, as a potential application in spintronics. Transport coefficients like ZT, PF, thermal conductivity, electrical conductivity and S have been researched via BoltzTraP that are recommended for thermoelectric applications.

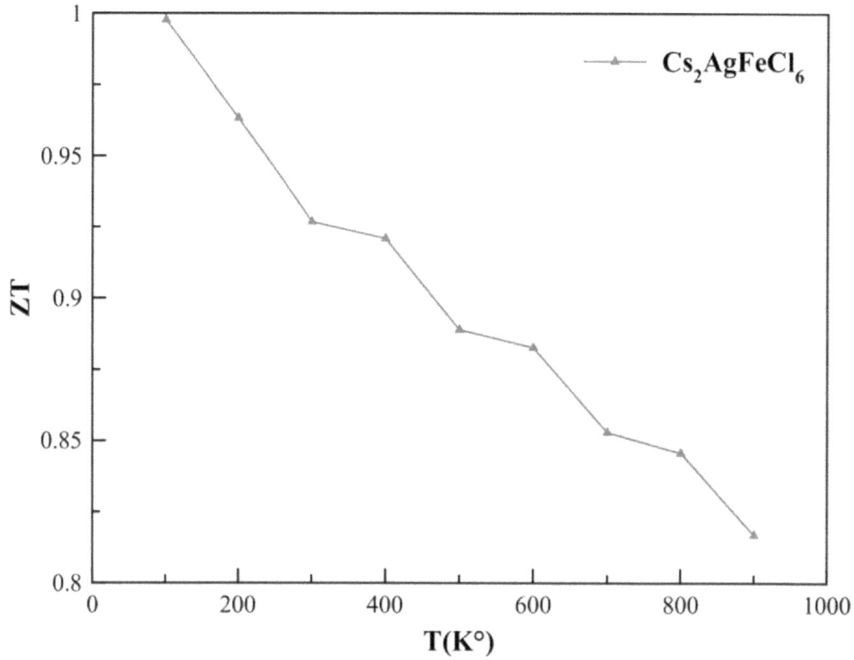

FIGURE 9.10 Variation of figure of merit (ZT) as a function of temperature for the $Cs_2AgFeCl_6$ compound.

KEYWORDS

- **FP-LAPW**
- **perovskite**
- **structural**
- **mechanical**
- **thermoelectric**

REFERENCES

1. Han, C.; Li, Z.; Dou, S. Recent Progress in Thermoelectric Materials. *Chin. Sci. Bull.* **2014**, *59* (18), 2073–2091. DOI: 10.1007/s11434-014-0237-2
2. Camargo, J. R.; de Oliveira, M. C. C. Heat Analysis and Thermodynamic Effects. In *BOD*; Ahsan, A., Ed.; InTech Europe, 2011; vol. 5, pp 327–346.
3. Ullah, R; Ali, M. A.; Haq, B. U.; Khan, A.; Mahmood, Q. Materials Science in Semiconductor Processing **2022,** *137,* 106218 DOI: 10.1016/j.mssp.2021.106218
4. Nolas, G. S.; Sharp, J. W.; Goldsmid, H. J. *Thermoelectrics: Basics Principles and New Materials Developments*; Springer Publications: Berlin, 2001.
5. Mahan, G. D.; Sofo, J. O. The Best Thermoelectric. *Proc. Natl. Acad. Sci.* **1996,** *93* (15), 7436–7439. DOI: 10.1073/pnas.93.15.7436
6. Ghasemi, M.; Hao, M.; Xiao, M.; Chen, P.; He, D.; Zhang, Y.; Wen, X. Lead-free Metal-halide Double Perovskites: from Optoelectronic Properties to Applications. *Nanophotonics* **2020,** *10* (8), 2181–2219. DOI: 10.1515/nanoph-2020-0548
7. Giustino, F.; Snaith, H. J. Toward Lead-Free Perovskite Solar Cells. *ACS Energy Lett.* **2016,** *1* (6), 1233–1240. DOI: 10.1021/acsenergylett.6b00499
8. Liang, G.-Q.; Zhang, J. A Machine Learning Model for Screening Thermodynamic Stable Lead-free Halide Double Perovskites. *Comput. Mater. Sci.* **2022,** *204,* 111172–111181 DOI: 10.1016/j.commatsci.2021.111172
9. Chung, I.; Lee, B.; He, J.; Chang, R. P. H.; Kanatzidis, M. G. All-solid-state dye-Sensitized Solar Cells with High Efficiency. *Nature* **2012,** *485* (7399), 486–489. doi:10.1038/nature11067
10. Lee, M. M.; Teuscher, J.; Miyasaka, T.; Murakami, T. N.; Snaith, H. J. Efficient Hybrid Solar Cells Based on Meso-Superstructured Organometal Halide Perovskites. *Science* **2012,** *338* (6107), 643–647. DOI: 10.1126/science.1228604
11. Burschka, J.; Pellet, N.; Moon, S.-J.; Humphry-Baker, R.; Gao, P.; Nazeeruddin, M. K.; Grätzel, M. Sequential Deposition as a Route to High-performance Perovskite-sensitized Solar Cells. *Nature* **2013,** *499* (7458), 316–319. DOI: 10.1038/nature12340

12. Kojima, A.; Teshima, K.; Shirai, Y.; Miyasaka, T. Organometal Halide Perovskites as Visible-Light Sensitizers for Photovoltaic Cells. *J. Am. Chem. Soc.* **2009,** *131* (17), 6050–6051. DOI: 10.1021/ja809598r
13. Etgar, L.; Gao, P.; Xue, Z.; Peng, Q.; Chandiran, A. K.; Liu, B.; ... Grätzel, M. Mesoscopic CH3NH3PbI3/TiO2 Heterojunction Solar Cells. *J. Am. Chem. Soc.* **2012,** *134* (42), 17396–17399. DOI: 10.1021/ja307789s
14. Jin, H.; Im, J.; Freeman, A. J. Topological Insulator Phase in Halide Perovskite Structures. *Phys. Rev. B* **2012,** *86* (12), 8713–8725. DOI: 10.1103/physrevb.86.121102
15. Takahashi, Y.; Obara, R.; Lin, Z.-Z.; Takahashi, Y.; Naito, T.; Inabe, T.; ... Terakura, K. Charge-transport in tin-Iodide Perovskite CH3NH3SnI3: Origin of High Conductivity. *Dalton Transac.* **2011,** *40* (20), 5563. DOI: 10.1039/c0dt01601b
16. Chen, Q.; De Marco, N.; Yang, Y. (Michael); Song, T.-B.; Chen, C.-C.; Zhao, H.; ... Yang, Y. Under the Spotlight: The Organic–inorganic Hybrid Halide Perovskite for Optoelectronic Applications. *Nano. Today* **2015,** *10* (3), 355–396. DOI:10.1016/j.nantod.2015.04.009
17. Khandy, S. A.; Gupta, D. C. Insight View of Double Perovskites Ba 2 XNbO 6 (X = Ho,Yb) for Spintronics and Thermoelectric Applications. *Int. J. Energy Res.* **2021,** *45* (9), 13338–13354. DOI: 10.1002/er.6662
18. Khandy, S. A.; Gupta, D. C. Intrinsic Magnetism and Thermoelectric Applicability of Novel Halide Perovskites Cs2GeMnX6 (X = Cl, Br): Route Towards Spintronics and Energy Harvesting Technologies. *Mater. Sci. Eng. B* **2021,** *265,* 114985. DOI: 10.1016/j.mseb.2020.114985
19. Xian, Y.; Yin, H.; Bao, Y.; Xiao, Y.; Yuan, S.; Rahman, N. U.; ... Fan, J. Engineered Electronic Structure and Carrier Dynamics in Emerging Cs2AgxNa1–xFeCl6 Perovskite Single Crystals. *J. Phys. Chem. Lett.* **2020,** *11* (22), 9535–9542. DOI: 10.1021/acs.jpclett.0c02963
20. Yin, H.; Xian, Y.; Zhang, Y.; Chen, W.; Wen, X.; Rahman, N. U.; ... Li, W. An Emerging Lead-Free Double-Perovskite Cs 2 AgFeCl 6:In Single Crystal. *Adv. Funct. Mater.* **2020,** *30* (43), 2002225. DOI: 10.1002/adfm.202002225
21. Li, Z.; Xu, Q.; Sun, Q.; Hou, Z.; Yin, W. J. 2018, arXiv preprint arXiv:1803.06042.
22. Hohenberg, P.; Kohn, W. Inhomogeneous Electron Gas. *Phys. Rev.* **1964,** *136,*B864.
23. Kohn, W.; Sham, L. J. Self-Consistent Equations Including Exchange and Correlation Effects. *Effects, Phys. Rev.* **1965,** *140,*A1133.
24. Blaha, P.; et al. *An Augmented Plane Wave Plus Local Orbitals Program for Calculating Crystal Properties*; Vienna University of Technology: Vienna, Austria, 2001.
25. Berber, M.; et al. Structural, Electronic, and Optical Properties of Quaternary Alloys Al0.50Ga0.50NxSb1-x: A First Principles Study. *Rev. Mex. Fis* **2020,** *66,* 790.
26. Tran, F.; Blaha, P. Accurate Band Gaps of Semiconductors and Insulators with a Semilocal Exchange-Correlation Potential. *Phys. Rev. Lett.* **2009,** *102* (22) 6478–6481. DOI: 10.1103/physrevlett.102.22640
27. Koller, D.; Tran, F.; Blaha, P. Merits and Limits of the Modified Becke-Johnson Exchange Potential. *Phys. Rev. B* **2011,** *83* (19) 4563–4572. DOI: 10.1103/physrevb.83.195134
28. Blaha, L. F.; Maafa, A.; Chahed, A.; Boukli, M. A. H.; Sayade, A. The First Principle Calculations of Structural, Magnetoelectronic, Elastic, Mechanical, and Thermoelectric Properties of Half-metallic Double Perovskite Oxide Sr2TiCoO6. *Rev. Mex. Fis* **2021,** *67,* 114.
29. Monkhorst, H. J.; Pack, J. D. Special points for Brillouin-zone Integrations. *Phys. Rev. B* **1976,** *13* (12), 5188–5192. DOI: 10.1103/physrevb.13.5188

30. Charpin, T. *A Package for Calculating Elastic Tensors of Cubic Phases Using WIEN*; Laboratory of Geometrix F-75252: Paris, France, 2001.
31. Madsen, G. K. H.;Singh, D. J. BoltzTraP. A Code for Calculating Band-structure Dependent Quantities. *Comput. Phys. Commun.* **2006,** *175,* 67.
32. Filip, M. R.; Hillman, S.; Haghighirad, A. A.; Snaith, H. J.; Giustino, F. Band Gaps of the Lead-Free Halide Double Perovskites Cs2BiAgCl6 and Cs2BiAgBr6 from Theory and Experiment. *J. Phys. Chem. Lett.* **2016,** *7* (13), 2579–2585.
33. Birch, F. The Effect of Pressure Upon the Elastic Parameters of Isotropic Solids, According to Murnaghan's Theory of Finite Strain. *J. Appl. Phys.* **1938,** *9,* 279. DOI: 10.1063/1.1710417
34. Li, Z.; Yang, M.; Park, J.-S.; Wei, S.-H.; Berry, J. J.; Zhu, K. Stabilizing Perovskite Structures by Tuning Tolerance Factor: Formation of Formamidinium and Cesium Lead Iodide Solid-State Alloys. *Chem. Mater.* **2015,** *28* (1), 284–292.
35. Shannon, R. D. Revised Effective Ionic Radii and Systematic Studies of Interatomic Distances in Halides and Chalcogenides. *Acta Crystallogr A* **1976,** *32* (5), 751–767.
36. Jamal, M. Cubic-elastic. 2019 [Online]. http://www.WIEN2k.at/reg_user/unsupported/cubic-elast/
37. Born, M. On the Stability of Crystal Lattices. *I. Math. Proc. Camb. Philos. Soc.* **1940,** *36* (02), 160. DOI: 10.1017/s0305004100017138
38. Aslam, F.; Ullah, H.; Hassan, M. Theoretical Investigation of Cs2InBiX6 (X = Cl, Br, I) Double Perovskite Halides Using First-Principle Calculations. *Mater. Sci. Eng.: B* **2021,** *274,* 115456–115466. DOI: 10.1016/j.mseb.2021.115456
39. Birch, F. The Effect of Pressure upon the Elastic Parameters of Isotropic Solids, According to Murnaghan's Theory of Finite Strain. *J. Appl. Phys.* **1938,** *9* (4), 279–288. DOI: 10.1063/1.1710417
40. Haid, S.; Bouadjemi, B.; Houari, M.; Matougui, M.; Lantri, T.; Bentata, S.; Aziz, Z. Investigation of DFT+U effect of Holmium Rare-earth on the Electronic, Magnetic and the Half-metallic Ferromagnetic Properties' of Double Perovskite Ba2HoReO6. *Solid State Commun.* **2019,** *294,* 29–35.
41. Braun, M.; Iváñez, I. Numerical Study of Damaged Micro-Lattice Blocks Subjected to Uniaxial Compressive Loading. *Extreme Mech. Lett.* **2020,** *39,* 100821–100833
42. Frantsevich, I. N.; Voronov, F. F.; Bokuta, S. A. *Elastic Constants and Elastic Moduli of Metals and Insulators Handbook*; Frantsevich, I. N., Ed.; Naukova Dumka: Kiev, 1982; vol. 60, 1990.
43. Kholil, M. I.; Bhuiyan, M. T. H. Electronic, Elastic, Vibrational and Superconducting Properties of a Ternary Superconductors LaIr*P* (*P* = P, As): Insights from DFT. *Solid State Commun.* **2020,** *322,* 114053–114062
44. Wu, M. S. A pressure-loaded Soft Functionally Gradient Spherical Capsule Under Finite Deformation. *Mech. Mater.* **2020,** *150,* 103573–103584
45. Fine, M. E.; Brown, L. D.; Marcus, H. L. Elastic Constants Versus Melting Temperature in Metals. *Scr. Metallurgica* **1984,** *18* (9), 951–956.
46. Khandy, S. A.; Gupta, D. C. Study of Ferromagnetism, Spin-polarization, Thermoelectrics and Thermodynamics of Layered Perovskite Ba2FeMnO6 Under Pressure and Temperature. *J. Phys. Chem. Solids* **2019,** *135,* 109079–109086.
47. Mir, S. A.; Seh, A. Q.; Gupta, D. C. New Ferromagnetic Half-metallic Perovskites for Spintronic Applications: BaMO3 (M = Mg and Ca). *RSC Adv.* **2020,** *10* (60), 36241–36252.

48. Singh, S.; Gupta, D. C. Lanthanum Based Quaternary Heusler alloys LaCoCrX (X= Al, Ga): Hunt for Half-Metallicity and High Thermoelectric Efficiency. *Results Phys.* **2019,** *13,* 102300.
49. Sofi, S. A.; Yousuf, S.; Gupta, D. C. Prediction of Robustness of Electronic, Magnetic and Thermoelectric Properties Under Pressure and Temperature Variation in Co2MnAs Alloy. *Comput. Condens. Matter* **2019,** *19,* e00375.
50. Bhat, T. M.; Gupta, D. C. Transport, Structural and Mechanical Properties of Quaternary FeVTiAl Alloy. *J. Electron. Mater.* **2016,** *45* (11), 6012–6018.
51. Riyadh A. Al-Samarai, Y. Al-Douri, *Friction and Wear in Metals*, Springer, Germany. 2024
52. Abraham George, Exp. Theo. NANOTECHNOLOGY 8 (2024) 1–9
53. M. Tadres, Exp. Theo. NANOTECHNOLOGY 8 (2024) 11–16

CHAPTER 10

Electrical Transport and Magnetic Properties of $La_{0.67}Sr_{0.33}Mn_{0.95}Sn_{0.05}O_3$ (LSMSO) Manganite Thin Film Grown by Pulsed Laser Deposition

TEJAS M. TANK[1], D. S. RANA[2], and SANKAR P. SANYAL[3]

[1]*Department of Physics, The Patidar Gin Science College, Bardoli, India*

[2]*Terahertz Physics Laboratory, Department of Physics,*
Indian Institute of Science Education and Research, Bhopal, India

[3]*Solid State Physics Laboratory, Department of Physics,*
Barkatullah University, Bhopal, India

ABSTRACT

$La_{0.67}Sr_{0.33}MnO_3$ is a promising material due to robust spin polarization with a ferromagnetic transition temperature above 300 K in the manganite family. As a potential material for a variety of spintronic, memory, and multifunctional devices, it has generated a great deal of interest. We have studied the effect of non-magnetic Sn^{4+} substitution to understand its effect on the structural, electrical, and magnetic properties of the $La_{0.67}Sr_{0.33}Mn_{0.95}Sn_{0.05}O_3$ manganite thin film onto a single crystalline $SrTiO_3$ (100) substrate prepared using pulsed laser deposition (PLD) technique. The grown film is single-phase and epitaxial, as confirmed by X-ray diffraction, X-ray reflectivity, and Reciprocal space mapping. Atomic force microscopy reveals that the film possesses low surface roughness and a good ordering of grains/domains. The electrical and magnetic measurements of this film reveal that both the metal–insulator transition (T_P) and ferromagnetic-paramagnetic transition temperatures (T_C)

Contemporary Advancements in Materials Technology: Novel Solutions to Complex Problems.
Shrikaant Kulkarni, Vipul Srivastava, & P. William (Eds.)

coincide. The electron-electron, electron-phonon, and electron-magnon processes all contribute to the scattering of carriers in electrical transport at low temperatures. The PLD-grown $La_{0.67}Sr_{0.33}Mn_{0.95}Sn_{0.05}O_3$ thin film shows a high magnetoresistance (MR) and temperature coefficient of resistance (TCR) value, which makes it especially suitable for next-generation spintronic and infrared bolometer sensors.

10.1 INTRODUCTION

The discovery of interesting phenomena like superconductivity, two-dimensional electron gas (2DEG), ferroelectricity, colossal magnetoresistance (CMR), and so forth has prompted significant attention in strongly correlated functional oxide perovskites.[1-4] Manganites, in particular, have attracted interest due to the observance of key phenomena such as colossal magnetoresistance (CMR) and putative applications in epitaxial thin film form. In the absence of prevailing grain boundary effects, the study of epitaxial thin films enriches the understanding of various physical properties of manganite materials. It is well-established that disorder and crystallographic defects influence a number of manganite parameters, including the metal-insulator transition temperature (T_P), ferromagnetic-paramagnetic transition temperature (T_C), magnetoresistance, anisotropic magnetoresistance, and so on.[5] Several groups have tried to substitute different types of dopants at the A-site (RE) and various magnetic and non-magnetic ions at the B-site (Mn) in ABO_3 type $RE_{1-x}A_xMnO_3$ manganite in order to investigate these properties; where RE = Rare earth trivalent metal ions (such as La, Nd, Pr, etc.); A = divalent alkaline earth ions (such as Ba, Ca, Sr, etc.). This suggests that the physical properties of CMR materials are significantly influenced by the A-site/B-site cation mismatch or size variance.[6–9] The strain at the film-substrate interface due to the lattice mismatch, causes manganite's thin film properties to differ from those of the bulk material. The annealing temperature and time, annealing conditions, film thickness, substrate mismatch, orientation, etc. affect the magnetic and electrical properties of the manganite films.[10–12] Both the grain boundary and the lattice strain effects are observed to create important dissimilarities between the bulk and thin film properties.[13] High-quality epitaxial manganite thin films are typically deposited using techniques such as chemical solution deposition (CSD), molecular beam epitaxy (MBE), RF sputtering, pulsed laser deposition (PLD), etc.[14–17]

Among the several known perovskite manganite, $La_{1-x}Sr_xMnO_3$ (LSMO) at $x = 0.33$ concentration of Sr, is the most promising material for room-temperature spintronic applications since it exhibits robust spin polarization at Curie temperature (T_C) 370 K.[18] External perturbations, including La-site/Mn-site substitutions, buffer layers, and substrate-induced strain, significantly impact the magneto-transport properties of manganite thin films.[18–21]The partial substitution of Mn ions with diamagnetic and tetravalent Sn ions is expected to provide a scope to understand structural, electrical, and magnetic properties in the form of a thin film. Since Sn ions lack a similar electronic configuration as manganese ions, their substitution could play a vital role in the physical properties. The ionic radii of Mn^{3+} and Mn^{4+} are 0.64 and 0.53, respectively, in contrast to 0.69 for Sn^{4+}, and hence due to their bigger size, Sn^{4+} cations cause local lattice distortions.[22] Therefore, in the present work, we have chosen to focus on the importance of structural effects on the thin film properties due to the substitution of nonmagnetic tin (Sn) at the Mn-site in $La_{0.67}Sr_{0.33}MnO_3$ manganite, for which no experimental investigation has been reported so far. The main objective of our studies has been to successfully prepare single-phase and single-crystalline $La_{0.67}Sr_{0.33}Mn_{0.95}Sn_{0.05}O_3$ (LSMSO) compounds and to systematically characterize their physical properties. In this chapter, we report the structural, electrical, and magnetic properties of 5% substitution of nonmagnetic tetravalent Sn in $La_{0.67}Sr_{0.33}Mn_{0.95}Sn_{0.05}O_3$ thin film grown on a single crystalline $SrTiO_3$ (100) substrate using the PLD technique.

10.2 EXPERIMENTAL PROCEDURE AND MATERIAL SYNTHESIS

The solid-state reaction method was used to synthesize a polycrystalline sample of LSMSO. Stoichiometric amounts of La_2O_3, $SrCO_3$, Mn_2O_3, and SnO_2 powders were mixed, pierced, and reacted for 24 h at 1100°C in air. The sample was ground, formed into a pellet, and sintered for 24 h at 1350°C. The PLD technique was used to deposit a single-crystalline epitaxial thin film of LSMSO, and a well-sintered 15 mm diameter pellet was used as a target during the deposition.

In PLD technique, inside a high vacuum chamber, a giant power-pulsed laser beam is used to grow high-quality films on a target with desired stoichiometry. PLD uses photon interaction to expel material from any target in the form of a plume. The vapor is collected on a substrate that is positioned close to the target. The real physical process of the material removal is relatively complex. The rapid explosion of the target surface due to superheating could be considered the cause of the material's ejection. As different from thermal evaporation, this provides vapor composition that depends on the vapor pressure of the element in the target material. The material produced by the laser-induced explosion has a plume with a stoichiometry exactly similar to the target. The laser-induced explosion produces a plume of material with stoichiometry similar to the target.

The following four subsequent stages constitute the entire thin film formation via the PLD technique:

1. Laser interaction with the target
2. The target's ablation material's escape dynamics
3. Deposition stage of the ablation materials on the substrate
4. A thin film's nucleation and expansion on the substrate surface

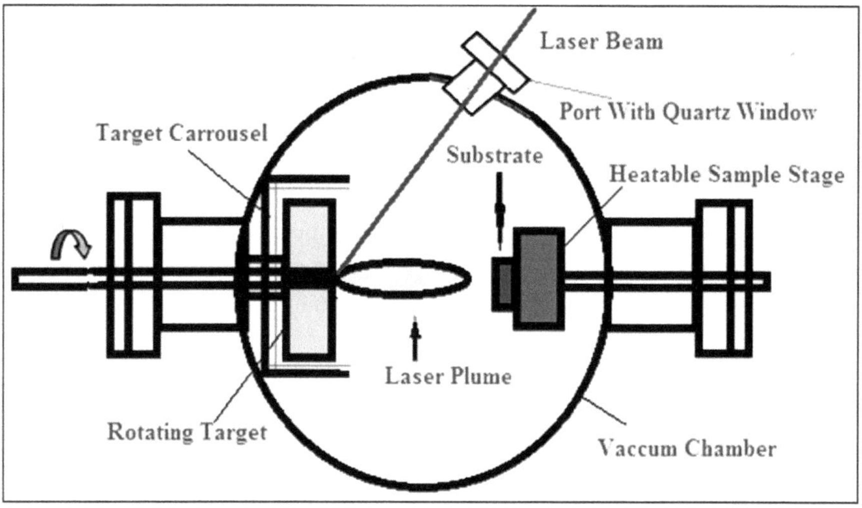

FIGURE 10.1 Schematic diagram for the pulsed laser deposition technique.

A hot pressed LSMSO target was ablated using a KrF excimer laser with a 248 nm wavelength, 300 mJ pulse energy, and a 4 Hz repetition rate in a vacuum chamber having a base pressure of 10^{-3} Pa. One side polished (100) $SrTiO_3$ (STO) single crystal substrate was positioned in front of the target distant of 52 mm and heated to 750°C. The substrate was kept at a temperature of 750°C with an O_2 pressure of 40 Pa during laser deposition, while the film was annealed at an O_2 pressure of 200 Pa at 750°C for 10 min during post-deposition. A typical growth rate was ~0.5 ~ 1.0 Å/pulse and obtained 40 nm thickness of the thin film. Then the film was cooled down to room temperature for 2 hours inside the deposition chamber. The process parameters for the thin film deposition are tabulated in Table 10.1.

The structure of the thin film and film thickness were analyzed using X-ray diffraction (XRD) X-ray reflectivity measurements respectively (PANalytical Empyrean). Reciprocal space mapping (RSM) was performed to confirm the thin film's coherent growth, epitaxial nature, and phase purity. The micro-structure and surface morphology data for the thin film was collected using Atomic Force Microscopy (AFM) (Nanoscope). The electrical transport of the film was measured in the temperature range from 2 K to 300 K by using the four-probe method without and with an applied magnetic field of 1 T and 8 T using PPMS (Quantum Design). Isotherms MR% measured at various temperatures up to 8 T applied magnetic field. Zero-field cool (ZFC) and field cool (FC) magnetization (M-T) measurements were carried out at 500 Oe field from 10 K–300 K using SQUID-VSM setup (Quantum Design). Isothermal magnetization (M-H) was performed at 10 K and 300 K.

TABLE 10.1 Thin Film Process Parameters During PLD Technique.

Target material	$\mathbf{La_{0.67}Sr_{0.33}Mn_{0.95}Sn_{0.05}O_3}$ **(LSMSO)**
Substrate	$SrTiO_3$ (STO) (100) Single Crystal Size: 5 mm × 5 mm
Laser source	KrF
Laser frequency	4 Hz
Laser energy	300 mJ
Pulse width	6 ns
A substrate to target distance	5.2 cm
Deposition temperature	750°C
Oxygen partial pressure	40 Pa
Oxygen annealing pressure	200 Pa

10.3 STRUCTURAL AND SURFACE MORPHOLOGICAL STUDIES

The LSMSO compound was formed as a perovskite pseudo-cubic structure without a preferred orientation, as demonstrated in the XRD pattern of the film, shown in Figure 10.2. Moreover, within the resolution of the XRD analysis, no obvious change in the crystalline structure was observed. This result is in good accordance with the result of the PLD-grown thin film of Ru-substituted LSMO reported by Krishnan et al.,[23] indicating that the Sn ions don't significantly alter the lattice. The XRD studies indicate that the sample is single-phase and single crystalline, suggesting a Sn substitute at the Mn site and possibly at the Mn^{4+} site resulting in charge neutrality, and no secondary phase was observed. It can be seen from the XRD pattern that only (100) groups of diffraction peaks of LSMSO and STO (a = 3.8745 Å) are apparent. The Rietveld refinement pattern of Sn substituted bulk target of $La_{0.67}Sr_{0.33}Mn_{0.95}Sn_{0.05}O_3$ was seen to possess a rhombohedral structure ($\bar{R}$3C space group),[24] but, its film grown on STO (100) substrate suggests the epitaxial growth of the film.

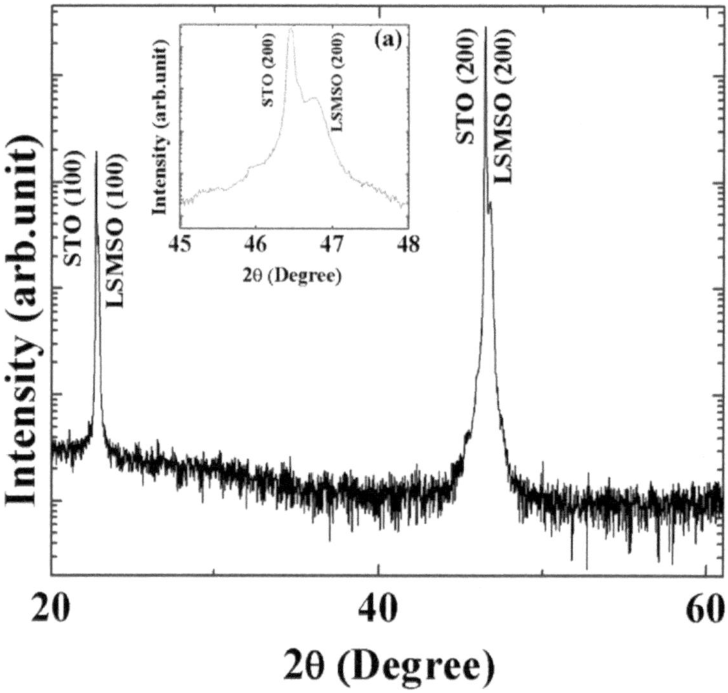

FIGURE 10.2 Room temperature XRD pattern of LSMSO thin film on the STO substrate. (a) Enlarged view of 200 peaks of STO and LSMSO.

Figure 10.3 depicts the thickness profile of LSMSO (100) /STO (100) thin film. The X-ray reflectivity measurement was carried out to derive film thicknesses by using the X-ray Diffractometer. The thickness of the thin film was determined to be around 40 nm from Figure 10.3. The strong peak structure reveals the minimum surface roughness and good quality of the thin film. The thin film can undergo a structural strain of some percent. Different strain states can be obtained in thin films by choosing substrates and/or buffer layers. As well as tuning and manipulating their functionalities. Epitaxial strains can also stabilize phases that are not stable under ambient conditions. For understanding the structural strain state of the thin film, RSM was carried out

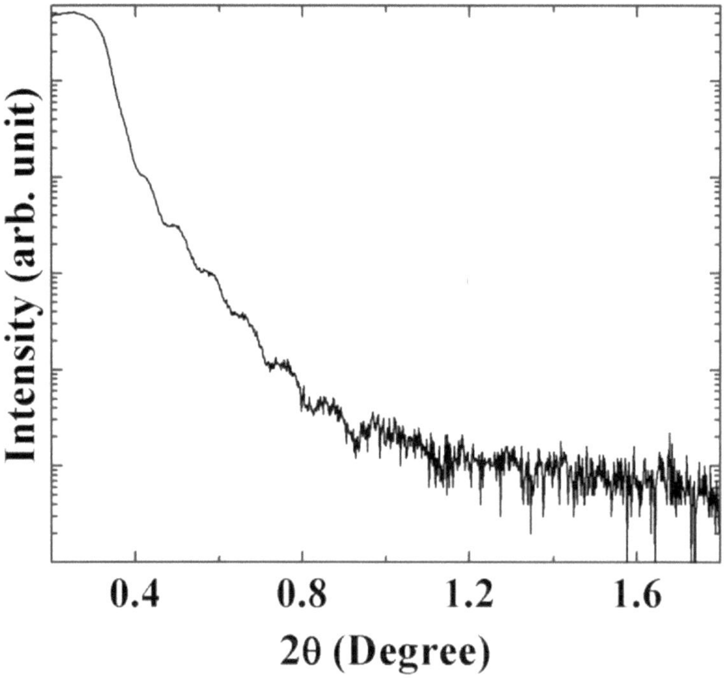

FIGURE 10.3 Room temperature X-ray reflectivity pattern of LSMSO thin film on the STO substrate.

using a high-resolution PIXcel^{3D} detector. Figure 10.4 shows the Reciprocal Space mapping (RSM) of LSMSO (100) /STO (100) thin film around the (301) crystallographic reflection, confirming the coherent growth of the thin film. By measuring the (301) reflections, the in-plane lattice parameters are determined, and the results indicate that the film is structurally strained to the underlying STO substrate. The Sn-substituted film's in-plane lattice parameter was found to be $a = 3.8515$ Å, indicating the *a*-axis preferred orientation along the [100] substrate crystallographic direction. Moreover, the relative positions of symmetric peaks of LSMSO and STO indicate the highly epitaxial growth of LSMSO onto the STO substrate. Lattice strain has been calculated using the relation, δ (%) = $[(d_{substrate} - d_{film})/d_{substrate}] \times 100$. The positive value of mismatch suggests the tensile strain in LSMSO/STO film is ~0.05% for 40 nm thin film. A deformation of the in-plane lattice parameters also produces an out-of-plane deformation of the cell as the strain (δ) is usually elastic. $\delta_{substrate} > \delta_{film}$, $\delta > 0$, for the in-plane tensile strain of the thin film, and the out-of-plane axis will be compressed. The strain is compressive for $\delta < 0$ ($d_{substrate} < d_{film}$), and the in-plane lattice is compressed while the out-of-plane axis will be widened. In addition, during the epitaxial growth of a thin film, the substrate being a single crystal, the grains are crystallographically oriented. The strain would enable this anisotropy to reflect in the electrical and magnetic properties. After optimizing several process conditions, we successfully deposited single-phase film having desired thickness of 40 nm, using a 248 nm KrF excimer laser.

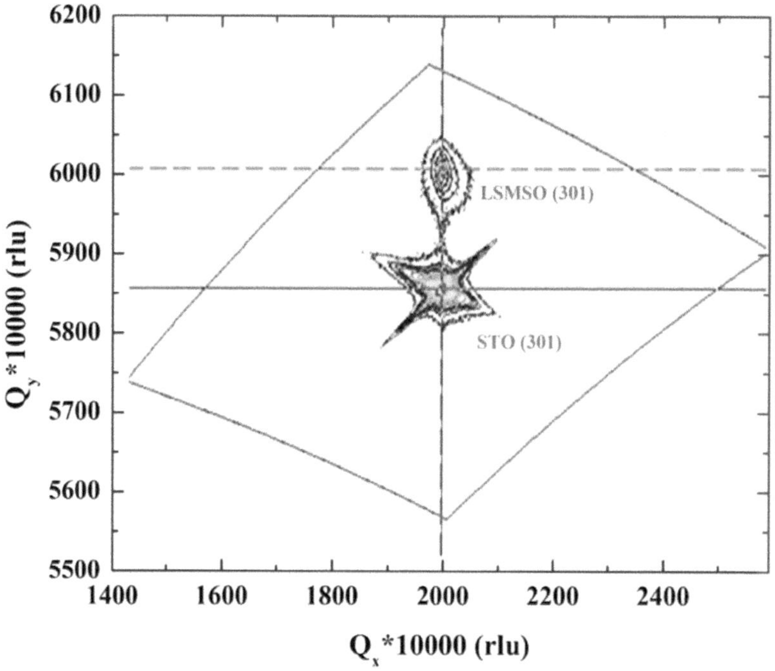

FIGURE 10.4 Room temperature RSM of LSMSO thin film on the STO substrate around (301) peak.

10.3.1 SURFACE MORPHOLOGY AND MICROSTRUCTURE

Figure 10.5 shows (a) 2D and (b) 3D images of AFM micrographs of LSMSO/STO film. The film surface probed by AFM reveals that it acquires well-developed grain growth, with a pronounced grain size indicating the better formation of an island-like growth of LSMSO film on the STO substrate. It can be confirmed from Figure 10.5 that the RMS (root mean square) surface roughness for LSMSO film is ~1.25 nm. AFM micrographs also indicate that the film possesses excellent grain and domain orientation, alignment, and ordering. As a result, electronic transport across grain interfaces would be improved as analyzed in detail in the next section.

TABLE 10.2 Values of RMS Roughness, Average Height, and Grain Size of LSMSO Thin Film.

RMS roughness (nm)	Average height (nm)	Maximum height (nm)	Grain size (nm)
1.25	8.25	18.15	45

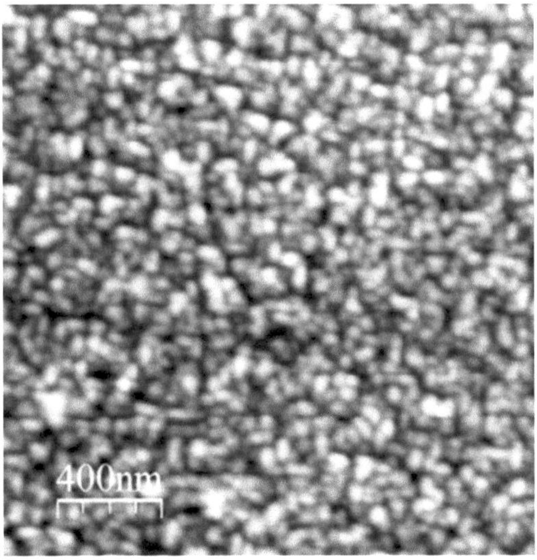

FIGURE 10.5(a) AFM 2D micrograph of PLD-grown LSMSO/STO thin film.

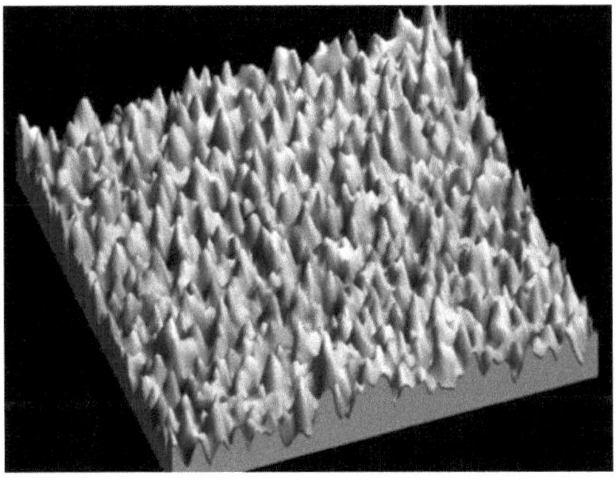

FIGURE 10.5(b) AFM 3D micrograph of PLD-grown LSMSO/STO thin film.

10.4 ELECTRICAL-TRANSPORT AND MAGNETO-TRANSPORT STUDIES

10.4.1 ELECTRICAL-TRANSPORT STUDIES

Figure 10.6 shows the variation of resistivity (ρ) as a function of temperature (T) for the LSMSO film on the STO substrate, in the temperature range of 2 K–300 K without and with applied magnetic fields. The metal-insulator transition temperature (T_P) for the LSMSO thin film is around 295 K for Sn^{4+} substitution under the 0 T field, while the LSMO thin film on the STO single crystal substrate shows T_P at around 315 K.[17] Due to the scattering of the carriers by the Mn-site, it has been observed that the resistivity of the manganite thin film at the T_P increases and is susceptible to the Mn-site substitution.[6] The value of $T_P \sim 297$ K under the applied magnetic field of 1 T behaves similarly but at a further high magnetic field of 8 T, T_P shifts towards the high-temperature region above 300 K.

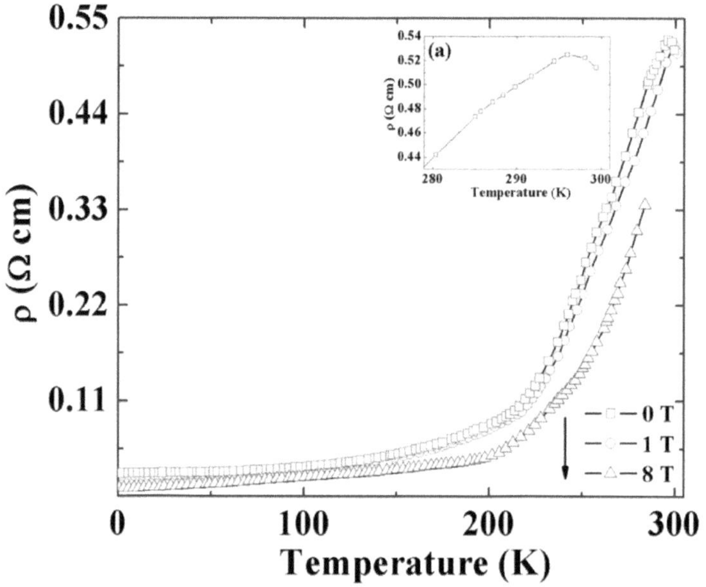

FIGURE 10.6 ρ vs. *T* plot for the LSMSO thin film without and with the magnetic field. (a) Enlarged part of T_P at 0 T.

The double exchange (DE) process, wherein a parallel alignment of the magnetic moments facilitates the transfer of electrons between Mn^{3+} and

Mn^{4+} ions, is compatible with the increase in resistivity near T_P and the metallic conducting behavior below $T_{P.}$ The results presented in Figure 10.6 may be ascribed to the crucial role played by the Sn-ion. This could be mainly attributable to the random potential of Sn^{4+}, leading to the localization of the carrier concentration. In addition, the decrease in the average Mn-O-Mn bond angle and the increase in Mn-O bond length have been caused by larger Sn^{4+} ions replacing smaller Mn^{4+} ions. This fact consequently reduces the single electron bandwidth, and leads to reduced T_P. This makes sense because the Mn^{4+} ion is replaced by Sn^{4+}, which disrupts the Mn^{3+}–O^{2-}–Mn^{4+} network. It is found that, as the magnetic field increases, the resistivity values decrease and the TP shifts toward the high-temperature region. Therefore, it implies that the spin-dependent scattering in bulk samples, impeded by the external magnetic field, might have been the origin of the resistivity minimum. Further, it can be noted that just a small distortion in the bond length or bond angle of the Mn-O-Mn significantly affects the transport and magneto-transport properties.[25, 26] The strain, which can be produced by, among others, the lattice mismatch between the STO single crystal substrate and LSMSO film as a result of substitution of the Sn ion at the Mn site, has a significant impact on the Mn-O-Mn bond angles. In order to understand the conduction process responsible for the electrical-transport in the metallic region, we have used an equation to analyze the resistivity data

$$\rho = \rho_0 + \rho_2 T^2 + \rho_{4.5} T^{4.5} \qquad (10.1)$$

where the term ρ_0 denotes residual resistivity and is temperature-independent; the second term ρ_2 implies electron-electron and electron-phonon interactions; and the third term with a coefficient of $\rho_{4.5}$ indicates electron-magnon scattering.[26] The fitted data are shown as a solid line in Figure 10.7 for the LSMSO thin film without and with a magnetic field. We observe that the electrical resistivity data below T_P best fits the equation (1), suggesting electron-electron and electron-magnon scattering processes as the causes of the electrical resistivity in this region. The fitted parameters ρ_0, ρ_2, and $\rho_{4.5}$ are listed in Table 10.3 for H = 0 T, 1 T, and 8 T. The table data suggests clearly that a combination of electron-electron, electron-phonon, and electron-magnon scattering processes mainly governs the metallic region. It can also be observed that all three parameters decrease with the application of the magnetic field. As the magnetic field increases, the value of ρ_0 reduces which reflects a strong disorder characteristic for Sn-substituted manganite compounds. Additionally, the domains expand when a magnetic field is applied, leading the value of ρ_0 to decrease. In contrast, the decrease in ρ_2 and $\rho_{4.5}$ could have been attributed to a reduction in electron spin fluctuations in the presence of the magnetic field.

The reduction in ρ_2 and $\rho_{4.5}$ parameters follows the same variation trend and was found to decrease with the magnetic field.

TABLE 10.3 The Values of the Parameters ρ_0, ρ_2, and $\rho_{4.5}$ Were Calculated by Fitting Equation (10.1) to the Metallic Region both with and Without a Magnetic Field.

Magnetic field (T)	ρ_0 ($10^{-4}\Omega$-cm)	ρ_2 ($10^{-4}\Omega$-cm K^2)	$\rho_{4.5}$ ($10^{-14}\Omega$-cm $K^{4.5}$)
0	0.345	4.125	4.654
1	0.240	3.054	4.305
8	0.175	2.187	1.853

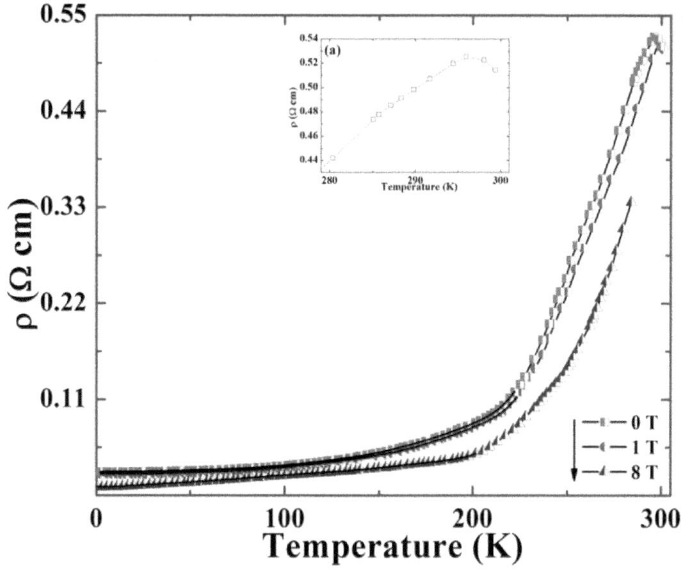

FIGURE 10.7 Fitted temperature dependence of resistivity plot of LSMSO film in a metallic region without and with a magnetic field.

10.4.2 FIELD DEPENDENCE OF MAGNETO-RESISTANCE

The resistivity as a function of magnetic field (H) at constant temperatures was measured to differentiate the intrinsic and granular parts of MR. The field was applied parallel to the film plane during the MR measurement. Figure 10.8 shows the MR% vs. H isotherms for the LSMSO/STO thin film at different temperatures. This figure indicates that the MR exhibits a sharp linear decline at all temperatures up to 8 T magnetic field. The MR (%) is calculated using the relation:

$$\text{MR (\%)} = \frac{\rho H - \rho 0}{\rho 0} \times 100 \tag{10.2}$$

Here, ρ_0 and ρ_H, denote the resistivity before and after the applied magnetic field respectively. It should be noted that Figure 10.8 displays two distinguishable parts, extrinsic MR or Low Field MR (LFMR) followed by $H \leq 1$ T and intrinsic MR, with the progression of the applied magnetic field ($H \geq 1$ T). Extrinsic MR is the result of a vital drop in resistance due to the spin-polarized tunneling effect.[27] This is caused by the enhancement of electron hopping, for example, as randomly oriented spins in the grains are aligned in the direction of a weak magnetic field.[28] The spins within the cores of the grains are further aligned in intrinsic MR, at a larger H, which favors a gradual decrease in resistance. Since no LFMR is observed in an entire magnetic field, the LSMSO thin film doesn't show a spin-polarized tunneling effect due to the grain boundary at a low field ($H < 1$ T). This behavior shows the single crystalline nature of the thin film. The Isotherms MR% values at 100 K, 240 K, and 300 K (corresponding to its T_P) up to 8 T magnetic fields are tabulated in Table 10.4. These results can be understood in terms of Sn substitution, which is affected by field-induced magneto-resistance.

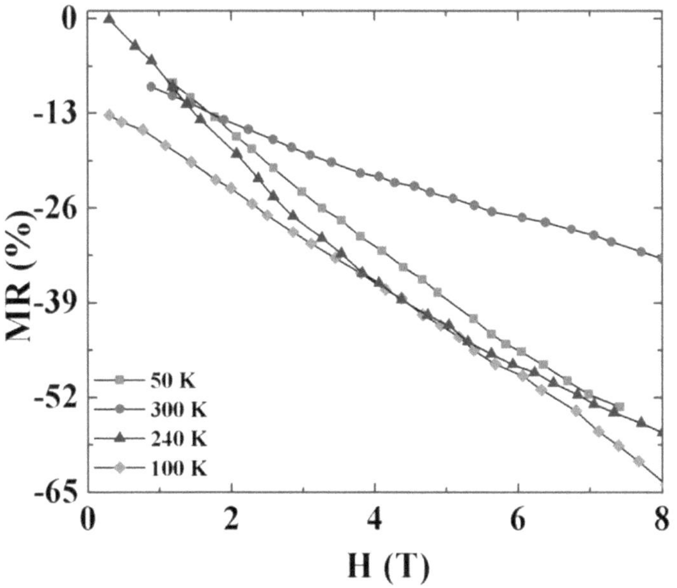

FIGURE 10.8 MR% vs. H isotherms for the LSMSO thin film at various temperatures.

TABLE 10.4 Isotherms MR (%) Values of LSMSO Thin Film at Various Temperatures.

Isotherm MR (%)			
50 K	100 K	240 K	300 K
52	65	59	33

10.4.3 TEMPERATURE COEFFICIENT OF RESISTANCE (TCR)

To explore the possibility of this PLD-grown LSMSO film for its bolometric application, the TCR which is a significant parameter, has to be examined properly. High bolometric device efficiency is implied by a high TCR. The temperature sensitivity of resistivity has been calculated from the ρ-*T* data. This parameter is quantified as the temperature coefficient of resistance and calculated from

$$TCR\% = \frac{\partial R}{\partial T} \times \frac{1}{R} 100\% \qquad (10.3)$$

In Figure 10.9, the TCR behavior of LSMSO film is shown as a function of temperature. The TCR vs. Temperature exhibits the same behavior as ρ

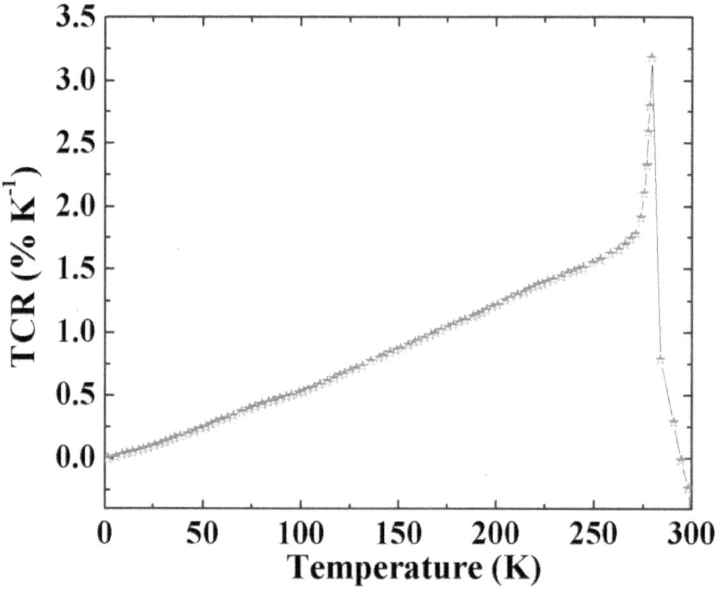

FIGURE 10.9 Variation of TCR% as a function of temperature for the LSMSO thin film.

vs. T with a shift from the paramagnetic-insulator state to the ferromagnetic-metallic state. The T_m is the temperature that indicates the highest TCR value. The devices must be cooled to the temperature T_m, at which the sample exhibits the highest TCR value, for application as an IR detector. It can be seen that the T_m value is about 10 K lower than T_P. For PLD-grown thin film, Todd et al.[29] have reported a T_m value 15 K lower than the T_P. The highest value of the TCR is 3.18 %/K at 285 K for LSMSO thin film. This feature is extremely beneficial for applications using bolometric sensors.

10.5 MAGNETIC PROPERTIES

10.5.1 MAGNETIZATION MEASUREMENT

To understand the magnetic phase transition of the LSMSO thin film, ZFC and FC magnetization measurements have been performed in the temperature range of 10 K to 300 K with an applied magnetic field of 500 Oe and the results are displayed in Figure 10.10. The temperature corresponding to a minimum in the derivative dM/dT of the M-T curve is the Curie temperature (T_C). The LSMSO thin film exhibits T_C at 276 K for 5% Sn content at the Mn site. Note the unusual doping effect of Sn which shows a very marginal decrease in T_c for the bulk LSMSO (5% Sn substitution) at 331 K (ΔTc~55 K)[24] This lower T_c value of thin film as compared to bulk is because, during thin film deposition, the effect of O_2 annealing time and O_2 partial pressure,

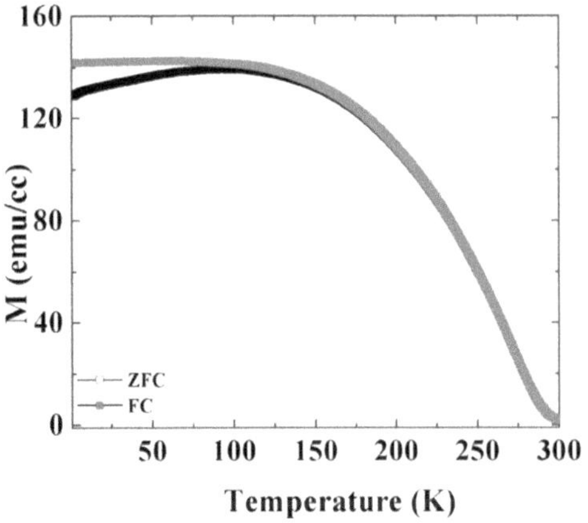

FIGURE 10.10 ZFC and FC magnetization of the LSMSO thin film as a function of temperature at 500 Oe.

strain, and structural distortions are responsible for such magnetic behavior and appear to account for the reduction in T_C. Since the Sn^{4+} cations cannot magnetically connect to manganese, neither ferromagnetism nor metallicity can be produced via these cations. The Sn substitution in thin film results in the 55 K difference between T_C and T_P and thus is more impactful than the Fe substitution, which has been seen in $(La_{0.67}Sr_{0.33})(Mn_{0.95}Fe_{0.05})O_3$ manganite (T_C–T_P = 65 K).[30] Furthermore, Sn^{4+} cations create local lattice deformation due to their larger size, which reduces electron hopping, resulting in higher suppression of T_C and T_P in this LSMSO thin film.

These results of both electrical transport and magnetic measurements indicate that the transition from FM metal to PM insulator in $La_{0.67}Sr_{0.33}Mn_{0.95}Sn_{0.05}O_3$ appears to be via.FM insulator phase. Figure 10.11 shows the temperature-dependent inverse of the susceptibility ($1/\chi$) of the PM region (above T_C). At relatively high temperatures, where the curve shows a distinct value of the paramagnetic Curie temperature, $\theta = 284$ K, it can be fitted by the Curie Weiss law, $\chi = C/(T - \theta)$, leading to $T_C = 276$ K. A comparison of Curie temperature (T_C) with that of $La_{0.67}Sr_{0.33}MnO_3$ on STO single crystal substrate shows that the reported T_C is around 332 K[31] and reveals that the 5% Sn substitution at the Mn site leads to decrease in T_C (276 K).

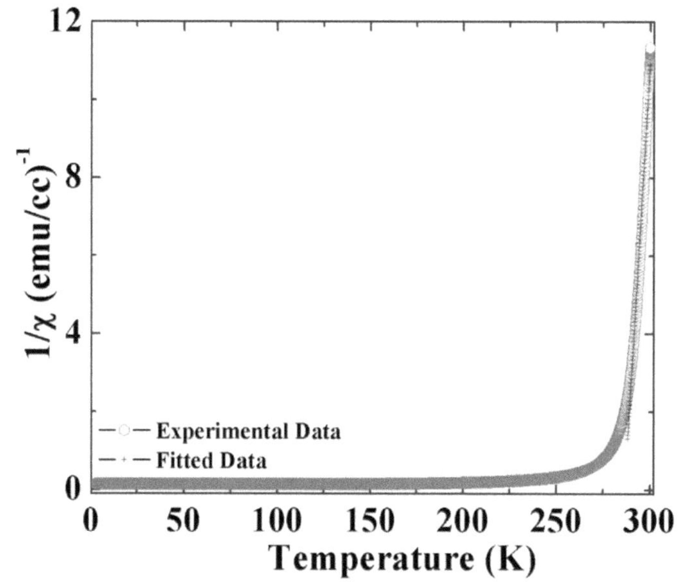

FIGURE 10.11 Inverse susceptibility (χ = M/H in H = 500 Oe) versus temperature for the LSMSO thin film.

10.5.2 FIELD DEPENDENCE OF MAGNETIZATION

Figure 10.12 (a,b) shows the magnetization (M) versus magnetic field (H) isotherms recorded for LSMSO thin film at 10 K and 300 K up to a field of 5 T. It shows a saturation magnetization (M_S) at a field of 5 Tesla. It can be seen that the film shows a weak ferromagnetic behavior with loop saturation around 2 T measured at a temperature of 10 K (Fig. 10.12 a). The ferromagnetism in $La_{0.67}Sr_{0.33}MnO_3$ originates from a double exchange interaction between the Mn^{3+} and Mn^{4+} ions. Both Mn^{3+} and Mn^{4+} have the electronic configurations $3d^4 4s^0 = (t_{2g}^3 e_g^1) 4s^0$ and $3d^3 4s^0 = (t_{2g}^3 e_g^0) 4s^0$, respectively, while the same for non-magnetic Sn^{4+} cation is $4d^{10} 5s^2 5p^2$. Hence no direct magnetic contribution to the system could be expected. Thus, the doping Sn^{4+} significantly alters the magnetic features of the $La_{0.67}Sr_{0.33}Mn_{0.95}Sn_{0.05}O_3$ thin film sample.

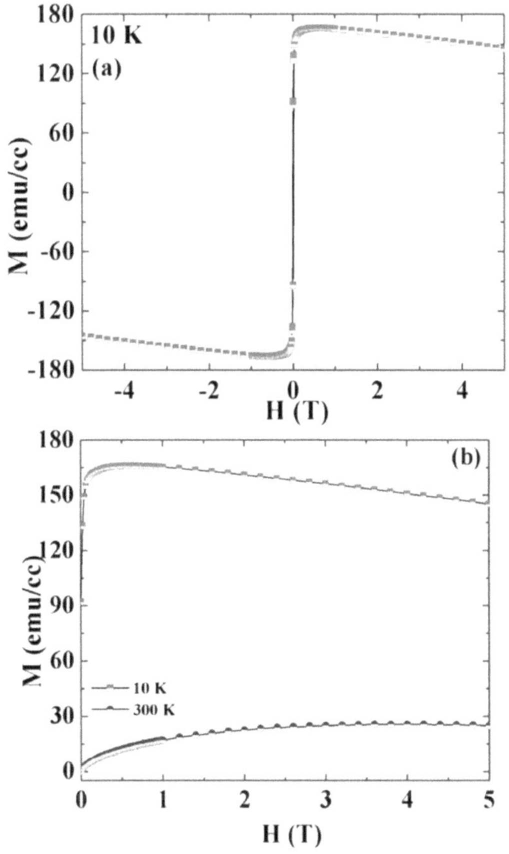

FIGURE 10.12 Magnetization vs. *H* (a) at 10 K (b) Isotherms for the LSMSO thin film.

10.6 CONCLUSIONS

In summary, we have grown the thin film of $La_{0.67}Sr_{0.33}Mn_{0.95}Sn_{0.05}O_3$ on a single crystalline $SrTiO_3$ (100) substrate using the PLD technique. By substituting the non-magnetic tetravalent Sn^{4+} ion at the Mn-site of the $La_{0.67}Sr_{0.33}Mn_{0.95}Sn_{0.05}O_3$ thin film, The structural, electrical, and magnetic properties have been studied. The crystalline structure of the thin film is not largely altered when Mn is substituted with 5% Sn. The thin film showed a phase transition from paramagnetic to ferromagnetic coupled with an insulator-metal transition. There is a slight difference between these two phase transitions because of the larger Sn^{4+} cations that create local lattice deformation that reduces the electron hopping, and results in higher suppression of T_C and T_P in this LSMSO thin film. The PLD-grown $La_{0.67}Sr_{0.33}Mn_{0.95}Sn_{0.05}O_3$ thin film shows a high MR and TCR value near room temperature, which makes it especially suitable for next-generation spintronic and IR bolometer devices.

ACKNOWLEDGMENT

One of the authors (TMT) acknowledges the award of a postdoctoral fellowship at IISER, Bhopal. TMT is also grateful to Dr. B. R. Naik, Principal, The Patidar Gin Science College, Bardoli, for motivating him to conduct research work at the institute.

KEYWORDS

- **manganites thin film**
- **Mn-site substitution**
- **magnetoresistance**
- **pulsed laser deposition**
- **bolometer sensor**

REFERENCES

1. Ohtomo, A.; Hwang, H. Y. A High-Mobility Electron Gas at the $LaAlO_3/SrTiO_3$ Hetero-interface. *Nature* **2004,** *427,* 423–427.
2. Dagotto, E. Complexity in Strongly Correlated Electronic Systems. *Science* **2005,** *309,* 257–262.

3. Catalan, G.; Scott, J. F. Physics and Applications of Bismuth Ferrite. *Adv. Mater.* **2009,** *21,* 2463–2485.
4. Baldini, M.; Muramatsu, T.; Sherafati, M.; Mao, H. K.; Malavasi, L.; Postorino, P.; Satpathy, S.; Struzhkin, V. V. Origin of Colossal Magnetoresistance in $LaMnO_3$ Manganite. *Proc. Natl. Acad. Sci.* **2015,** *112,* 10869–10872.
5. Renard, J.-P.; Haghiri-Gosnet A-M. CMR Manganites: Physics, Thin Films and Devices. *J. Phys. D. Appl. Phys.* **2003,** *36,* R127–R150.
6. Raveau, B. The Crucial Role of Mixed Valence in the Magnetoresistance Properties of Manganites and Cobaltites. *Philos. Trans. R. Soc. A Math. Phys. Eng. Sci.* **2008,** *366,* 83–92.
7. Tokura, Y. Critical Features of Colossal Magnetoresistive Manganites. *Reports Prog. Phys.* **2006,** *69,* 797–851.
8. Chen, X.; Dong, S.; Wang, K.; Liu, J. M.; Dagotto, E. Nonmagnetic B-Site Impurity-Induced Ferromagnetic Tendency in CE-Type Manganites. *Phys. Rev. B* **2009,** *79,* 2–8.
9. Tank, T. M.; Bhargava, D.; Sridharan, V.; Samatham, S. S.; Ganesan, V.; Sanyal, S. P. Influence of Mn Site Substitution on Electrical Resistivity and Magnetoresistance Properties of Rare Earth Manganite. *Adv. Mater. Res.* **2014,** *1047,* 123–129.
10. Kim, Y.; Ryu, S.; Jeen, H. Strain-Effected Physical Properties of Ferromagnetic Insulating $La_{0.88}Sr_{0.12}MnO_3$ Thin Films. *RSC Adv.* **2019,** *9,* 2645–2649.
11. Annese, E.; Mori, T. J. A.; Schio, P.; Bonfim, R. P. F.; Rache Salles, B.; Peña Serna, J. D.; Cezar, J. C. Magnetic Properties of $La_{0.67}Sr_{0.33}MnO_3$ Epitaxial Thin Films Grown on $SrTiO_3$(001). *J. Magn. Magn. Mater.* **2020,** *507,* 166812.
12. Burema, A. A.; van Rijn, J. J. L.; Banerjee, T.; Temperature-Dependent out-of-Plane Anisotropy in Compressively Strained $La_{0.67}Sr_{0.33}MnO_3$ Thin Films. *J. Magn. Magn. Mater.* **2022,** *549,* 168910-168910-5.
13. Hwang, H. Y.; Cheong, S. W.; Ong, N. P.; Batlogg, B. Spin-Polarized Intergrain Tunneling in $La_{2/3}Sr_{1/3}MnO_3$. *Phys. Rev. Lett.* **1996,** *77,* 2041–2044.
14. Tank, T. M.; Thaker, C. M.; Chhatrala, R. S.; Ganesan, V.; Sanyal, S. P.; Enhancement of Temperature and Field Coefficient of Resistance in CSD Grown Nanostructure $La_{0.7}Ca_{0.3}MnO_3$ Thin Films. *J. Nano Res.* **2013,** *24,* 155–162.
15. Chaluvadi, S. K.; Polewczyk, V.; Petrov, A. Y.; Vinai, G.; Braglia, L.; Diez, J. M.; Pierron, V.; Perna, P.; Mechin, L.; Torelli, P.; Orgiani, P. Electronic Properties of Fully Strained $La_{1-x}Sr_xMnO_3$ Thin Films Grown by Molecular Beam Epitaxy ($0.15 \leq x \leq 0.45$). *ACS Omega* **2022,** *7,* 14571–14578.
16. Wong, Y. J.; Lau, L. N.; Lim, K. P.; Hon, X. T.; Daud, N. A. A.; Kechik, M. M. A.; Chen, S. K.; Shabdin, M. K. Bin; Shaari, A. H.; Miryala, M. Characterisations of La-Sr-Mn-O (LSMO) Thin Film Fabricated by RF Sputtering. *Coatings* **2023,** *13,* 541-541-11.
17. Majumdar, S.; Van Dijken, S. Pulsed Laser Deposition of $La_{1-x}Sr_xMnO_3$: Thin-film Properties and Spintronic Applications. *J. Phys. D. Appl. Phys.* **2014,** *47,* 034010-034010-15.
18. Chaluvadi, S. K.; Ajejas, F.; Orgiani, P.; Rousseau, O.; Vinai, G.; Petrov, A. Y.; Torelli, P.; Pautrat, A.; Camarero, J.; Perna, P.; Mechin, L. Room Temperature Biaxial Magnetic Anisotropy in $La_{0.67}Sr_{0.33}MnO_3$thin Films on $SrTiO_3$ Buffered MgO (001) Substrates for Spintronic Applications. *Appl. Phys. Lett.* **2018,** *113,* 052403-052403-5.
19. Zhan, J. M.; Li, P. G.; Liu, H.; Tao, S. L.; Ma, H.; Shen, J. Q.; Pan, M. J.; Zhang, Z. J.; Wang, S. L.; Yuan, G. L.; Carrier Tuning The Metal-insulator Transition of Epitaxial $La_{0.67}Sr_{0.33}MnO_3$ Thin Film on Nb doped $SrTiO_3$ Substrate. *AIP Adv.* **2016,** *6,* 045001–045001-6.

20. Schopf, J.; Thampi, A.; Milde, P.; Ivaneyko, D.; Kondovych, S.; Kononenko, D. Y.; Eng, L. M.; Jin, L.; Yang, L.; Wysocki, L.; van Loosdrecht, P. H. M.; Richter, K.; Yershov, K. V.; Wolf, D.; Lubk A.; Lindfors-Vrejoiu, I. Neel Skyrmion Bubbles in $La_{0.7}Sr_{0.3}Mn_{1-x}Ru_xO_3$Multilayers. *Nano Lett.* **2023,** doi.org/10.1021/acs.nanolett.3c00693.
21. Othmani, I.; Regaieg, Y.; Ayadi, F.; Cheikhrouhou-Koubaa, W.; Koubaa, M.; Nachbaur, V.; Abdelhedi, M.; Cheikhrouhou, A. Effect of Lithium Substitution on the Structural, Magnetic, and Magnetocaloric Properties of $La_{0.7}Sr_{0.3-x}Li_xMnO_3$ ($0 \leq x \leq 0.15$). *J. Supercond. Nov. Magn.* **2023,** doi.org/10.1007/s10948-023-06554-z.
22. Shanon, R. D. Revised Effective Ionic Radii and Systematic Studies of Interatomie Distances in Halides and Chaleogenides. *Acta Cryst.* **1976,** *A32,* 751-767.
23. Krishnan, K. M.; Lu, H. L. Role of Stoichiometry and Structure in Colossal Magnetoresistive $La_{1-x}Sr_xMn_{1-y}Ru_yO_3$. *Phys. Rev. B* **1999,** *60,* 14793-14803.
24. Tank, T. M.; Das, S.; Rana, D. S.; Thaker, C. M.; Sanyal, S. P. Magnetic field Driven Anomalous Electrical Behavior of $La_{0.67}Sr_{0.33}Mn_{1-x}Sn_xO_3$. *Manuscript Submitted to Mat. Res. Bull.* **2023.**
25. Ju, H. L.; Gopalakrishnan, J.; Peng, J. L.; Li, Q.; Xiong, G. C.; Venkatesan, T.; Greene, R. L. Dependence of Giant Magnetoresistance on Oxygen Stoichiometry and Magnetization in Polycrystalline $La_{0.67}Ba_{0.33}MnO_3$. *Phys. Rev. B* **1995,** *51,* 6143–6146.
26. Tank, T. M.; Shelke, V.; Das, S.; Rana, D. S.; Thaker, C. M.; Samatham, S. S.; Ganesan, V.; Sanyal, S. P. Low Temperature Transport Anomaly in Cr Substituted ($La_{0.67}Sr_{0.33}$) MnO_3 Manganites. *J. Magn. Magn. Mater.* **2017,** *432,* 581–590.
27. Lau, L. N.; Lim, K. P.; Ishak, A. N.; Kechik, M. M. A.; Chen, S. K.; Ibrahim, N. B.; Miryala, M.; Murakami, M.; Shaari, A. H.; The Physical Properties of Submicron and Nano-Grained $La_{0.7}Sr_{0.3}MnO_3$ and $Nd_{0.7}Sr_{0.3}MnO_3$ Synthesised by Sol–Gel and Solid-State Reaction Methods. *Coatings*, **2021,** *11,* 361-361-15.
28. Lim, K. P.; Albert Gan, H. M.; Chen, S. K.; Talib, Z. A.; Halim, S. A.; Ng, S. W.; Chin, H. W. Lattice Strain Effect in Structural, Magnetoresistance and Electrical Properties of $La_{0.67}Sr_{0.33}MnO_3$ Bulk and Thin Film System. *Mater. Sci. Forum* **2016,** *846,* 635–640.
29. Todd, M. A.; Donohue, P. P.; Wright, P. J.; Crosbie, M. J.; Lane, P. A.; Jo, M. H.; Pang, B. S. H.; Blamire, M. G. Colossal Magnetoresistive Manganite Thin-Films for Infrared Detection and Imaging. *Ann. der Phys.* **2004,** *13,* 48–51.
30. Zhou, Y.; Zhu, X.; Shengli, L. Structure, Magnetic, Electrical Transport and Magnetoresistance Properties of $La_{0.67}Sr_{0.33}Mn_{1-x}Fe_xO_3$($x = 0$–$0.15$) Doped Manganite Coatings. *Ceramics Int.* **2017,** *43,* 3679-3687.
31. Dallocchio, M.; Boileau A.; Mercey, B.; David, A.; Lüders, U.; Froissart, S.; Larose, X.; Berini, B.; Dumont, Y.; Pautra, A. Tunable Magnetic and Magnetotransport Properties in Locally Epitaxial $La_{0.67}Sr_{0.33}MnO_3$ Thin Films on Polycrystalline $SrTiO_3$ by Control of Grain Size. *J. Phys. D: Appl. Phys.* **2022,** *55,* 235303.

CHAPTER 11

Effect of Composition on the Structural and Elastic Properties of Boron-based III–V Ternary Alloys Using First-Principles Calculations

ARUN NEGI and NEHA MUNJAL

Department of Physics, Lovely Faculty of Science and Technology, Lovely Professional University, Phagwara, India

ABSTRACT

In the projected plan of the study, we have investigated ternary semiconductor compounds for their structural and elastic properties with varying concentrations of constituents based on First-Principles calculations. The computational methods will be in accordance with Density Functional Theory and the coding will be done on the CRYSTAL coding software which is a coding software based on the ab-initio methods. The ternary systems investigated in this study are BN_xP_{1-x}, BN_xAs_{1-x}, and BP_xAs_{1-x} in the NaCl (B1) and the zinc blende (B3) phases. In the study, the lattice constants, the bulk modulus, and the pressure derivative of the bulk modulus for all the three systems have been determined in the B1 and B3 phases. Apart from these properties the elastic constants have also been determined in the B3 phase. Properties such as Poisson's ratio, Young's Modulus, and shear modulus which are the functions of the elastic constants have been determined based on the elastic constants for the B3 phase. The outcomes of the study have been discussed in the Results and Conclusions section. The energy vs volume plots have also been plotted for all the intermediate compounds in both the B1 and the B3 phase. The energy vs volume plots have also been discussed in the Results section.

Contemporary Advancements in Materials Technology: Novel Solutions to Complex Problems.
Shrikaant Kulkarni, Vipul Srivastava, & P. William (Eds.)

11.1 INTRODUCTION

Semiconductor alloys find applications in a vast area of technological fields, with the manufacturing of electronic and electro-optical devices being the fields in which they are mostly used.[1] However, they are not just limited to these fields; they still find applications in various technologically important areas. The most reliable and convenient way of altering the properties of semiconductors is by alloying.[2] Currently, we are investigating the III–V semiconductor ternary systems obtained by slowly adding slight bit of a material in the compound and thus forming a ternary alloy, which is an easy way of forming material with intermediate properties with respect to the two end products. The III–V semiconductors find uses in high-performance optoelectronic devices which is due to their superior electronic properties such as direct band gap, low exciton binding energy, and high electron mobility.[3] Even though these materials have a lot of properties and applications, the III–V materials are still less developed as compared to the II–IV and VI–IV semiconductor compounds which can be attributed to the fact that they are hard to crystallize in the solution phase due to more covalent bonding environment.[3] Since this issue has been removed by the introduction of modern thin-film growth techniques, which facilitates the development of artificial crystalline structures which naturally do not occur, this issue is no longer a barrier to the study of the nitride compounds now. This can be seen by the successes in the studies related to the group III nitrides in the last decade.[4] Other areas for which these materials can become a suitable option are the development of 2D materials having moderate band gaps and high carrier mobility for which the boron phosphide layer has been shown to be a contender.[5] These properties and applications are the cause of interest in the study of the boron-V semiconductor compounds.

In this study, we have determined the structural and elastic properties such as lattice constants, bulk modulus, elastic constants, and other properties that can be derived from the elastic constants such as the shear modulus, Young's modulus of some of the ternary semiconductor compounds BN_xP_{1-x}, BN_xAs_{1-x}, BP_xAs_{1-x}. Furthermore, the energy vs volume diagrams have been plotted.

11.2 EXPERIMENTAL METHODS AND MATERIALS

In this study, we employ the first-principles linear combination of atomic orbitals (LCAO) method to investigate the structural properties of some ternary semiconductor compounds, specifically BN_xP_{1-x}, BN_xAs_{1-x}, and BP_xAs_{1-x}.

The LCAO method has demonstrated its effectiveness in numerous cases. Our calculations are based on solving the Kohn–Sham Hamiltonian equation within the density functional theory, utilizing the CRYSTAL code.

To construct the local functions for B, P As, and N, Gaussian basis sets are utilized. Structural optimization of $BNxP_{1-x}$, BN_xAs_{1-x}, and $BP_xAs_{1-x.}$ As is performed by minimizing the total energy with respect to the volume of the cell for each composition. The resulting parameters are then compared with the previous findings.

The properties that are based on the elastic constant such as anisotropy factor (A), the Reuss shear modulus (G_R), Voigt shear modulus (G_v), the Hill shear modulus (G_H), the Young's modulus (E), and Poisson's ratio (v) are calculated with the use of the following formulas:

$$\text{Bulk modulus: } B = \frac{C_{11} + 2C_{12}}{3}$$

$$\text{Poisson's ratio: } v = \frac{3B - 2G_H}{2(3B + G_H)}$$

$$\text{Anisotropy factor: } A = \frac{2C_{44}}{C_{11} - C_{12}}$$

$$\text{Voigt shear modulus: } G_v = \frac{C_{11} - C_{12} + C_{44}}{5}$$

$$\text{Reuss shear modulus: } G_R = \frac{5C_{44}(C_{11} - C_{12})}{4C_{44} + 3(C_{11} - C_{12})}$$

$$\text{Hill shear modulus: } G_H = \frac{G_V + G_R}{2}$$

$$\text{Young's modulus (E): } E = \frac{9BG_H}{G_H + 3B}$$

The energy has been calculated using the Birch Murnaghan equation that is given by

$$E(V) = E_0 + \frac{9V_0B_0}{16}\left[\left(\frac{v_0}{v}\right)^{\frac{2}{3}} - 1\right]^3\left[6 - 4\left(\frac{V_0}{V}\right)^{\frac{2}{3}}\right]B_0'\left[\left(\frac{v_0}{v}\right)^{\frac{2}{3}} - 1\right]^2$$

The energy has been determined using the theoretical model. The two energies calculated are then plotted and compared in the energy–volume diagrams in the Results and Discussion section.

11.3 RESULTS AND DISCUSSION

The above table shows the values of the lattice constants, the bulk modulus, and the pressure derivative of the bulk modulus for the compounds BN_xP_{1-x}, BN_xAs_{1-x}, and BP_xAs_{1-x} in the B1 phase.

For the BN_xP_{1-x} sample in the B1 phase the lattice constant has a value of 4.21 Å when there is 75% concentration of phosphorus and 25% concentration of nitrogen while the bulk modulus takes a value of 215 GPa. When the concentration becomes even at 50% each for nitrogen and phosphor the value of the lattice constant reduces to 3.97 Å while the bulk modulus increases to 298 GPa. When the concentration is again changed to 25% phosphor and 75% nitrogen, the values for the lattice constant reduce to 3.82 Å and the value of the bulk modulus increases to 340 GPa. In general, for BN_xP_{1-x}, the lattice constant decreases while the value of bulk modulus increases with an increase in the concentration of nitrogen in the sample.

For the BN_xAs_{1-x} sample in the B1 phase the lattice constant takes the values of 4.34 Å at 25% nitrogen and 75% arsenide, 4.02 Å at a concentration of 50% both nitrogen and arsenide, and 3.74 Å at 75% nitrogen and 25% arsenide. The bulk modulus takes the values of 239 GPa at 25% nitrogen and 75% arsenide, 294 GPa at a concentration of 50% both nitrogen and arsenide, and 329 GPa at 75% nitrogen and 25% arsenide. Again, we get a decreasing trend in the values of the lattice constant and an increasing trend in the values of the bulk modulus with the increasing concentration of nitrogen.

For the BP_xAs_{1-x} sample in the B1 phase the lattice constant is 4.51 Å and the bulk modulus is 144 GPa at 25% phosphorus and 75% arsenide. At 50% phosphorus and 50% arsenide the lattice constant is 4.47 Å and the bulk modulus is 148 GPa. At 75% phosphorus and 25% arsenide the lattice constant is 4.35 Å and the bulk modulus is 152 GPa.

The values of the partial derivative of the bulk modulus with respect to pressure in the NaCl(B1) phase for all the given above samples are also given in Table 11.1.

For the BN_xP_{1-x} sample in the B3 phase the lattice constant takes the values of 4.52 Å for BP, 4.37 Å at 25% nitrogen and 75% phosphorus, 4.21 Å at a concentration of 50% both nitrogen and phosphorus, and 3.98 Å at 75% nitrogen and 25% phosphorus and 3.61 Å for BN. The bulk modulus takes the values of 188 GPa at 25% nitrogen and 75% phosphorus, 216 GPa at a concentration of 50% both nitrogen and phosphorus, and 269 GPa at 75% nitrogen and 25% phosphorus. We get a decreasing trend in the values of lattice constant and an increasing trend in the values of the bulk modulus with the increasing concentration of nitrogen.

TABLE 11.1 The Lattice Constant, the Bulk Modulus, and the Pressure Derivative of the Bulk Modulus for BN_xP_{1-x}, BN_xAs_{1-x}, BP_xAs_{1-x} in B1 Phase.

		BNxP1-x		
	X	**a (Å)**	**B (GPa)**	**B'**
BP	0			
$BN_{0.25}P_{0.75}$	0.25	4.21	215	4.12
$BN_{0.5}P_{0.5}$	0.5	3.97	298	4.13
$BN_{0.75}P_{0.25}$	0.75	3.82	340	4.14
BN	1			
		BNxAs1−x		
	X	**a (Å)**	**B (GPa)**	**B'**
BAs	0			
$BN_{0.25}As_{0.75}$	0.25	4.34	239	4.24
$BN_{0.5}As_{0.5}$	0.5	4.02	294	4.2
$BN_{0.75}As_{0.25}$	0.75	3.74	329	4.18
BN	1			
		BPxAs1−x		
	x	**a(Å)**	**B (GPa)**	**B'**
BAs	0			
$BP_{0.25}As_{0.75}$	0.25	4.51	144	4.23
$BP_{0.5}As_{0.5}$	0.5	4.47	148	4.16
$BP_{0.75}As_{0.25}$	0.75	4.35	152	4.14
BP	1			

For the BN_xAs_{1-x} sample the lattice constant takes the values of 4.73 Å for BAs, 4.65 Å at 25% nitrogen and 75% arsenide, 4.41 Å at a concentration of 50% both nitrogen and arsenide, 4.12 Å at 75% nitrogen and 25% arsenide, and 3.61 Å for BN. The bulk modulus takes the values of 158 GPa at 25% nitrogen and 75% arsenide, 192 GPa at a concentration of 50% both nitrogen and arsenide, and 256 GPa at 75% nitrogen and 25% arsenide. We get a decreasing trend in the values of the lattice constant and an increasing trend in the values of the bulk modulus with the increasing concentration of nitrogen.

For the BP_xAs_{1-x} sample the lattice constant takes the values of 4.73 Å for BAs, 4.71 Å at 25% phosphorus and 75% arsenide, 4.69 Å at a concentration of 50% both phosphorus and arsenide, and 4.64 Å at 75% phosphorus and 25% arsenide, 4.53 Å for BP. The bulk modulus takes the values of 141 GPa at 25% phosphorus and 75% arsenide, 141 GPa at a concentration of

50% both phosphorus and arsenide, and 157 GPa at 75% phosphorus and 25% arsenide. Again, we get a decreasing trend in the values of the lattice constant and an increasing trend in the values of the bulk modulus with the increasing concentration of nitrogen.

The values of the partial derivative of the bulk modulus with respect to pressure in zinc blende (B3) phase for all the given above samples are also given in Table 11.2.

TABLE 11.2 The Lattice Constant, Bulk Modulus, and the Pressure Derivative of the Bulk Modulus in B3 Phase for BN_xP_{1-x}, BN_xAs_{1-x}, BP_xAs_{1-x} in B3 Phase.

		BN_xP_{1-x}		
	x	**a**	**B (GPa)**	**B'**
BP	0	4.53		
$BN_{0.25}P_{0.75}$	0.25	4.37	188	3.89
$BN_{0.5}P_{0.5}$	0.5	4.21	216	3.92
$BN_{0.75}P_{0.25}$	0.75	3.98	269	3.98
BN	1	3.61		
		BN_xAs_{1-x}		
	x	**a**	**B (GPa)**	**B'**
BAs	0	4.73		
$BN_{0.25}As_{0.75}$	0.25	4.65	158	4.13
$BN_{0.5}As_{0.5}$	0.5	4.41	192	4.1
$BN_{0.75}As_{0.25}$	0.75	4.12	256	4.08
BN	1	3.61		
		BP_xAs_{1-x}		
	x	**a**	**B (GPa)**	**B'**
BAs	0	4.73		
$BP_{0.25}As_{0.75}$	0.25	4.71	141	4.1
$BP_{0.5}As_{0.5}$	0.5	4.69	149	4.04
$BP_{0.75}As_{0.25}$	0.75	4.64	157	3.97
BP	1	4.53		

The elastic properties in the B3 phase of boron-based III–V compounds (BN_xP_{1-x}, BN_xAs_{1-x}, BP_xAs_{1-x})

The above table represents the elastic constants C_{11}, C_{12}, C_{44} and some of the properties dependent on these elastic constants, such as the anisotropy factor (A), the Reuss shear modulus (G_R), Voigt shear modulus (G_v), the Hill

TABLE 11.3 The Elastic Constants C11, C12, C44, and Dependent Properties for BNxP1-x, BNxAs1-x, BPxAs1-x in B3 Phase.

	C11	C12	C44	A	G_R	G_v	G_H	B	E	v
					BNxP1-x					
BP	310	96	156	1.46	131.84	74	102.92	167.33	256.24	0.24
$BN_{0.25}P_{0.75}$	402	102	254	1.69	198.85	110.8	154.82	202	369.96	0.19
$BN_{0.5}P_{0.5}$	481	112	324	1.76	248.76	138.6	193.68	235	455.82	0.18
$BN_{0.75}P_{0.25}$	601	119	398	1.65	315.73	176	245.86	279.67	570.43	0.16
BN	779	174	467	1.54	383.57	214.4	298.98	375.67	708.89	0.19
					BNxAs1-x					
BAs	292	72	147	1.34	129.57	73.4	101.48	145.33	246.97	0.22
$BN_{0.25}As_{0.75}$	358	87	278	2.05	195.68	109.8	152.74	177.33	356.01	0.17
$BN_{0.5}As_{0.5}$	456	94	346	1.91	253.55	141.6	197.57	214.67	453.57	0.15
$BN_{0.75}As_{0.25}$	550	124	401	1.88	296.37	165.4	230.88	266	537.22	0.16
BN	779	174	467	1.54	383.57	214.4	298.98	375.67	708.89	0.19
					BPxAs1-x					
BAs	292	72	147	1.34	129.57	73.4	101.48	145.33	246.97	0.22
$BN_{0.25}As_{0.75}$	294	78	149	1.38	129.36	73	101.18	150	247.82	0.22
$BN_{0.5}As_{0.5}$	297	84	153	1.44	130.25	73.2	101.73	155	250.40	0.23
$BN_{0.75}As_{0.25}$	306	91	154	1.43	131.28	73.8	102.54	162.67	254.21	0.24
BN	310	96	156	1.46	131.85	74	102.92	167.33	256.23	0.24

shear modulus (G_H), the Young's modulus (E), Poisson's ratio (v), and bulk modulus (B, as derived with the help of the elastic constants) for BN_xP_{1-x}, BN_xAs_{1-x}, BP_xAs_{1-x} in the zinc blende (B3) structure.

Energy vs volume diagrams

1. **Energy vs volume diagram for BN_xP_{1-x} in B1 phase:**
   a. **At a concentration of x = 0.25:** The minima is obtained at the volume of 18.65 Å³ and the value shown is –9.12

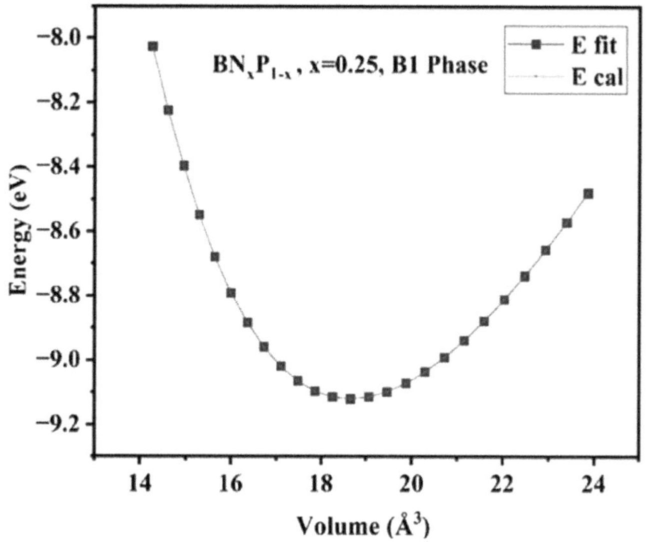

FIGURE 11.1 Energy vs volume diagram for BN_xP_{1-x} in B1 phase at 0.25.

   b. **At the concentration of x = 0.50:** The minima is obtained at a volume of 15.64 Å³ with a value of –8.6 eV.
   c. **At the concentration of x = 0.75:** The minima is obtained at the volume of 13.93 Å³ with a value of –7.5 eV.

2. **Energy vs volume diagram for BN_xAs_{1-x} in B1 phase:**
   a. **At a concentration of x = 0.25:** The minima are obtained at the volume of 20.43 Å³ and the value shown is –10.12 eV.
   b. **At the concentration of x = 0.50:** The minima are obtained at a volume of 16.24 Å³ with a value of –8.9 eV.

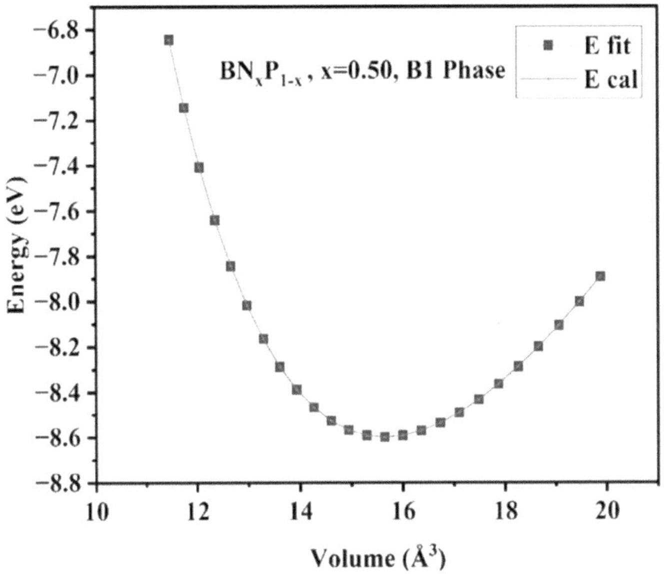

FIGURE 11.2 Energy vs volume diagram for BN_xP_{1-x} in B1 phase at 0.50

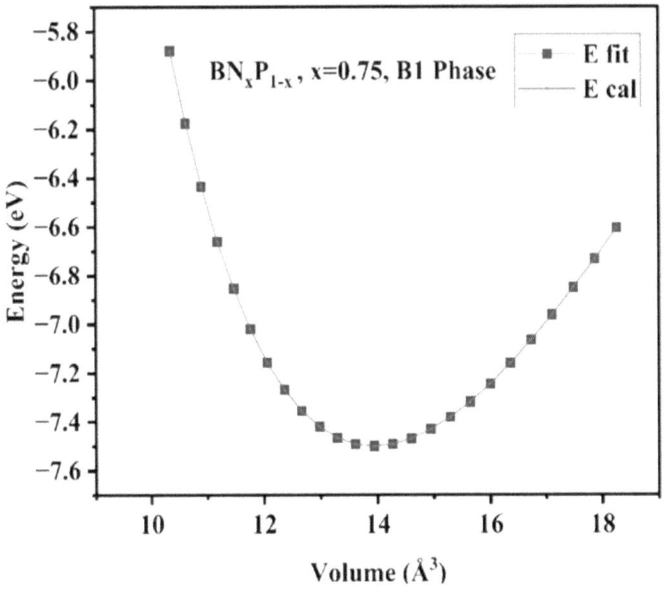

FIGURE 11.3 The energy vs volume diagram for BNxP1-x in B1 phase at 0.75.

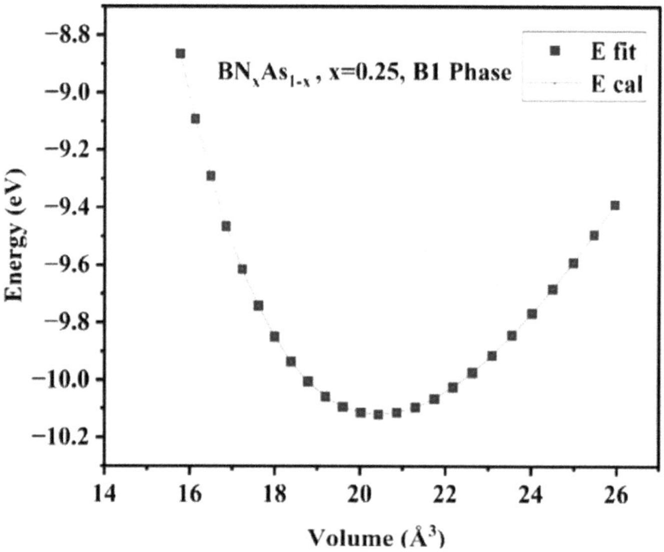

FIGURE 11.4 The energy vs volume diagram for BN_xAs_{1-x} in B1 phase at $x = 0.25$.

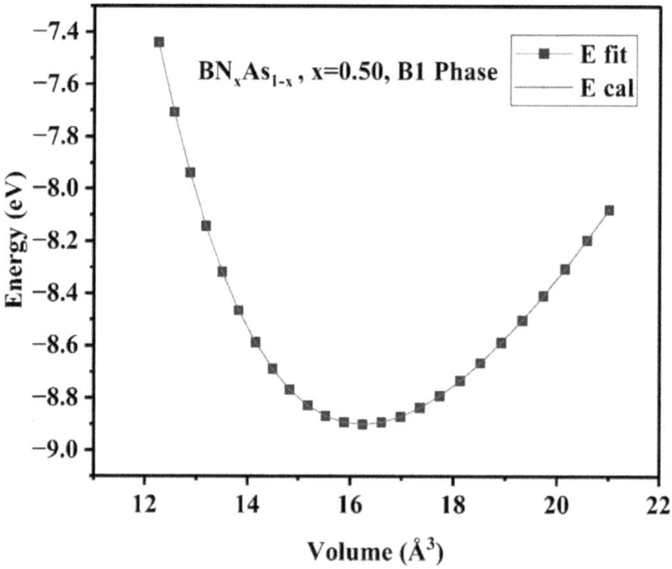

FIGURE 11.5 The energy vs Volume diagram for BN_xAs_{1-x} in B1 phaset x=0.50.

c. **At the concentration of x = 0.75:** The minima are obtained at the volume of 13.078 Å^3 with a value of –6.7 eV.

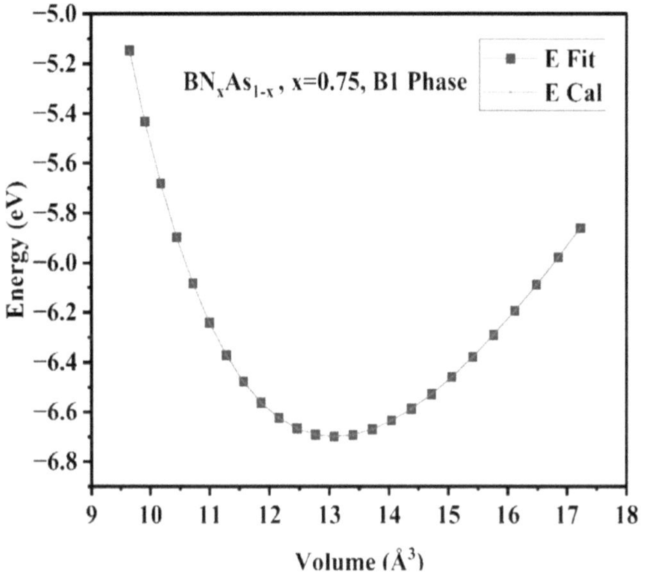

FIGURE 11.6 The energy vs volume diagram for BN_xAs_{1-x} in B1 phase at $x = 0.75$.

3. **Energy vs volume diagram for BPxAs1−x in B1 phase:**
   a. **At a concentration of $x = 0.25$:** The minima are obtained at the volume of 22.93 Å^3 and the value shown is –11.8 eV.

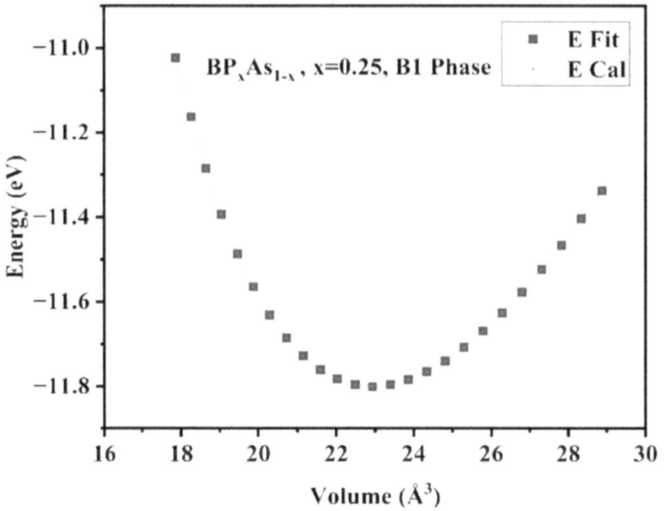

FIGURE 11.7 The energy vs volume diagram for BP_xAs_{1-x} in B1 phase at $x = 0.25$.

b. **At the concentration of $x = 0.50$:** The minima are obtained at a volume of 22.328 Å³ with a value of –11.4 eV.

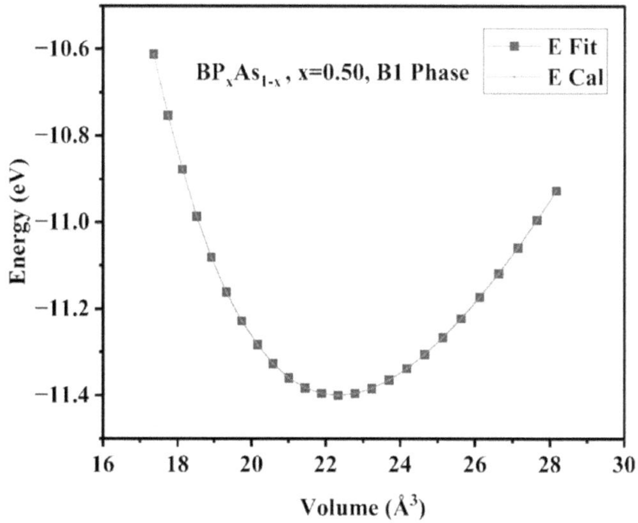

FIGURE 11.8 The energy vs volume diagram for BP_xAs_{1-x} in B1 phase at $x = 0.50$.

c. **At the concentration of $x = 0.75$:** The minima are obtained at the volume of 20.578 Å³ with a value of –10.6 eV.

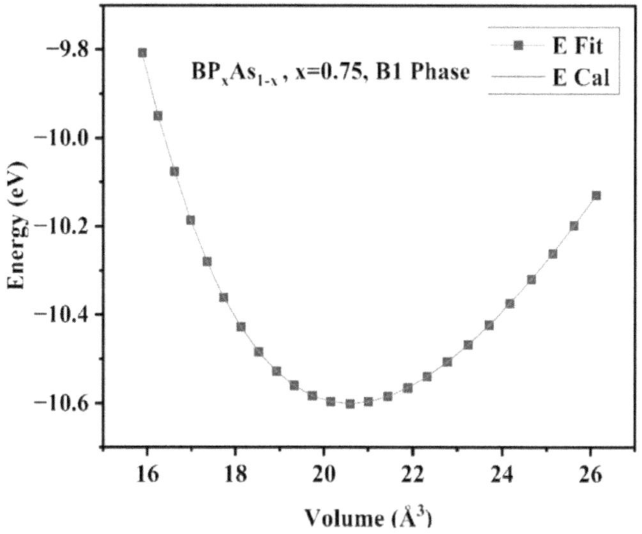

FIGURE 11.9 The energy vs volume diagram for BPxAs1−x in B1 phase at $x = 0.75$.

4. **Energy vs volume diagram for BN_xP_{1-x} in B3 phase:**
   a. **At a concentration of $x = 0.25$:** The minima are obtained at the volume of 20.86 Å³ and the value shown is –9.8 eV.

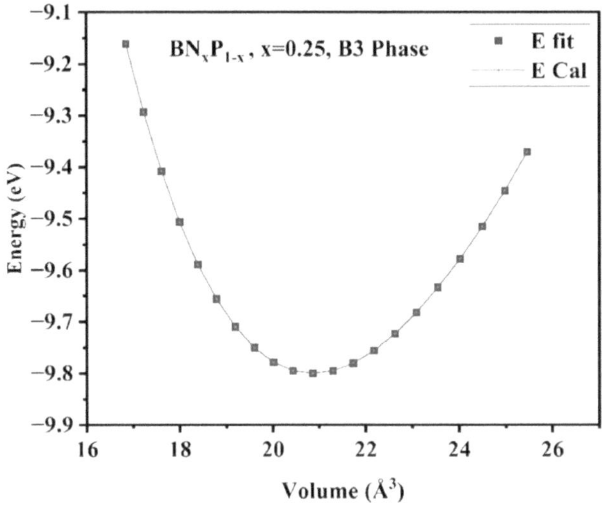

FIGURE 11.10 The energy vs Volume diagram for BN_xP_{1-x} in B3 phase at x= 0.25.

   b. **At the concentration of $x = 0.50$:** The minima are obtained at a volume of 18.65 Å³ with a value of –9.4 eV.

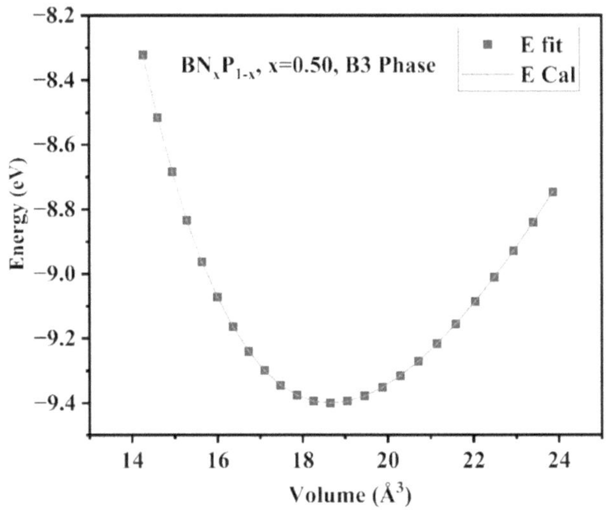

FIGURE 11.11 The energy vs Volume diagram for BNxP1-x in B3 phase at x= 0.50.

c. **At the concentration of $x = 0.75$:** The minima are obtained at the volume of 15.76 Å^3 with a value of –8.8 eV.

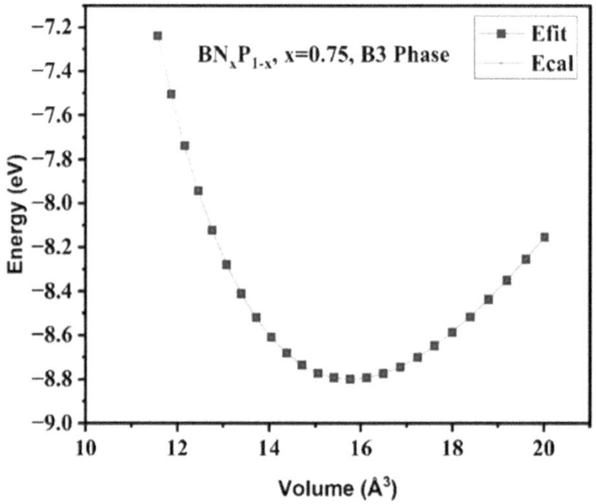

FIGURE 11.12 The energy vs Volume diagram for BNxP1-x in B3 phase at x= 0.75.

5. **Energy vs volume diagram for BN_xAs_{1-x} in B3 phase:**
   a. **At a concentration of $x = 0.25$:** The minima is obtained at the volume of 25.136 Å^3 and the value shown is –11.23 eV.

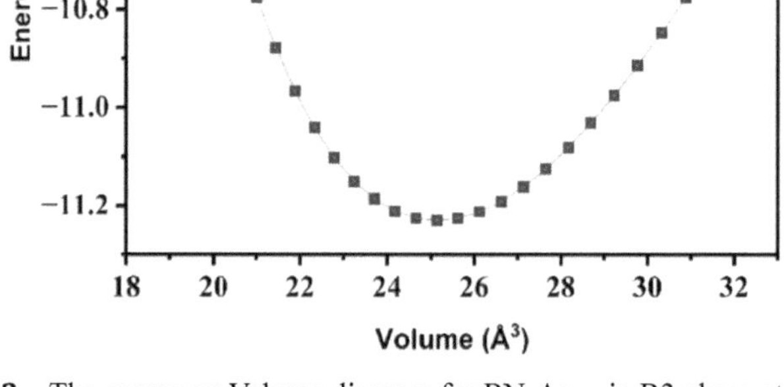

FIGURE 11.13 The energy vs Volume diagram for BN_xAs_{1-x} in B3 phase at x= 0.25.

b. **At the concentration of $x = 0.50$:** The minima are obtained at a volume of 21.44 Å^3 with a value of –10.12 eV.

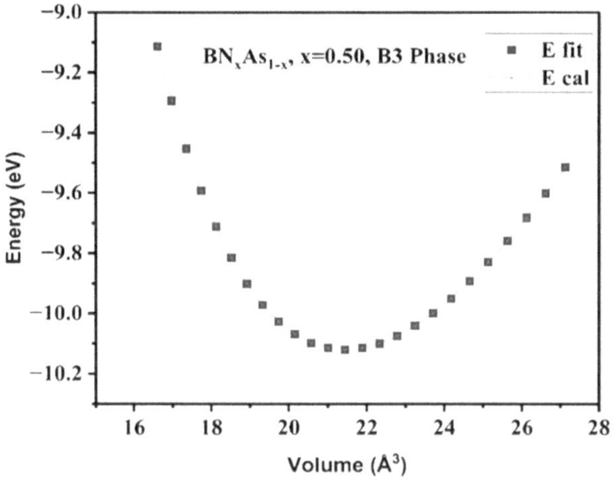

FIGURE 11.14 The energy vs Volume diagram BN_xAs_{1-x} in B3 phase at x= 0.50.

c. **At the concentration of $x = 0.75$:** The minima are obtained at the volume of 17.48 Å^3 with a value of –8.98 eV.

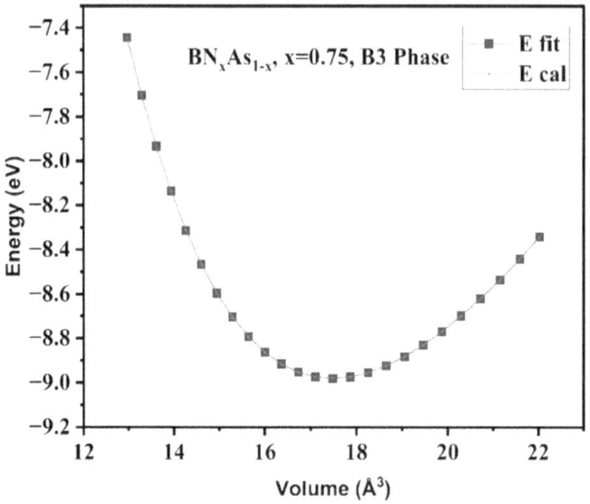

FIGURE 11.15 The energy vs volume diagram BN_xAs_{1-x} in B3 phase at = 0.75.

6. **Energy vs volume diagram for BP_xAs_{1-x}**
   a. **At a concentration of $x = 0.25$:** The minima are obtained at the volume of 26.12 Å³ and the value shown is –12.1 eV.

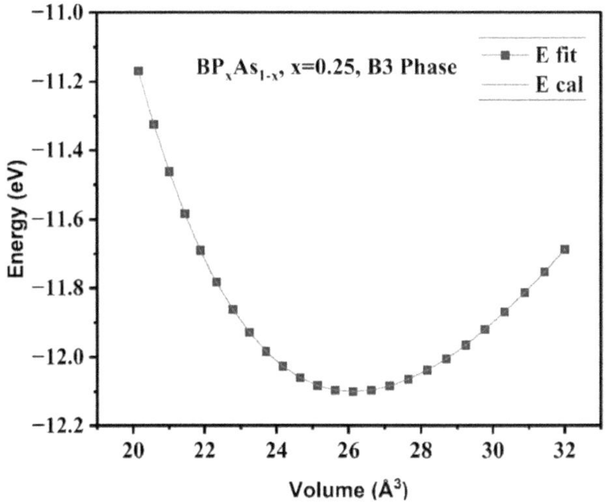

FIGURE 11.16 The energy vs volume diagram for BP_xAs_{1-x} in B3 phase at $x = 0.25$.

   b. **At the concentration of $x = 0.50$:** The minima are obtained at a volume of 25.79 Å³ with a value of –11.56 eV.

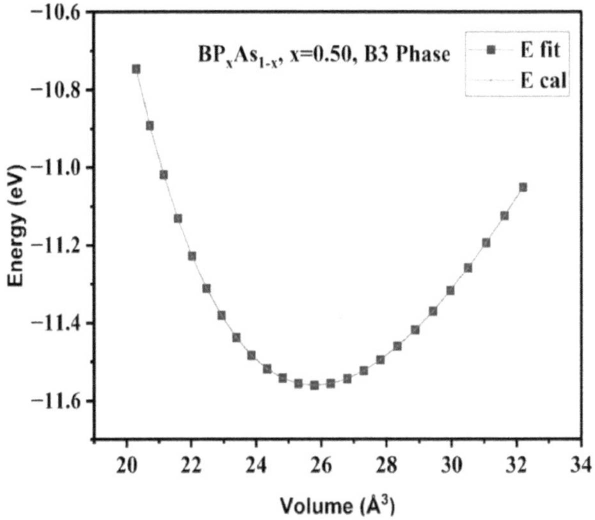

FIGURE 11.17 The energy vs volume diagram for BP_xAs_{1-x} in B3 phase at $x = 0.50$.

c. **At the concentration of $x = 0.75$:** The minima are obtained at the volume of 24.97 Å³ with a value of –10.9 eV.

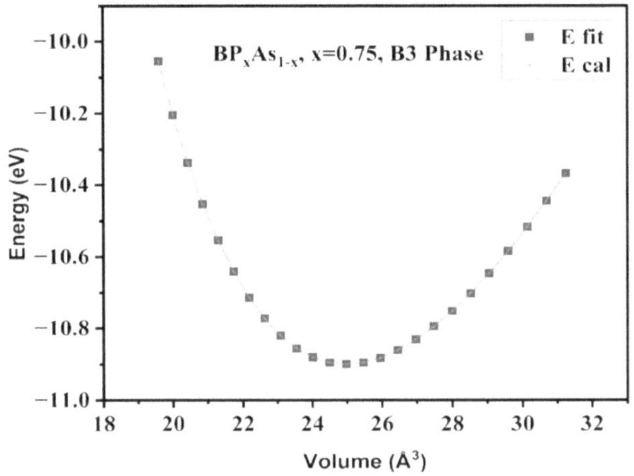

FIGURE 11.18 The energy vs volume diagram for BP_xAs_{1-x} in B3 phase at $x = 0.75$.

All the values of the energies calculated on the basis of the computational model are in good agreement with the values calculated based on the values calculated using the Birch–Murnaghan equation.

11.4 CONCLUSION

The lattice constant and the bulk modulus for the boron-based Group III–V compounds BN_xP_{1-x}, BN_xAs_{1-x}, and BP_xAs_{1-x} have been determined in the NaCl (B1) and the Zinc Blende (B3) structure. In this way, we were also able to see the effect of variation of the concentration on the values of the lattice constant and the bulk modulus for the above-said compounds and phases. Furthermore, the elastic constants C_{11}, C_{12}, and C_{44} have also been determined in the study along with the properties which are a function of elastic constants such as the anisotropy factor (A), The Reuss shear modulus (G_R), Voigt shear modulus (G_v), the Hill shear modulus (G_H), Young's modulus (E), and Poisson's ratio (ν) have also been calculated with it. Furthermore, we have also plotted and discussed the energy vs volume diagram in the B1 and the B3 phase for all the intermediate compounds in the study.

KEYWORDS

- **lattice constants**
- **elastic constants**
- **density functional theory**
- **NaCl(B1) structure**
- **zinc blende(B3) structure**

REFERENCES

1. Jain, C.; Willis, J. R.; Bullogh, R. *Adv. Phys.* **1990,** *39,* 127.
2. El Haj Hassan, F.; Akbarzadeh, H. *Mater. Sci. Eng. B* **2005,** *121,* 170–177.
3. Liu, Z.; Ma, W.; Ye, X. Chapter 2—Shape Control in the Synthesis of Colloidal Semiconductor Nanocrystals, Anisotropic Particle Assemblies. Elsevier, 2018; pp 37–54, ISBN 9780128040690, https://doi.org/10.1016/B978-0-12-804069-0.00002-2.
4. Wang, S. Q.; Ye, H. Q. First-principles Study on Elastic Properties and Phase Stability of III–V Compounds. *Physica Status Solidi (b)* **2003,** *240* (1), 45–54. doi:10.1002/pssb. 200301861
5. Wang, S.-F.; Xiaojun, W. First-Principles Study on Electronic and Optical Properties of Graphene-Like Boron Phosphide Sheets. *Chin. J. Chem. Phys.* **2015,** *28,* 588–594. 10.1063/1674-0068/28/cjcp1505100.
6. Chen, L.; et al. Ceramics International, https:// doi.rg/10.1016/j.ceramint.2020.11.090.
7. Chen, L. Ceramics International, https:// doi.org/10.1016/j.ceramint.2020.11.090.
8. Greenaway, A. L.; et al. Ternary Nitride Materials: Fundamentals and Emerging Device Applications, https://arxiv.org/pdf/2010.08058.pdf
9. Dadsetani, M.; Kianisadr, B.; Nejatipour, H. First Principles Investigation of the Optical Properties of BN_xP_{1-x} $(0 \leq x \leq 1)$ Boron Ternary Alloys. *J. Electron. Mater.* **2015,** *44,* 2699–2711.
10. Zhou, L.; *et al. J. Phys. D: Appl. Phys.* **2013,** *46,* 365301.
11. Anurag, S.; Bhoopendra, D. Structural and Elastic Properties of ZrN and HfN: Ab initio Study. *Canadian J. Phy.* 2014.
12. Kohn, W.; Sham, L. J. *Phys. Rev. Rev.* **1965,** *140,* A1133.
13. Hohenberg, P.; Kohn, W. *Phys. Rev.* **1964,** *136,* B864.
14. Toth, L. E. Transition Metal Carbides and Nitrides. Academic Press: New York/London, 1971.
15. Vegard, L. *Z. Phys. A: Hadrons Nucl.* **1921,** *5,* 17–26.
16. Wooten, F. Optical Properties of Solids. Academic Press: New York, 1972.
17. Perdew, J. P.; Burke, K.; Ernzerhof, M. *Phys. Rev. Lett.* **1996,** *77,* 3865.
18. Blaha, P.; Schwarz, K.; Madsen, G. K. H.; Kavanicka, D.; Luitz, J. *Wien2 k, An Augmented Plane Wave Plus Local Orbitals Program for Calculating Crystal Properties*; Wien, Austria, ISBN 3-9501031-1-2, 2001.
19. Murnaghan, F. D. *Proc. Natl. Acad. Sci. U.S.A.* **1944,** *30,* 244.

CHAPTER 12

Combined Investigation of Various Physical Properties of Fe-Based Quaternary Heusler Compounds

NAVDEEP KAUR[1], RAMESH SHARMA[2], and VIPUL SRIVASTAVA[1]

[1]*Department of Physics, School of Chemical Engineering and Physical Sciences, Lovely Professional University, Phagwara, Punjab, India*

[2]*Department of Applied Science, Feroze Gandhi Institute of Engineering and Technology, Raebareli, Uttar Pradesh, India*

ABSTRACT

Heusler compounds that have strengthened their uses in various thermoelectric materials have become the topic of research in these days. Therefore, in order to find efficient materials, we have investigated two quaternary Heusler compounds, FeCrTiZ (Z = Sn, Sb) concerning their various physical properties through the density functional theory. First, we have checked their structural stability via E–V curves. We have considered three possible types in two magnetic phases (ferromagnetic and nonmagnetic) for the purpose of optimization and found type-1 of ferromagnetic as a stable one. Due to their ferromagnetic nature, we further investigated their spin-polarized electronic and magnetic properties through two different potentials. These give the half-metallic (HM) nature of the compounds. We used the IRelast package within the WIEN2k code to find the strengthening and toughening behavior of the compounds through their elastic and mechanical stabilities. Further to investigate thermodynamical stability, we have shown the variation of some thermodynamical parameters. Moreover, we used the BoltzTrap code to check the thermoelectric competence of the compounds.

Contemporary Advancements in Materials Technology: Novel Solutions to Complex Problems.
Shrikaant Kulkarni, Vipul Srivastava, & P. William (Eds.)

12.1 INTRODUCTION

Ferromagnetism and high polarization at the Fermi level[1] have become extraordinary properties of Heusler compounds (HCs) to discuss. Due to these properties, HCs have made their ways to be used as source materials in spintronics, magnetoresistive materials, thermoelectric generators, and memory devices.[2–5] Therefore, the probes for efficient materials accelerated and such probes signified the solid-state devices known as thermoelectrics, which can be used as converters for power generation by converting waste heat occurring from automobiles into electricity or vice versa. These devices work specifically through the Seebeck effect.[6] This effect is outlined as an electromotive force that generates potential difference across the two different conducting materials through temperature gradient.[6,7] Typically, an interest in these sustainable thermoelectric materials gets increased due to the imperative need of clean energy. Also, working without any moving parts or environmentally unsafe fluids makes them reliable and effective. Consequently, HCs have become a matter of enormous significance owing to their several possibilities in various applications.[8,9]

The very first full-Heusler compounds, namely Cu_2MnAl and Cu_2MnSn, were investigated by Friedrich Heusler in 1903,[10, 11] which initially showed ferromagnetic nature even when all elements in the compounds were nonmagnetic. In addition to this, these materials exhibit another notable property known as half-metallicity, in which one spin demonstrates metallic character while another spin demonstrates semiconducting or insulating behavior and first appeared in the NiMnSb half-Heusler compound.[1]

The HCs are inhibited by three elements with three compositions that are generalized as XYZ, X_2YZ, and XX'YZ. Here, the first one is representing half-Heusler compounds, the second one is indicating full-Heusler compounds, and the latter one is representing quaternary Heulser compounds. The X, Y positions are usually seized by transition elements, Z positions are commonly acquired by "s" or "p" elements of the main group, and "X" in the quaternary Heuslers refers to another transition element.[12] Generally, all half-Heusler compounds hold MgAgAs ($C1_b$) crystalline structure with the F-43m (216) space group. The full HCs hold Cu_2MnAl ($L2_1$) or Hg_2CuTi (XA) crystalline structure. The $L2_1$ structure lies under Fm-3m (225) and XA lies under the F-43m (216) space group.

However, the quaternary Heusler compounds that have been investigated in the present study are generally crystallized in the LiMgPdSn (Y-type) structure.[13] This Y-type has further three types, specifically type-1, type-2, and type-3. Due to the assigned atomic positions these types are different from each other.

It has been projected that the quaternary Heuslers dependent electrical gadgets would dissipate less power[14] and hence are significantly more interesting as compared to other forms of HCs. Several quaternary Heusler compounds have already been synthesized theoretically as well as experimentally.[14–18] Alijani et al.[14] synthesized CoFeMnZ (Z = *p* block elements) quaternary Heuslers and compared their results with ab-initio calculations. The compounds were anticipated to exhibit characteristics of being HM ferromagnetics with elevated Curie temperatures. Shi et al.[19] used first-principle calculations to investigate CoFeRGa (R = Mn, Ti, Nb, V, Cr, Cu) quaternary Heuslers. They reported them in a cubic structure of Y-type with different electronic properties and also found their mechanical and dynamic stability. Shakil et al[20] investigated the FeTiVZ and FeZrVZ (Z=Ga, In) compounds and determined their stable phase with HM nature and mechanical stability. Along with these several quaternary Heuslers have been reported with their futuristic applications.[21–25] Also, the potential uses of HCs in various fields prompt experimental and theoretical research.[26–28]

In this order the motivation of this work is to look after the physical behavior of two (FeCrTiSn and FeCrTiSb) quaternary Heusler compounds. We considered GGA and mBJ potentials to evaluate the electronic and magnetic nature of compounds, which are then followed by their mechanical, thermodynamic, and thermoelectric properties. So far, there is no literature available on these compounds except for their crystal structure data.[29,30] Hence, the further paper is divided into three sections. The second section explains the methodology applied for the calculation purposes. A third section is dedicated to the obtained results, followed by a discussion on various physical properties and in Section 4, we finally conclude our work.

12.2 METHODOLOGY

The full-potential linearized augmented plane-wave (FP-LAPW) method,[31] supported by density functional theory (DFT)[32] integrated in the WIEN2k code,[33] was employed to inspect FeCrTiZ (Z = Sn, Sb) compounds. DFT is a kind of theory that uses electron density over many body electronic wave functions[34] that helps to determine all ground-state properties through the time-independent Schrodinger equation.[32] Therefore, we used two potentials for exchange–correlation (XC) interaction. First, we utilized generalized gradient approximation of Perdew Burke and Ernzerhof (PBE-GGA)[35] in which energy of XC depends upon gradient of electron densities. Secondly, we used modified Becke–Johnson (mBJ) introduced by Taran and Blaha,[36] which has been proven to be one of the best methods for accurately determining band gap.[37] To determine the matrix size, we take a plane wave basis set (R_{MT}* K_{Max}) as 7. A term R_{MT} is the radius of nonoverlapping muffin tin (MT) spheres, which is marked out as one region of the space being divided by the FP-LAPW method. The second region is known as the interstitial region. These two regions contain basis sets that get expanded into different wave functions. The potential and nonspherical contribution to the charge density inside these MT spheres is assigned as L_{max} = 10. Moreover, potential and charge density in the interstitial region are Fourier expanded with G_{max}=12 $(a.u.)^{-1}$. The R_{MT} values to retain the charge leakage were set to 2.16 for Fe, 2.18 for Cr, 2.2 for Ti, 2.26 for Sn, and 2.27 for Sb. Further, during the self-consistent calculations, we ensured that the convergence criteria in terms of energy and charge were met. For this, the convergences criteria for energy and charge are fixed as 10^{-4} Ry and 10^{-3} a.u.^{3}. The Brillouin zone integration has been performed with a 1000 k-points.

The structural stability of the compounds has been verified through the GGA potential. To estimate the mechanical stability, elastic constants have been calculated through IRelast package.[38] Then the thermodynamic stability of the compounds has been explored by the quasi-harmonic Debye approximation (QHA) that is employed in Gibbs2 package[39–41] QHA minimized the Gibbs function concerning volume (V) at constant pressure (P) and temperature (T).

$$\left(\frac{\partial G^*(V;P,T)}{\partial V}\right)_{P,T} = 0, \tag{12.1}$$

where Gibbs function is

$$G^*(V;P,T) = E(V) + PV + A_{vib}\left(\Theta_D(V);T\right) \tag{12.2}$$

where, $E(V)$ is a static energy of the compounds, PV is a condition of constant hydrostatic pressure, A_{vib}, and Θ_D are defined as vibrational Helmholtz free energy, and Debye temperature. Different thermoelectric parameters have been calculated through a code, namely BoltzTrap code based on the semi-classical Boltzmann transport theory.[42]

12.3 RESULTS AND DISCUSSION

The subsequent sections go over the physical parameters of FeCrTiZ (Z = Sn, Sb).

12.3.1 STRUCTURAL PROPERTIES

FeCrTiSn and FeCrTiSb are cubic quaternary Heusler compounds, henceforth both subsist in Y-type structure under the F-43m (216) space group. However, based on distinct Wyckoff positions of four atoms, there are three potential nonequivalent superstructures.[14]

TABLE 12.1 Various Occupation Sites for the Possible Types of Y-type Crystal Structure.

Types	A(0,0,0)	B(0.25,0.25,0.25)	C(0.5,0.5,0.5)	D(0.75,0.75,0.75)
Type 1	Sn/Sb	Cr	Ti	Fe
Type 2	Sn/Sb	Ti	Cr	Fe
Type-3	Cr	Sn/Sb	Ti	Fe

As a result, Table 12.1 displays the occurrence of possible types upon altering atomic locations. The different positions are represented in Figure 12.1 or two compounds, the equilibrium energy for the Type-1 structure is found to be more negative as compared to other two types, known as Type-2 and Type-3. This signifies the stability of Type-1 structure as shown in Figure 12.2 and therefore we considered Type-1 structure for further investigation. In such a context, the ground-state parameters are obtained through the fitting of *E-V* data into the Birch Murnaghan equation of state (EOS)[43] given below and are tabulated below (Table 12.2).

$$E(V) = E_0 + \frac{B}{B'(B'-1)}\left[V\left(\frac{V_0}{V}\right)^{B'} - V_0\right] + \frac{B}{B'}(V - V_0) \tag{12.3}$$

Further, we considered nonmagnetic (NM) and ferromagnetic (FM) phases for a stable type whose equilibrium parameters are in Table 12.2. The *E–V* data of these two magnetic phases infer on the FM phase as a stable one, as depicted in Table 12.2. Therefore, Type-1 structure with ferromagnetic phase exhibits the lowest energy which is equal to –18713.277 Ry (FeCrTiSn) and –19322.280 Ry (FeCrTiSb). We also estimated the formation energy (E_{form}) and cohesive energies (E_{coh}) for the investigated HCs by using the following equations.

$$E_{for} = \frac{E^{Total}_{\mathrm{FeCrTiZ}} - \left[E^{bulk}_{Fe} + E^{bulk}_{Cr} + E^{bulk}_{Ti} + E^{bulk}_{Z}\right]}{4} \tag{12.4}$$

$$E_{coh} = \frac{\left[E^{iso}_{Fe} + E^{iso}_{Cr} + E^{iso}_{Ti} + E^{iso}_{Z}\right] - E_{\mathrm{FeCrTiZ}}}{4} \tag{12.5}$$

where E_i^{bulk}, (i = Fe, Cr, Ti, Z) are the bulk energies of the individual elements, E_i^{iso}, (i = Fe, Cr, Ti, Z) symbolize energies of isolated atoms, and $E_{FeCrTiZ}^{Total}$ is an equilibrium total energy of the HCs. High values of cohesive energies (5.93 for FeCrTiSn and 5.81 for FeCrTiSb) represent the strong atomic bondings with high mechanical stability. The possibility for synthesizing these compounds in the lab has been revealed by the predicted negative formation energies (–0.11 for FeCrTiSn and –0.119 eV/atom for FeCrTiSb).

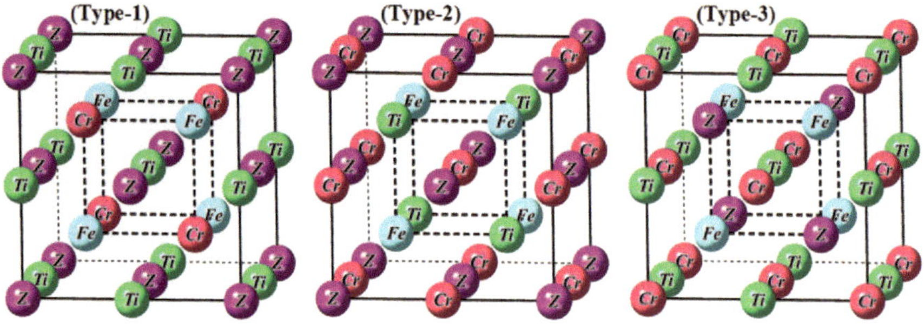

FIGURE 12.1 Crystal structures of different types of Y-type structure based on different atomic positions.

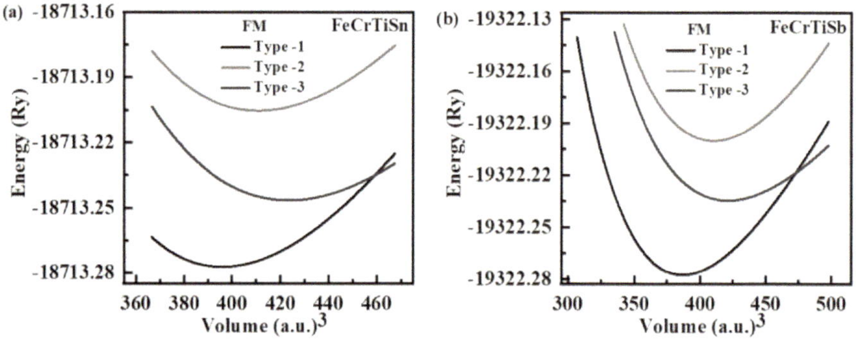

FIGURE 12.2 Energy versus volume curve for ferromagnetic phase of (a) FeCrTiSn and (b) FeCrTiSb.

12.3.2 ELECTRONIC AND MAGNETIC PROPERTIES

To thoroughly explore and classify the electronic behavior, band structure is of utmost importance. Therefore, we estimated the electronic band profile for FeCrTiZ (Z = Sn, Sb) through GGA and mBJ potentials and it is displayed in Figure 12.3 concerning Brillouin zone symmetry.

TABLE 12.2 Calculated Equilibrium Parameters for FeCrTiZ (Z=Sn, Sb) Cubic Quaternary Heuslers within Their Possible Structures using the PBE-GGA Method.

Compounds	Types	Magnetic phases	a_0 (Å)	B (GPa)	B′	E_0 (Ry)
FeCrTiSn	Type-1	FM	6.16	160.54	4.49	–18713.277
		NM	6.13	187.69	4.96	–18713.240
	Type-2	FM	6.24	139.73	3.85	–18713.205
		NM	6.16	174.71	4.82	–18713.177
	Type-3	FM	6.3	129.01	4.2	–18713.246
		NM	6.19	169.46	4.6	–18713.154
	Type-1 [14, 30]	–	6.14	–	–	–
FeCrTiSb	Type-1	FM	6.10	173.37	5.64	–19322.280
		NM	6.17	178.79	3.82	–19322.177
	Type-2	FM	6.24	136.91	3.03	–19322.201
		NM	6.15	176.3	3.95	–19322.181
	Type-3	FM	6.29	89.88	6.99	–19322.233
		NM	6.17	178.81	3.82	–19322.178
	Type-1 [14,30]	–	6.092	–	–	–

Here, GGA showed a purely metallic behavior of FeCrTiSn, and FeCrTiSb in spin-down state whereas spin-up of two compounds showed a nearly semiconducting nature. Although GGA is well known for its ability to determine a material's electrical and magnetic profiles, in some cases it reveals a very small band gap or even eliminates it in some materials when compared to experimental research.[44] Therefore we applied mBJ exchange potential, whose accuracy is equivalent to highly expensive GW computations.[33] On applying mBJ potential, electronic states get shifted with the dependency of V_{XC} on the kinetic-energy density.[37] This originates the indirect band gap in the up spin of both investigated compounds which leads to the semiconducting nature of FeCrTiZ (Z = Sn, Sb). An indirect band gap in these two materials is caused by the valence band (VB) maxima and conduction band (CB) minima which are at the Γ and X locations, correspondingly. In FeCrTiSn, VB maxima (at Γ point), as well as CB minima (at X point), have occurred from a combination of Fe:*d* and Cr:*d* states, whereas for FeCrTiSb, VB maxima at Γ have occurred from cr:*d* and CB minima at X point have taken place due to Fe:*d*. The evaluated band gap with their symmetry direction is listed in Table 12.3. Hence, the two investigated compounds have come out as HM materials with 100% spin polarization at the Fermi level (E_f = 0 eV) in one spin. Moreover, the electronic states far from E_f are due to the *p* of Sn and Sb elements.

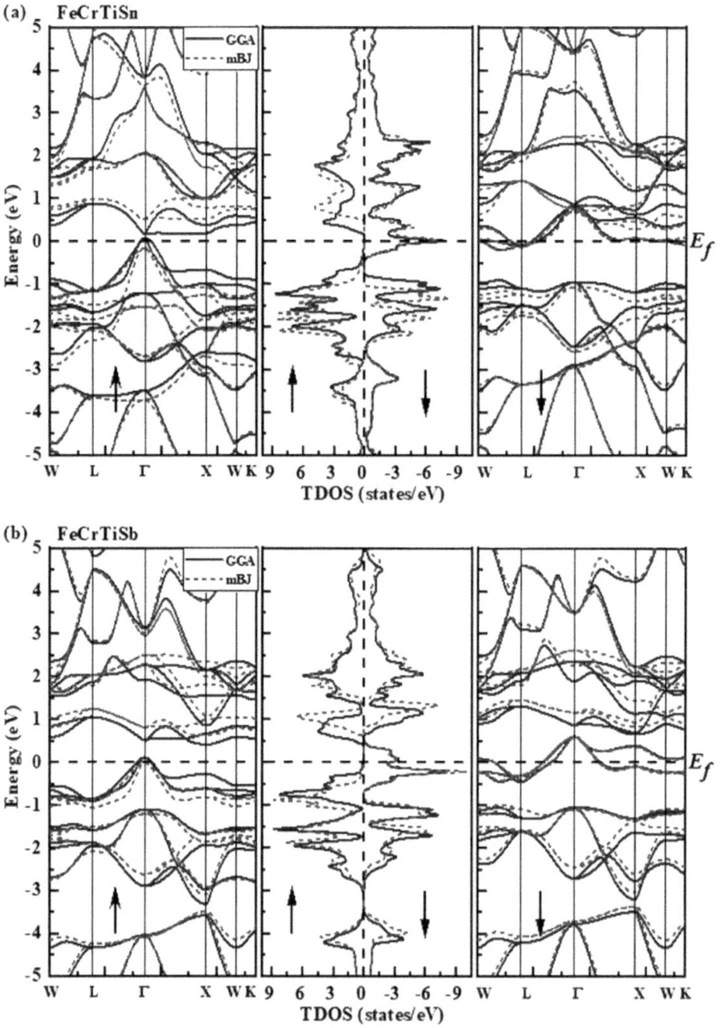

FIGURE 12.3 Spin-polarized band structure and total density of states (TDOS) for (a) FeCrTiSn and (b) FeCrTiSb quaternary Heusler compounds. Solid blue lines are corresponding to GGA potential and dashed red lines are relating to mBJ potential.

TABLE 12.3 Magnitude, Location, and Nature of Obtained Half-Metallic Gap of FeCrTiZ (Z = Sn, Sb).

Compounds	Fermi energy	Band gap (eV)	Location	Nature
FeCrTiSn	0.65291	0.662	Indirect	Γ-X
FeCrTiSb	0.774312	0.604	Indirect	Γ-X

Further to elaborate and categorize the nature of electronic states, elemental and orbital density of states (DOS) with mBJ potential are in Figure 12.4. These DOS plots are signifying the HM nature of the investigated HCs. In spin-up states, the semiconducting gap with zero density of states is clearly visible between the negative (VB) and positive (CB) energies, whereas in spin-down states the presence of electronic states led to its metallic nature and hence results in the HM nature of FeCrTiZ (Z = Sn, Sb). The elemental DOS for FeCrTiZ (Z=Sn, Sb) are marking the main contribution of transition elements in the total DOS and are showing very less contribution p-block elements (Sn, Sb). Near E_f, the hybridization of d-block elements is more active over Sn of FeCrTiSn and Sb of FeCrTiSb, which have a lesser contribution to the total DOS.

Furthermore, Figure 12.4(c, d) ascertains the presence of orbital DOS, for which we consider the p and d states of the elements. For both materials, p states have very less DOS near E_f and occupy bands at lower energies in the VB mainly within –3 eV and –4 eV. However, d states of Fe and Cr are mainly occupying the bands near E_f in the VB as well as the CB for both the compounds. Moreover, Figure 12.4(c) evidences that in FeCrTiSn, bands at VB maxima and CB minima are occupied by the combination of Fe-'d' and Cr-'d' that led to a band gap of 0.662 eV (spin-up). Figure 12.4(d) is evincing the band gap of 0.645 eV in spin-up states of FeCrTiSb, due to the occupation of VB maxima by Cr-'d' and CB minima by Fe-'d'.

The calculations for the structural and electronic properties of FeCrTiZ have revealed a ferromagnetic phase of the mentioned materials. In support of this, we have investigated the magnetic moment for FeCrTiZ (Z = Sn, Sb). Our results reveal that FeCrTiSn and FeCrTiSb compounds are ferromagnetically stable with their calculated total magnetic moments (M_{Tot}) within GGA potential. We have also estimated the M_{Tot} of both the compounds within mBJ potential as 2 μ_B (FeCrTiSn) and 1 μ_B (FeCrTiSb). The calculated M_{Tot} and partial magnetic moments of FeCrTiZ (Z = Sn, Sb) Heusler compounds in a stable type-1 within GGA and mBJ potentials are given in Table 12.4. An integer value of M_{Tot} illustrates the HM nature of FeCrTiZ (Z=Sn, Sb). The magnetic moments of individual elements are substantiating the high contribution of Cr element in M_{Tot}. Furthermore, the negative values of Ti and Z (=Sn, Sb) elements are indicating their ferromagnetic coupling with the remaining atoms of both the compounds.

The calculated values of M_{Tot} are also following the Pauling–Slater rule.[45-48] Hence, in the current case the results of the electronic structure and integer values of M_{Tot}, both have revealed the HM character of FeCrTiZ (Z = Sn, Sb).

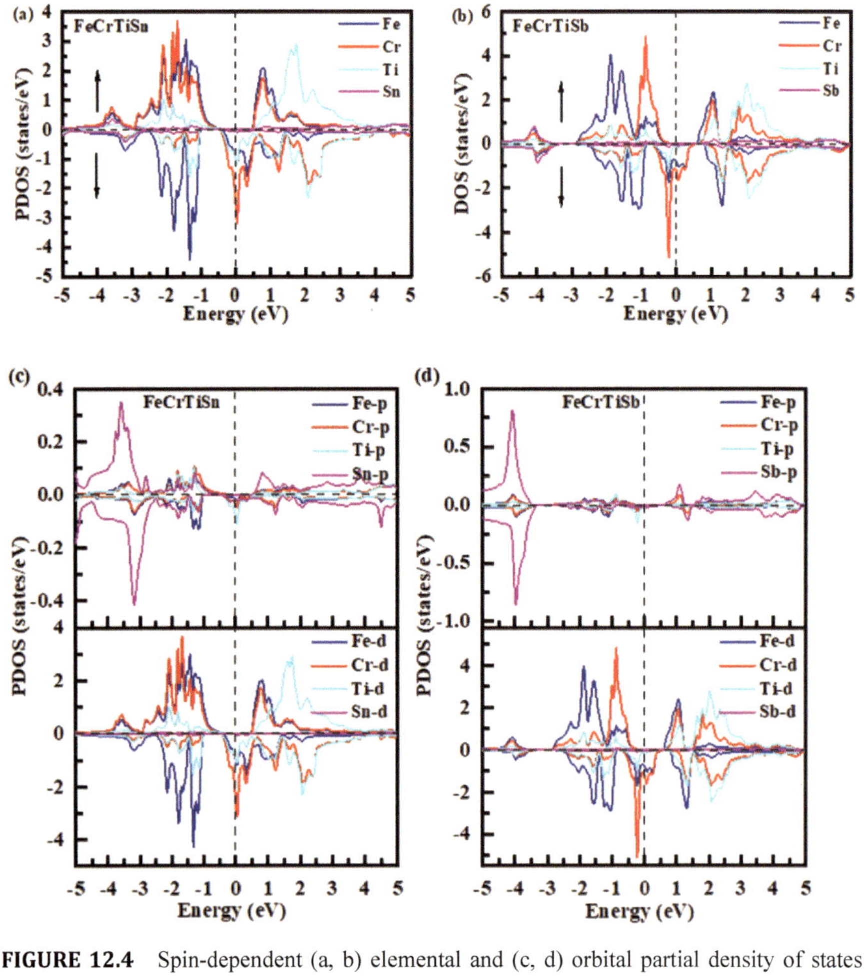

FIGURE 12.4 Spin-dependent (a, b) elemental and (c, d) orbital partial density of states (PDOS) within mBJ potential for FeCrTiSn and FeCrTiSb.

TABLE 12.4 Total and Individual Magnetic Moment in Bohrmagneton (μ_B) Within GGA and mBJ Potentials for FeCrTiZ (Z = Sn, Sb).

Compounds	Methods	M_{Fe}	M_{Cr}	M_{Ti}	M_Z	M_{Int}	M_{Tot}
FeCrTiSn	GGA	0.492	1.873	–0.378	–0.021	0.032	1.99
	mBJ	0.443	2.005	–0.44	–0.032	0.033	2.00
	Pauling–Slater	–	–	–	–	–	2.00
FeCrTiSb	GGA	0.49	0.735	–0.221	-0.009	0.003	0.99
	mBJ	0.536	0.823	–0.312	–0.017	–0.029	1.00
	Pauling–Slater	–	–	–	–	–	1.00

12.3.3 ELASTIC AND MECHANICAL PROPERTIES

Elastic constants and moduli assist to determine crystalline solids under external stress of various natures. These fundamental parameters are related to the lattice stability and bonding strength of the compounds[49] and can be estimated from the equilibrium energy calculations. As a result, we used the IRelast program[38] to estimate various parameters and this is shown in Table 12.5. However, the compounds under investigation have a cubic structure; therefore, three elastic constants, designated as C_{ij} (ij = 11, 12, 44), have been computed. We further estimated various parameters to find the mechanical stability by using different formulas.[50]

TABLE 12.5 Calculated Elastic Constants and Mechanical Parameters for FeCrTiZ (Z = Sn, Sb) Compounds.

Properties	FeCrTiSn	FeCrTiSb
C_{11} (GPa)	229.84	259.41
C_{12} (GPa)	100.19	141.12
C_{44} (GPa)	118.47	178.24
B (GPa)	143.41	180.55
G_V (GPa)	97.01	130.6
G_R (GPa)	89.01	98.72
G (GPa)	93.01	114.66
Y (GPa)	229.43	283.89
N	0.23	0.24
CP (GPa)	−18.28	−37.12
A	1.83	3.01
B/G	1.54	1.57
v_l (m/s)	5866.5	6414.48
v_t (m/s)	3459.78	3761.56
v_m (m/s)	3833.64	4170.21
θ_D (K)	466.35	512.54
T_m (K)	1911±300	2086±300

The shear modulus (G) measures the resistance of materials against plastic deformation, whereas bulk modulus (B) analyzes the compressibility inside the substance. A ratio of these two estimates the brittle/ductile nature, which is formulated as B/G and is given by Pugh.[51] The measured values of B/G for FeCrTiSn and FeCrTiSb demonstrate their brittle nature because they are less than the crucial value (1.75) of B/G. In addition to these negative values of Cauchy pressure ($C_{12} - C_{44}$) also affirm the brittle nature of the compounds. Another distinct parameter is Young's modulus (Y) that gauges resistance of materials to deformation caused by applied force and describes the stiffness among substances. The higher Y directs toward the more stiffed material and for the investigated materials it is obtained as 229.43 (FeCrTiSn) and 283.89 GPa (FeCrTiSb).

Further, Poisson's ratio (v) assists to recognize the bonding within solid materials and an average value of v is ranged between 0.25 and 0.5, which enumerates the stability of crystals. However, the critical values 0.1, 0.25, and 0.33 of v characterize the bonding nature as covalent, ionic, and metallic, respectively. Since our reported values are close to 0.25, this indicates the ionic bonding within the FeCrTiSn and FeCrTiSb compounds. Elastic anisotropy (A) is associated with micro cracks in materials. The materials are isotropic for $A \leq 1$ and typify as anisotropic materials for $A > 1$.[52] The higher value of A for both the compounds has revealed their anisotropic behavior. Moreover, the melting temperature (T_m) for FeCrTiSn and FeCrTiSb has been calculated as 1911 K and 2086 K, in a row.

12.3.4 THERMODYNAMIC PROPERTIES

The thermodynamic behavior is significant for designing the materials and facilitates in understanding the chemical bond and cohesion of the materials. The thermodynamic properties are capable of specifying the state of the system[53–56] and offer assistance to determine the heat absorption capacity of the materials that help to understand the phase transition against temperature.[57] Therefore, it turns out to be significant to envisage the behavior of HCs concerning the various thermodynamic variables because of their applications in device fabrication. Thus, the impact of pressure (P) and temperature (T) on various thermodynamic parameters by using the quasi-harmonic approximation (QHA) and Gibbs2 program[39–41] has been investigated.

To evaluate the resistance of material under the influence of P and T, we have projected bulk modulus (B) of materials. The trend of B is shown in Figure 12.5(a, b). One can see that B is merely decreasing at all different temperatures. FeCrTiSn is decreasing from 129.15 GPa (100 K) to 74.63 GPa (1200 K) and FeCrTiSb is descending from 171.46 GPa (100 K) to 128.12 GPa (1200 K) at 0 GPa. But it increases noticeably by pressure from 0 to 40 GPa and an increasing rate under P is superior to the decreasing rate under T. However, at 300 K and 40 GPa, B gives a value of about 322.19 and

350.56 GPa for FeCrTiSn and FeCrTiSb, respectively. As a result, FeCrTiSb compound is discovered to be tougher than FeCrTiSn compound.

Further, to inquire about thermal conductivity of the materials based on Debye temperature (θ_D), we have presented a variation of θ_D concerning temperature and pressure in Figure 12.5(c, d). The decreasing trend of θ_D concerning T is stronger at lower values of P. However, it continuously increases with the rise in P, which illustrates the strong impact of P on θ_D. It gives a value of about 658.9 and 675.4 K for FeCrTiSn and FeCrTiSb, respectively, at 300 K and 40 GPa. Additionally, these high values of θ_D for both materials predicted their nature as hard materials and refer to the good thermal conductivity of these materials.[58, 59]

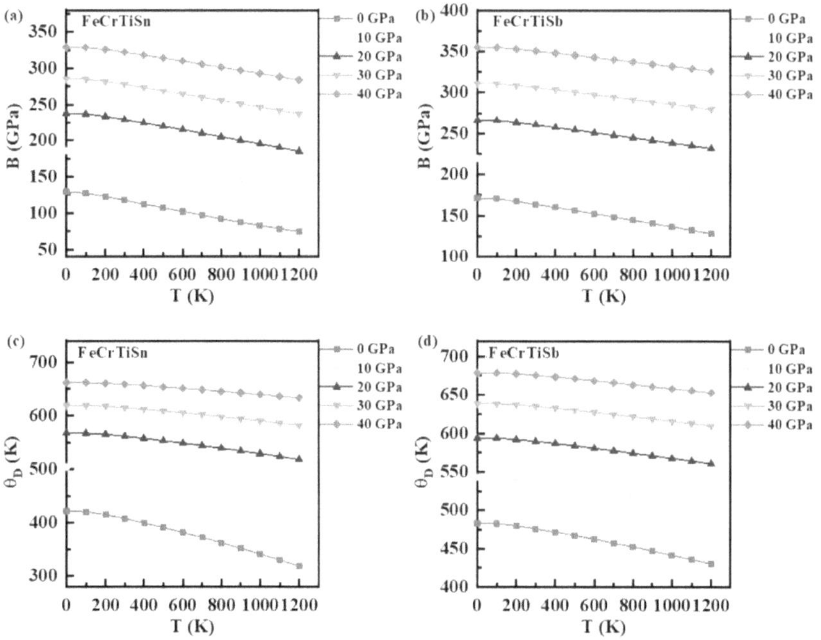

FIGURE 12.5 Temperature and pressure reliance of thermodynamical parameters such as bulk modulus (B), and Debye temperature (θ_D) for FeCrTiSn and FeCrTiSb compounds.

12.3.5 THERMOELECTRIC PROPERTIES

To find the efficiency of FeCrTiZ (Z = Sn, Sb) compounds in thermoelectric (TE) materials, we estimated TE properties of both compounds by calculating figure of merit (ZT). It acts as a pathway to recite the significance of various materials in TE generators. It depends on Seebeck coefficient (S), thermal conductivity (κ/τ), electrical conductivity (σ/τ), and power factor (PF). Here, τ is a relaxation time and has been kept constant to ~10^{-14} throughout our calculations.

In Figure 12.6(a, b) we have evaluated S for FeCrTiZ (Z = Sn, Sb) for the up and down spin-electrons. The S leads to assessing an induced thermoelectric voltage at different temperatures. The variation of S is presented in Figure 12.6(a, b). The graph is showing a sharp decreasing trend of S in the spin-up case till 300 K, followed by a steady drop on rising temperature and exposes small values of S in the spin-down state.

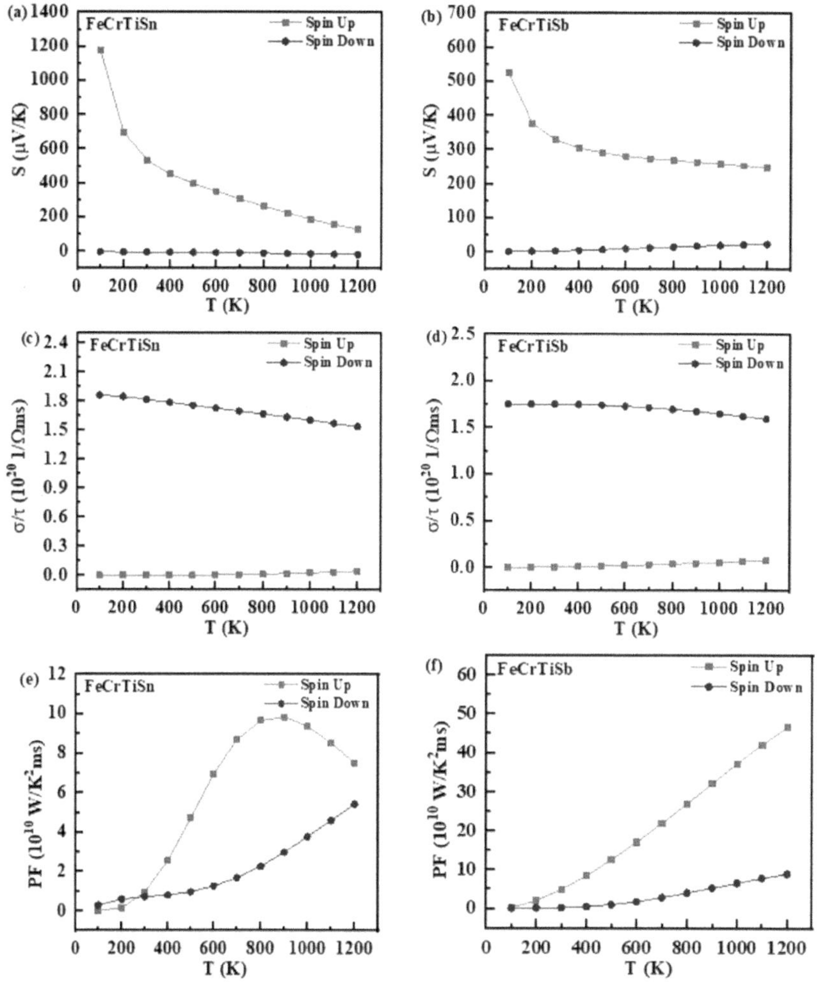

FIGURE 12.6 *(Continued)*

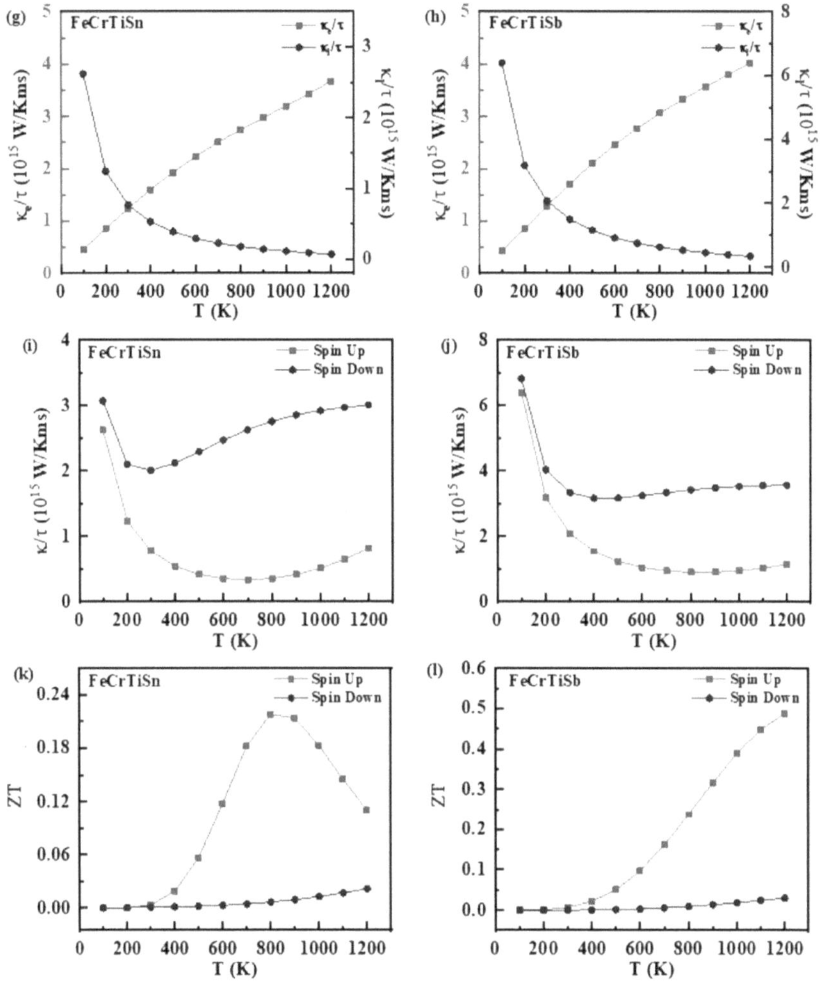

FIGURE 12.6 The variation of thermoelectric parameters, Seebeck coefficient (S), electrical conductivity (σ/τ), power factor (PF), thermal conductivity (κ/τ), and figure of merit (ZT) for FeCrTiSn and FeCrTiSb compounds.

The positive values of S at most signify the occupancy of holes as charge carriers. Furthermore, the two spins manifest diverse nature of variation with a huge difference in their values. The values of S for FeCrTiSn and FeCrTiSb at 300 K within the spin-up case are estimated as 533.01 and 328.9 in μV/K, respectively.

The variation of σ/τ is presented in Figure 12.6(c, d). In the spin-up case σ/τ is increasing somewhat on increasing temperature; however, it shows a reverse effect in another spin. The higher values of σ/τ in spin-down states are pointing toward the presence of DOS at E_f. Furthermore, an unlike behavior of σ/τ in two spins characterizes the HM nature of both the compounds. It gives values as $0.04 \times 10^{20}/\Omega ms$ ($1.54 \times 10^{20}/\Omega ms$) and $0.07 \times 10^{20}/\Omega ms$ (1.6×10^{20} $/\Omega ms$) in up (down) spin for FeCrTiSn and FeCrTiSb, respectively, at 1200 K.

Through the parameters defined above, we further calculated PF of materials by using the equation:

$$PF = S^2 \frac{\sigma}{\tau} \tag{12.6}$$

This parameter is related to the generated power and its variation with temperature is presented in Figure 12.6(e, f). The PF for both materials is increasing with temperature. However, an ascent is stronger in spin-up states. At 1200 K, PF is accounted as 7.49×10^{10} (46.69×10^{10}) and 5.41×10^{10} (9.01×10^{10}) for FeCrTiSn (FeCrTiSb) within spin-up and spin-down, correspondingly, in W/K^2ms. A high PF of FeCrTiSb indicates its significance in TE materials.

The decisive parameter to portray ZT, thermal conductivity (κ/τ) is determined through the united contribution of lattice (κ_l/τ) and electronic (κ_e/τ) thermal conductivity. Here, κ_l/τ is computed through the Slack equation[60,61] and κ_e/τ is described through the BoltzTrap code.[42] Hence, the variation of two thermal conductivities versus temperature is shown in Figure 12.6(g, h) for both the compounds. According to this, total κ_e/τ is increasing linearly and κ_l/τ is decaying exponentially with temperature. This decreasing trend is highlighting the dominance of phonon scattering and is giving the lowest value as 0.07×10^{15} W/Kms (0.34×10^{15} W/Kms) for FeCrTiSn (FeCrTiSb) at 1200 K. Small values of κ_e/τ at low temperatures suggest the significance of κ_l/τ to the κ/τ; moreover, at high temperatures, κ_e/τ took the control. The united contribution of κ_e/τ, κ_l/τ points up a variation of κ/τ in Figure 12.6(i, j) on account of increasing temperature that proposed a similar trend for both the envisaged compounds. The κ/τ value for FeCrTiZ (Z = Sn, Sb) at 300 K is computed as 0.77×10^{15} W/Kms (2.01×10^{15} W/Kms) and 2.09×10^{15} W/Kms (3.34×10^{15} W/Kms) in a sequence for spin-up (spin-down) states. However, it gives a minimum value at 700 K for FeCrTiSn and at 800 K for FeCrtisb, in spin-up states.

The defined parameters have assisted our calculations to define ZT. This evaluates competence of TE materials. Its highest value is estimated as 0.22 at 800 K for FeCrTiSn and as 0.49 at 1200 K for FeCrTiSb in the spin-up case as demonstrated in Figure 12.6(k, l). These results concluded that

FeCrTiSb is a better TE material as it gives a higher value of ZT. However, the increasing trend of ZT is significant in terms of TE material and inherently gives direction to the experimental study.

12.4 CONCLUSIONS

In this report, we have applied the density functional theory to investigate FeCrTiZ (Z=Sn, Sb) compounds. For the two compounds, we have used three different types of "LiMgPdSn structure" generally known as "Y-type" with the F-43m (216) space group. Structural optimization has been accomplished through the Birch–Murnaghan EOS, which resulted in the stability of type-1 structure. In electronic properties, FeCrTiZ (Z = Sn, Sb) compounds exhibit nearly HM nature in the GGA potential; however, they exhibit unambiguous HM nature in the mBJ potential. According to this, both the FeCrTiSn and FeCrTiSb compounds are semiconducting and metallic in their respective up and down spin states. In magnetic properties, FeCrTiZ (Z = Sn, Sb) compounds give integer values to M_{Tot} that also symbolize their HM nature. Further, elastic and mechanical properties objectify the brittle nature of FeCrTiZ (Z = Sn, Sb) with ionic bonding and anisotropic behavior. The thermodynamical parameters through the Gibbs 2 program have shown the variations of bulk modulus and Debye temperature within a certain range of temperature and pressure. Further, figure of merit (ZT) for each compound has been examined by using the BoltzTrap code. The highest value of ZT is estimated as 0.22 at 800 K for FeCrTiSn and as 0.49 at 1200 K for FeCrTiSb in the spin-up states. As a result, we infer that the outcomes that give high values of ZT are significant.

KEYWORDS

- **Heusler compounds**
- **ferromagnetic**
- **half-metallic**
- **thermoelectric properties**

REFERENCES

1. De Groot, R. A.; Mueller, F. M.; Van Engen, P. G.; Buschow, K. H. J. New Class of Materials: Half-Metallic Ferromagnets. *Phys. Rev. Lett.* **1983,** *50* (25), 2024.
2. Cahaya, A. B.; Tretiakov, O. A.; Bauer, G. E. Spin Seebeck Power Generators. *Appl. Phys. Lett.* **2014,** *104* (4), 042402.
3. Binasch, G.; Grünberg, P.; Saurenbach, F.; Zinn, W. Enhanced Magnetoresistance in Layered Magnetic Structures with Antiferromagnetic Interlayer Exchange. *Phys. Rev. B* **1989,** *39* (7), 4828.
4. Moodera, J. S.; Kinder, L. R.; Wong, T. M.; Meservey, R. Large Magnetoresistance at Room Temperature in Ferromagnetic thin Film Tunnel Junctions. *Phys. Rev. Lett.* **1995,** *74* (16), 3273.
5. Žutić, I.; Fabian, J.; Sarma, S. D. Spintronics: Fundamentals and Applications. *Rev. Modern Phys.* **2004,** *76* (2), 323.
6. Seebeck, T. J. Magnetic Polarization of Metals and Minerals. *Abh. Dtsch. Akad. Wiss. Berlin* **1822,** *265,* 1822–1823.
7. Mondal, S.; Mazumdar, C.; Ranganathan, R.; Alleno, E.; Sreeparvathy, P. C.; Kanchana, V.; Vaitheeswaran, G. Ferromagnetically Correlated Clusters in Semimetallic Ru_2NbAl Heusler Compound and its Thermoelectric Properties. *Phys. Rev. B* **2018,** *98* (20), 205130.
8. Tritt, T. M.; Subramanian, M. A. Thermoelectric Materials, Phenomena, and Applications: A Bird's Eye View. *MRS Bull.* **2006,** *31* (3), 188–198.
9. Graf, T.; Felser, C.; Parkin, S. S. Simple Rules for the Understanding of Heusler Compounds. *Prog. Solid State Chem.* **2011,** *39* (1), 1–50.
10. Heusler, F. Über Magnetische Manganlegierungen. *Verh Deutsch. Phys. Ges.* **1903,** *5* (12), 219.
11. Heusler, F.; Starck, W.; Haupt, E. Magnetisch-chemische Studien. *Verh Deutsch. Phys. Ges.* **1903,** *5,* 219–232.
12. Felser, C.; Hirohata, A.; Ed.; Heusler Alloys Properties, Grows, Applications. Springer Series in Material Science, Springer: Verlag, Berlin, 2016; p 222.
13. Eberz, U.; Seelentag, W.; Schuster, H. U. Zur Kenntnis Farbiger Ternärer und Quaternärer Zintl-Phasen/Coloured Ternary and Quaternary Zintl-Phases. *Z. Naturforsch. B* **1980,** *35* (11), 1341–1343.
14. Alijani, V.; Ouardi, S.; Fecher, G. H.; Winterlik, J.; Naghavi, S. S.; Kozina, X.; Stryganyuk, G.; Felser, C.; Ikenaga, E.; Yamashita, Y.; Ueda, S.; Kobayashi, K. Electronic, Structural, and Magnetic Properties of the Half-Metallic Ferromagnetic Quaternary Heusler Compounds CoFeMnZ (Z= Al, Ga, Si, Ge). *Phys. Rev. B* **2011,** *84* (22), 224416.
15. Rani, D.; Bainsla, L.; Suresh, K. G.; Alam, A. Experimental and Theoretical Investigation on the Possible Half-metallic Behaviour of Equiatomic Quaternary Heusler Alloys: CoRuMnGe and CoRuVZ (Z= Al, Ga). *J. Magn. Magn. Mater.* **2019,** *492,* 165662.
16. Şarlı, N.; Ak, F.; Özdemir, E. G.; Saatçi, B.; Merdan, Z. Key Role of Central Antimony in Magnetization of $Ni_{0.5}Co_{1.5}MnSb$ Quaternary Heusler Alloy Revealed by Comparison Between Theory and Experiment. *Phys. B: Condens. Matter* **2019,** *560,* 46–50.
17. Stephen, G. M.; Buda, G.; Jamer, M. E.; Lane, C.; Kaprzyk, S.; Barbiellini, B.; Graf, D.; Lewis, L.H.; Bansil, A.; Heiman, D. Structural and Electronic Properties of the Spin-filter Material CrVTiAl with Disorder. *J. Appl. Phys.* **2019,** *125* (12), 123903.

18. Venkateswara, Y.; Gupta, S.; Samatham, S. S.; Varma, M. R.; Suresh, K. G.; Alam, A. Competing Magnetic and Spin-gapless Semiconducting Behavior in Fully Compensated Ferrimagnetic CrVTiAl: Theory and experiment. *Phys. Rev. B* **2018,** *97* (5), 054407.
19. Shi, B.; Li, J.; Zhang, C.; Zhai, W.; Jiang, S.; Wang, W.; Chen, D.; Yan, Y.; Zhang, G.; Liu, P. F. First-principles Investigation on the Transport Properties of Quaternary CoFeRGa (R= Ti, V, Cr, Mn, Cu, and Nb) Heusler Compounds. *Phys. Chem. Chem. Phys.* **2020,** *22* (40), 23185–23194.
20. Shakil, M.; Arshad, H.; Aziz, S.; Gillani, S. S. A.; Rizwan, M.; Zafar, M. Determination of Phase Stability, Half Metallicity, Mechanical and Thermal Behavior of Fe Based Quaternary Heusler Alloys. *J. Alloys Compd.* **2021,** *856,* 157370.
21. Feng, Y.; Chen, H.; Yuan, H.; Zhou, Y.; Chen, X. The Effect of Disorder on Electronic and Magnetic Properties of Quaternary Heusler alloy CoFeMnSi with LiMgPbSb-type structure. *J. Magn. Magn. Mater.* **2015,** *378,* 7–15.
22. Seh, A. Q.; Gupta, D. C. Quaternary Heusler Alloys a Future Perspective for Revolutionizing Conventional Semiconductor Technology. *J. Alloys Compd.* **2021,** *871,* 159560.
23. Singh, J.; Kaur, K.; Bhat, M. A.; Sharopov, U. B.; Dhiman, S.; Goyal, M.; Verma, S.S.; Khandy, S. A. First-principles Calculations on the Electronic Structure and Thermoelectric Properties of Quaternary Heusler Compounds: LiScPtSi and LiScPdGe. *Mater. Today Commun.* **2022,** *32,* 103961.
24. Alqurashi, H.; Haleoot, R.; Hamad, B. First-principles Investigations of the Electronic, Magnetic and Thermoelectric Properties of VTiRhZ (Z= Al, Ga, In) Quaternary Heusler alloys. *Mater. Chem. Phys.* **2022,** *278,* 125685.
25. Andharia, E.; Alqurashi, H.; Hamad, B. Lattice Dynamics, Mechanical Properties, Electronic Structure and Magnetic Properties of Equiatomic Quaternary Heusler Alloys CrTiCoZ (Z= Al, Si) Using First Principles Calculations. *Materials* **2022,** *15* (9), 3128.
26. Webster, P. J.; Ziebeck, K. R. A. Alloys and Compounds of D-elements with Main Group Elements. Part 2. *Landolt-Bornstein, New Series, Group III* **1988,** *19,* 75–184.
27. Ziebeck, K.; Neumann, K. U. In Magnetic Properties of Metals. Wijn, H. P. J., Ed.; Landolt-Börnstein, New Series, Group III, pt. c Springer: Berlin, 2001; *32,* 64.
28. Pierre, J.; Skolozdra, R. V.; Tobola, J.; Kaprzyk, S.; Hordequin, C.; Kouacou, M. A.; Karla, I.; Currat, R.; Lelievre-Berna, E. Properties on Request in semi-Heusler Phases. *J. Alloys Compd.* **1997,** *262,* 101–107.
29. Jain, A.; Ong, S. P.; Hautier, G.; Chen, W.; Richards, W. D.; Dacek, S.; Cholia, S.; Gunter, D.; Skinner, D.; Ceder, G.; Persson, K. A. Commentary: The Materials Project: A materials Genome Approach to Accelerating Materials Innovation. *APL Mater.* **2013,** *1* (1), 011002.
30. Kirklin, S.; Saal, J. E.; Meredig, B.; Thompson, A.; Doak, J. W.; Aykol, M.; Rühl, S.; Wolverton, C. The Open Quantum Materials Database (OQMD): Assessing the Accuracy of DFT Formation Energies. *npj Comput. Mater.* **2015,** *1* (1), 1–15.
31. Wimmer, E.; Krakauer, H.; Weinert, M.; Freeman, A. J. Full-potential Self-consistent Linearized-augmented-plane-wave Method for Calculating the Electronic Structure of Molecules and Surfaces: O_2 molecule. *Phys. Rev. B* **1981,** *24* (2), 864.
32. Sholl, D. S.; Steckel, J. A. Density Functional Theory: A Practical Introduction. John Wiley & Sons, 2011.
33. Blaha, P.; Schwarz, K.; Madsen, G. K.; Kvasnicka, D.; Luitz, J. wien2k. An Augmented Plane Wave+ Local Orbitals Program for Calculating Crystal Properties, **2001,** 60.
34. Slater, J. C. A Simplification of the Hartree-Fock Method. *Phys. Rev.* **1951,** *81* (3), 385.

35. Perdew, J. P.; Burke, K.; Ernzerhof, M. Generalized Gradient Approximation Made Simple. *Phys. Rev. Lett.* **1996,** *77* (18), 3865.
36. Tran, F.; Blaha, P. Accurate Band Gaps of Semiconductors and Insulators with a Semilocal Exchange-Correlation Potential. *Phys. Rev. Lett.* **2009,** *102* (22), 226401.
37. Tran, F.; Blaha, P.; Schwarz, K. Band Gap Calculations with Becke–Johnson Exchange Potential. *J. Phys.: Condens. Matter* **2007,** *19* (19), 196208.
38. Jamal, M.; Bilal, M.; Ahmad, I.; Jalali-Asadabadi, S. IRelast Package. *J. Alloys Compd.* **2018,** *735,* 569–579.
39. Blanco, M. A.; Francisco, E.; Luana, V. GIBBS: Isothermal-isobaric Thermodynamics of Solids from Energy Curves Using a Quasi-harmonic Debye Model. *Comput. Phys. Commun.* **2004,** *158* (1), 57–72.
40. Otero-de-la-Roza, A.; Luaña, V. Gibbs2: A New Version of the Quasi-harmonic Model Code. I. Robust Treatment of the Static Data. *Comput. Phys. Commun.* **2011,** *182* (8), 1708–1720.
41. Otero-de-la-Roza, A.; Abbasi-Pérez, D.; Luaña, V. Gibbs2: A New Version of the Quasiharmonic Model Code. II. Models for Solid-state Thermodynamics, Features and Implementation. *Comput. Phys. Commun.* **2011,** *182* (10), 2232–2248.
42. Madsen, G. K.; Singh, D. J. BoltzTraP. A Code for Calculating Band-structure Dependent Quantities. *Comput. Phys. Commun.* **2006,** *175* (1), 67–71.
43. Murnaghan, F. D. The Compressibility of Media Under Extreme Pressures. *Proc. Natl. Acad. Sci.* **1944,** *30* (9), 244–247.
44. Perdew, J. P. Density Functional Theory and the Band Gap Problem. *Int. J. Quantum Chem.* **1985,** *28* (S19), 497–523.
45. Pauling, L. The Nature of the Interatomic Forces in Metals. *Phys. Rev.* **1938,** *54* (11), 899.
46. Slater, J. C. The Ferromagnetism of Nickel. *Phys. Rev.* **1936,** *49* (7), 537.
47. Slater, J. C. The Ferromagnetism of Nickel. II. Temperature Effects. *Phys. Rev.* **1936,** *49* (12), 931.
48. Galanakis, I.; Dederichs, P. H.; Papanikolaou, N. Slater-Pauling Behavior and Origin of the Half-Metallicity of the Full-Heusler Alloys. *Phys. Rev. B* **2002,** *66* (17), 174429.
49. Bouhemadou, A.; Khenata, R.; Zegrar, F.; Sahnoun, M.; Baltache, H.; Reshak, A. H. Ab Initio Study of Structural, Electronic, Elastic and High Pressure Properties of Barium Chalcogenides. *Comput. Mater. Sci.* **2006,** *38* (2), 263–270.
50. Srivastava, V.; Kaur, N.; Wang, X.; Mushtaq, M.; Dar, S. A. First-principles Study on Structural, Electronic, Magnetic, Elastic, Mechanical and Thermodynamic Properties of Mn_2PtCo Heusler Alloy. *Int. J. Energy Res.* **2021,** *45* (7), 11305–11319.
51. Pugh, S. F. XCII. Relations Between the Elastic Moduli and the Plastic Properties of Polycrystalline Pure Metals. *London, Edinburgh, Dublin Philos. Mag. J. Sci.* **1954,** *45* (367), 823–843.
52. Tvergaard, V.; Hutchinson, J. W. Microcracking in Ceramics Induced by Thermal Expansion or Elastic Anisotropy. *J. Am. Ceramic Soc.* **1988,** *71* (3), 157–166.
53. Kaur, T.; Kumar, S.; Bhat, B. H.; Want, B.; Srivastava, A. K. Effect on Dielectric, Magnetic, Optical and Structural Properties of Nd–Co Substituted Barium Hexaferrite Nanoparticles. *Appl. Phys. A* **2015,** *119,* 1531–1540.
54. Singh, P.; Singh, A.; Quraishi, M. A. Thiopyrimidine Derivatives as New and Effective Corrosion Inhibitors for Mild Steel in Hydrochloric Acid: Electrochemical and Quantum Chemical Studies. *J. Taiwan Inst. Chem. Eng.* **2016,** *60,* 588–601.

55. Lin, Y.; Singh, A.; Ebenso, E. E.; Wu, Y.; Zhu, C.; Zhu, H. Effect of Poly (methyl methacrylate-co-N-vinyl-2-pyrrolidone) Polymer on J55 Steel Corrosion in 3.5% NaCl Solution Saturated with CO_2. *J. Taiwan Inst. Chem. Eng.* **2015,** *46,* 214–222.
56. Bashir, S.; Sharma, V.; Lgaz, H.; Chung, I. M.; Singh, A.; Kumar, A. The Inhibition Action of Analgin on the Corrosion of Mild Steel in Acidic Medium: A Combined Theoretical and Experimental Approach. *J. Mol. Liq.* **2018,** *263,* 454–462.
57. Wen, Z.; Zhao, Y.; Hou, H.; Wang, B.; Han, P. The Mechanical and Thermodynamic Properties of Heusler Compounds Ni_2XAl (X= Sc, Ti, V) under Pressure and Temperature: A First-principles Study. *Mater. Design* **2017,** *114,* 398–403.
58. Hu, W. C.; Liu, Y.; Li, D. J.; Zeng, X. Q.; Xu, C. S. First-principles Study of Structural and Electronic Properties of C14-type Laves Phase Al_2Zr and Al_2Hf. *Comput. Mater. Sci.* **2014,** *83,* 27–34.
59. Zhao, Y.; Hou, H.; Zhao, Y.; Han, P. First-Principles Study of the Nickel–Silicon Binary Compounds Under Pressure. *J. Alloys Compd.* **2015,** *640,* 233–239.
60. Slack, G. A. The Thermal Conductivity of Nonmetallic Crystals. *Solid State Phys.* **1979,** *34,* 1–71.
61. Slack, G. A. Nonmetallic Crystals with High Thermal Conductivity. *J. Phys. Chem. Solids* **1973**, *34* (2), 321–335.

CHAPTER 13

One-Step Method for the Synthesis of Single-Phase $BiFeO_3$ Compound

DEEPIKA TRIPATHI[1], H. S. TIWARI[1], and VILAS SHELKE[2]

[1]*Department of Pure and Applied Physics, Guru Ghasidas Vishwavidyalaya, Koni, Bilaspur, Chhattisgarh, India*

[2]*Rechargion Energy Pvt. Ltd, Pune, Maharashtra, India*

ABSTRACT

The present study of $BiFeO_3$ (BFO) is prepared by the one-step method. To prepare the material BFO pure phase, we developed and optimized the introverted method for synthesis. The ceramic $BiFeO_3$ is very difficult to synthesize single phase, due to its volatile nature and formation of secondary phases like mullnite and sillenite. The synthesis parameters like sintering temperature, time were changed; quenching sequence was used to stabilize without impure phase. We synthesized several polycrystalline samples of $BiFeO_3$ to optimize parameters like temperature and time in the range of 850–870°C, 5–30 min. The samples were prepared under an optimized condition which shows a rhombohedral structure without any impurity. The structural characterization was carried out by using the Powder method, X-ray diffraction (XRD). The room temperature magnetization measurement showed antiferromagnetic behavior due to canting sublattice by an antisymmetric spin coupling. It was observed magnetic moment by using SQUID magnetometer and its microstructure behavior of these samples was studied using scanning electron microscopy (SEM) technique.

Contemporary Advancements in Materials Technology: Novel Solutions to Complex Problems.
Shrikaant Kulkarni, Vipul Srivastava, & P. William (Eds.)

13.1 INTRODUCTION

An unambiguous ordering of electric and magnetic dipoles takes place through two different and diversified mechanisms as they follow different spatial symmetry and time hitch aspects. Such materials are known as ferroelectric, ferromagnetic etc. irrespective of their frequent ferro-free compositions. A simultaneous presence and coupling between the so-called ferroic order parameters was observed in a new class of materials known as "multiferroics."[1–3]

The search for magneto-electric materials began in earnest of 1950s in the group of Smolenski.[4,5] H. Schimd was the first to coin the term "multiferroic" to indicate the presence of more than one ferroic ordering.[6] Ferro/antiferromagnetic is a group of materials having coupled ferroelectric and order parameter that result in synchronized ferroelectricity and magnetism in single-phase material. Ferromagnetic materials show a transition from a high-temperature paramagnetic phase to a low-temperature structure which possess a spontaneous, nonvolatile magnetization below a certain temperature called Curie temperature (T_c). The Curie–Weiss localized-moment theory of ferromagnetism can explain many properties of ferromagnetic and antiferromagnetic materials.[7, 8] Transition metal compounds often possess antiferromagnetic ordering. Multiferroic materials are prominent not only because of these properties but also they show magneto-electric coupling that leads to additional functionality.[9–11]

Among all the multiferroics, the prototype $BiFeO_3$ is a model for multiferroic materials. It received great attention due to its excellent characteristics and multifunctional application for devices. Controlling of electric and magnetic ordering through cross coupling may provide possibility of new types of memory devices. Such a type of coupling was explained by Maxwell in the 19th century in electromagnetic waves.[12] However, in solid materials electron charge is responsible for polarization and magnetization is governed by mainly orbital and spin motions. The spins are frustrated and their ordering may provide unique characteristics.[13, 14] The magnetic nature of $BiFeO_3$ is an antiferromagnetic type having a modulated spin structure. A spiral spin arrangement leads to cancellation of macroscopic magnetization and also inhibits scrutiny of linear magneto electric effect.

$BiFeO_3$ exhibits a high value of ferroelectric polarization with a very high ferroelectric Curie temperature (1100 K). Besides it is G-type antiferromagnetic with canted spin representing a spin cycloid structure of 62 nm length. Due to G-type canted antiferromagnetic structure, electric field can produce magnetization and control magnetization. Similarly, the magnetic field can produce magnetization and control polarization. In BFO ferroelectric plane and antiferromagnetic plane are perpendicularly coupled to each other so polarization can rotate the magnetic plane. The large relative ionic displacements toward polar direction can impart high polarization values.[14–16] One of the essential concerns about BFO, polarization switching is prevented by its low resistivity. So the leakage current may be by adapting different strategies like

(i) Improving microstructure and reducing structural defects[9, 10, 17]
(ii) By suitable doping in pure BFO.[11]

As we know Fe2+ is more stable than Fe3+, since half-filled orbitals are more stable, therefore it would help to improve the resistivity of BFO material. $BiFeO_3$ has potential for applications in information storage, spintronic and sensor devices and science of this material is truly fascinating for technological applications. But the changeable ferroelectric and ferromagnetic ordering in same phase the material leads to supplementary functionality in which ferroelectric polarization can be used to change magnetization and ferromagnetic ordering can be used to switch the electric polarization. One of the main applications which is associated with this material is 10^{12} hertz frequency radiation. Another important characteristic of bulk material ($BiFeO_3$) is highly sensitive to acetone, which opens the possibility in semiconductor gas sensor. BFO material properties like were most important driving forces for many experimental as well as theoretical studies.[18, 19] In theoretical studies we calculate ferroelectric polarization above the range of 100 μC/cm. In the invention of extra phenomena like photovoltaic effect, metal–insulator transition, BFO played a crucial role[20–22]

1. lead-free composition
2. high transition temperature
3. large polarization values

Extensive experimental studies were done on thin films and bulk ceramic; research work is still being over-involved by the formation of second phase during the synthesis of little resistivity in material. The task of formation of pure phase BFO is immensely difficult compared to other iso-structural perovskite oxides like $LaMnO_3$, $SrTiO_3$, $BaTiO_3$, and their variants. The constituent oxides like Bi_2O_3 and Fe_2O_3 have adverse thermodynamic characters as Bi_2O_3 is volatile and Fe_2O_3 is less reactive at low temperature. Therefore, the temperature range for phase formation is narrow. The combination of Bi_2O_3 and Fe_2O_3 compounds gives a single phase ternary compound which is probable within minimum set of parameters. The problem remains the same with intermediary phases like $Bi_2Fe_4O_9$ (sillenite), $Bi_2Fe_4O_9$ (mullite), which coexist.[23] Its great significance and the trouble during synthesizing difficulties persuade several strategies to get pure-phase $BiFeO_3$ compounds.

Bismuth ferrite ($BiFeO_3$) opened a new era for multifunctional material due to simultaneous presence of electric/magnetic ordering and photovoltaic retort sustainable over a wide range of temperature. The lead-free compositions, high value of ferroelectric polarization, and above room temperature functionalities make it attractive material for a variety of applications. Multiferroic properties of $BiFeO_3$ are very well linked to its structure and also strain generated in a thin-film form was presumed initially. Nevertheless, the bulk compound can represent the properties as a whole material without extrinsic factors. However, synthesis of single-phase material has been difficult due to diverse nature of constituent elements, particularly volatile nature of bismuth oxide and formation of impurity phases. The two compounds $Bi_2Fe_4O_9$ (mullite) and $Bi_{25}FeO_{36}$ (sillenite) can coexist with $BiFeO_3$ according to Gibb's phase rule.[23] Therefore, it is ardently required to design a synthesis strategy to get rid of these phases and obtain single-phase BFO compound. Rapid liquid phase sintering[15] and HNO_3 leaching[16] are the two commonly used methods for BFO synthesis. These methods have inherent drawbacks like poor control on sample quality and external impurities arising from rapid heating or HNO_3 usage. As an alternative, we developed a one-step solid-state method for the synthesis of a single-phase BFO compound. We observed that the initial particle size of a few microns, sintering temperature of 870°C, and a sintering duration of 30 min form a crucial set of synthesis parameters. Any deviation from these values resulted in impurity formation. The magnetic behavior and microstructure of representative samples was also studied to examine the quality of samples.

13.2 EXPERIMENTAL METHODS AND MATERIALS

Stoichiometric amounts of Bi_2O_3 and Fe_2O_3 powder were thoroughly ground in agate mortar for 10 hours using acetone as a medium. With the help of hydraulic pressure the homogenous mixture was pressed into pellets of 12 mm diameter by applying 7 ton pressure; then it was sintered in separate batches at 850, 860, and 870°C for 5, 15, and 30 min duration followed by quenching to room temperature.

For characterization purposes, the pellets were shaped in different forms. A finely ground powder is preferred for X-ray diffraction study with the help of advance equipped diffractometer with Cu Kα radiation (λ=1.54 Å) source and an eye detector with a step size of 0.02° was used in the range 10° ≤ 2θ ≤ 60° for phase identification and structure determination.. A rectangular-shaped pellet is used for magnetic measurement. A fractured surface of pellet can be used for scanning electron microscopy. The magnetic property measurements of the samples were conducted using SQUID magnetometer at room temperature. A magnetic field up to 50 KOe was applied using super-conducting magnet. The microstructures of the samples were examined by using scanning electron microscopy (SEM) technique.

13.3 RESULTS AND DISCUSSION

13.3.1 STRUCTURAL ANALYSIS

The phase purity of the BFO sample depends on sintering temperature and sintering time. Both the thermo-dynamical parameters play an important role in phase formation. The XRD patterns of the samples sintered at 850°C for 5, 15, and 30 min are shown in Figure 13.1. Several peaks corresponding to unreacted chemicals and intermediate phases were seen in these patterns. We observed that Bi_2O_3 and Fe_2O_3 have not reacted under these conditions. As we observed that at 850°C the reaction does not take place, we increased the temperature to 860°C keeping the time durations in similar trails. The XRD patterns for the samples sintered at a temperature of 860°C for 60 min and 120 min as shown in Figure 13.2. The partial reaction took place for 5 and 15 min samples, comparatively 30 min samples and exhibited a better phase formation. As demonstrated by the XRD patterns for samples sintered for 5, 15, and 30 min in Figure 3.3(a-c), we increased the temperature to 870°C in order to boost the reaction rate. But Figure 3.3(c) makes it quite evident that it produced a single phase BFO at 870°C. Using the R3c space group, all of the peaks were indexed to the rhombohedral lattice type (JCPDS file 20-0169). The calculated values of the lattice parameters were a = 5.48 Å c = 7.00 Å, and V = 182.04 $Å^3$. These values match with those reported for single-phase BFO synthesized by other methods.[15, 16, 23] It means 870°C, 30 min is optimized set of synthesis parameters for single-phase BFO formation. We also tried for higher temperature (880°C) or longer duration (90 min).

However, the XRD patterns for these samples showed reentrant impurity phases.

It is noteworthy to mention here that Bi_2O_3 compound decomposes at 820°C, but Fe_2O_3 needs a higher temperature. The reaction between the two is initiated at the grain boundaries and then it proceeds through diffusion. The diffusion process reduces the reaction kinematics, but a longer heating duration can cause significant "Bi" loss. Therefore, it is necessary to restrict the duration to 30 min and enhance the reaction rate by achieving grain size as small as possible. For this reason extensive grinding of starting chemicals is necessary. Higher temperature or long duration can also initiate decomposition of the product phase.

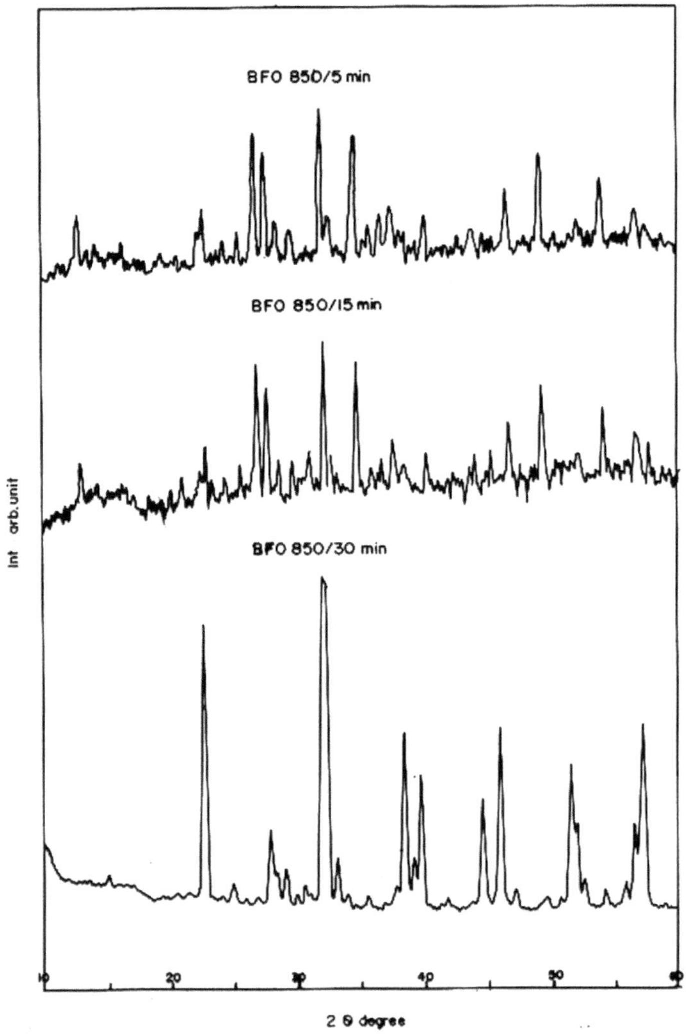

FIGURE 13.1 XRD of $BiFeO_3$ sintered at 850°C with different sintering times: (i) 5 min, (ii) 15 min, (iii) 30 min.

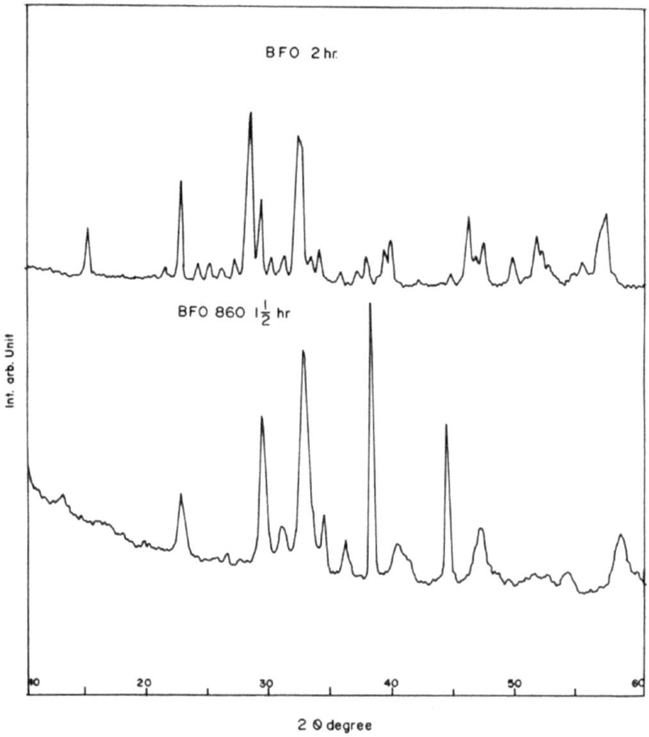

FIGURE 13.2 XRD of $BiFeO_3$ at 860°C with different sintering times: (i) 60 min, (ii) 120 min.

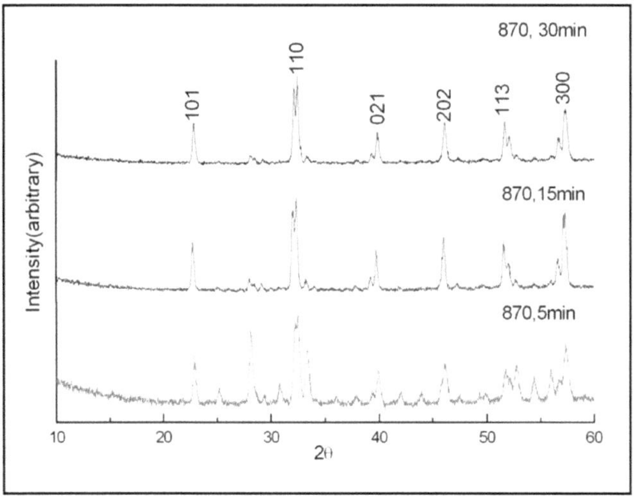

FIGURE 13.3 XRD of $BiFeO_3$ at 870°C with different sintering times: (i) 5 min, (ii) 15 min, (iii) 30 min.

At higher temperature Bi_2O_3 can decompose with primary loss of Bi which may result in residual Fe_2O_3 impurity. On the other hand at below temperature 820°C Bi_2O_3 decomposes and intermediate phase started formation ($BiFe_4O_9$ and $Bi_{25}FeO_3$); the three phases can coexist with zero degree of freedom as the number of components and oxidation state of Fe remain invariant according to Gibb's phase rule. In temperature range 820°C to 930°C the impurity phases can react to form perovskite $BiFeO_3$ phase and $BiFeO_3$ cannot decompose back to other phases. Although the temperature is a major thermodynamic variable, time duration and particle dimension are very decisive parameters. A longer duration can give sufficient time for substantial loss of Bi and a shorter duration can keep unreacted due to low ionic mobility. Therefore it is essentially required to reduce particle size, so proceed to a complete reaction in a shorter period. Our study shows that quick firing is not important as the sample takes a very smaller amount of time to reach 870°C (optimized value) from 820°C (disintegration temperature). A divergence from 870°C and duration of 30 min resulted in the structure of parasitic phases.

The predictable heating was followed by kinetic cooling to stay away from narrow temperature departure and to decrease the chance of other phase patterns. The unadventurous sintering was followed by kinetic cooling to keep away from narrow temperature divergence and to decrease the possibility of other phase patterns.

13.3.2 MICROSTRUCTURAL BEHAVIOR

The microstructure of the samples was revealed through scanning electron microscopy. Figures 13.4 (a) and (b) show the microstructure of samples sintered at 860°C for 15, 30 min, respectively. In these samples the incomplete grain growth is clearly visible, which also indicates that 860°C is not a sufficient temperature to synthesize a good quality sample.

The SEM images in Figure 13.5 (a) and (b) are for the samples sintered at 870°C for 15 and 30 min, respectively. An entire granule growth with inter-diffused granule boundaries can be observed in Figure 13.5 (b). It is remarkable that small particle size has a vital role in single-phase BFO formation, as the synthesis duration is limited to 30 min. No extra measures like intermediary grinding were used to improve the reaction rate in this single-step method. The reaction had to continue through the interface of product phase within the duration of 30 min. In this small duration, the grain growth is quite rapid achieving an average grain size of around 5 μm.

FIGURE 13.4 SEM images of samples sintered at 860°C for (a) 15, (b) 30 min.

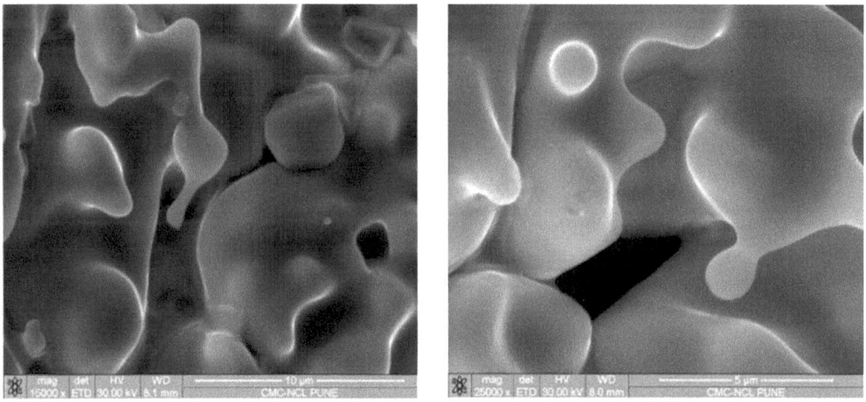

FIGURE 13.5 SEM images of samples sintered at 870°C for a) 15, b) 30 min.

13.3.3 MAGNETIC BEHAVIOR

We measured magnetization as a function of applied magnetic field using superconducting quantum interference device (SQUID) magneto-meter at 300 K. The measurement was carried out on the ceramic sample sintered at 870°C for 30 min. A typical M–H plot is shown in Figure 13.6, which indicates a linearly increasing magnetic moment with an applied magnetic field. It did not show saturation behavior up to the maximum applied field of 50 KOe. The low magnetic moment, characteristic of antiferromagnetic behavior was observed. A slightly higher moment may be expected from canting of spins. However, a high value of magnetic moment can also result from impurity phases and/or Fe^{3+} states in oxygen-deficient samples. The low value of magnetization in our sample is an indirect evidence for impurity-free good-quality sample.

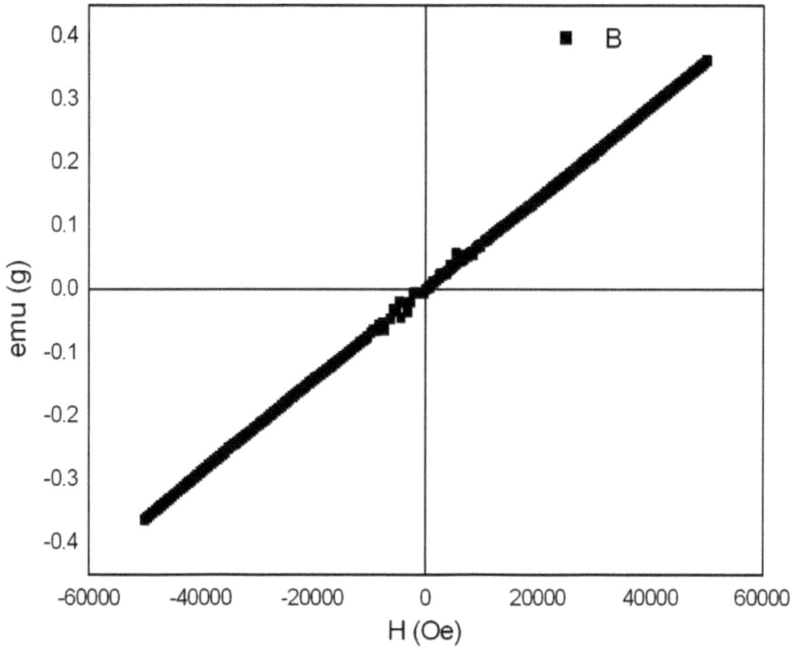

FIGURE 13.6 A typical M–H behavior of BFO sample.

13.4 CONCLUSION

We have synthesized a phase pure $BiFeO_3$ sample by the one-step solid-state reaction method followed by the quenching technique. Sintering temperature and time affected the phase, microstructural properties of the samples. An increase in sintering temperature improved the bulk density of the sample. Magnetic measurement showed that $BiFeO_3$ is antiferromagnetic at room temperature.

KEYWORDS

- **bismuth**
- **ferrite**
- **$BiFeO_3$**
- **XRD**
- **SQUID**
- **SEM**
- **antiferromagnetism**
- **quenching**

REFERENCES

1. Wang, K. F.; Liu, J. M.; Ren, Z. F. *Adv. Phys.* **2009,** *58,* 321.
2. Schmid, H.; *J. Phys.: Cond. Matt.* **2008,** *20,* 434201.
3. Engbaolin, F.; Hao, X.; Zhao, X. *J. Mater. Sci.* **2009,** *55,* 452.
4. Smolenski, G. A.; Isupov, V. A.; Agronovskaj, A. I. *Sov. Phys. Solid State* **1959,** *1,* 150.
5. Smolenski, G. A.; Chupis, I. E. *Sov. Phys. Usp* **1982,** *25,* 475.
6. Schmid, H. *Ferroelectrics* **1994,** *162,* 317
7. Kittel, C. Introduction to Solid State Physics. 7th ed.; Wiley, 1996.
8. Dekker, A. J. Solid State Physics. Oscar Pub.: New Delhi, India.
9. Scott, J. F.; Magn, J. *Magn. Mater.* **2009,** *321,* 1689.
10. Zhang, S. T.; Chen, Y. F.; Ming, N. *Appl. Phys. Lett.* **2005,** *87,* 262907.
11. Xu, J.; Wang, G.; Wang, H.; Ding, D.; He, Y. *Adv. Mater.* **2009,** *63,* 855.
12. Jackson, J. D. Electrodynamics. 3rd Ed.
13. Cheng, J. R.; Liand, N.; Cross, L. *J. Appl. Phys.* **2003,** *94,* 5153.
14. Catlon, G.; Scott, J. F. *Adv. Matt.* **2009,** *21,* 2463.
15. Kumar, M. M.; Palkar, V. R.; Srinivas, K.; Suryanarayan, S. V. *Appl. Phys. Lett.* **2000,** *76,* 2764.
16. Wang, Y. P.; Zhou, L.; Zhang, M. F, Chen, X. Y, Liu, J. M.; Liu, Z. G.; *J. Appl. Phys. Lett.* **2004,** *84,* 1731.
17. Shelke, V.; Harshan, V.; Kotru, S.; Gupta, A. *J. Appl. Phys.* **2009,** *106,* 104114.
18. Catalan, G.; Scott, J. F. *Adv. Mat.* **2009,** *21,* 2463.
19. Martin, L. W. *Dalton Trans.* **2012,** *39,* 1083.
20. Choi, T.; Lee, S.; Choi, Y. J.; KiryuKhin, V. *Science* **2009,** *324,* 63.
21. Zhang, Y.; Xie, Z. *Nanoscale* **2012,** *4,* 2926.
22. Palai, R.; Katyar, R. S.; Schmid, H.; Robertson, J.; Catalan, G.; Scott, J. F. *Phys. Rev. B* **2008,** *77,* 014110.
23. Selbech, S. M.; Einarsrud, M.; Grande, T. *Chem. Mater.* **2009,** *21,* 169.

Index

D

E

F

H

I

L

M

P

U

V

W

X

Y

Z